Geolocation of RF Signals

First Edition

Ilir Progri

Geolocation of RF Signals

Principles and Simulations

Springer

Ilir Progri
Giftet Inc.
118 Heywood St
01604 Worcester Massachusetts
USA

ISBN 978-1-4419-7951-3 e-ISBN 978-1-4419-7952-0
DOI 10.1007/978-1-4419-7952-0
Springer New York Dordrecht Heidelberg London

Printed on acid-free paper

Springer is part of Springer Science+Business Media (www.springer.com)

Preface

Geolocation of RF Signals: Principles and Simulations offers an overview of the best practices and innovative techniques in the art and science of geolocation over the last 20 years. It covers all research and development aspects including theoretical analysis, RF signals, geolocation techniques, key block diagrams, and practical principle simulation examples in the frequency band from 100 MHz to 18 GHz or even 60 GHz. Dr. Progri reveals the research and development process by demonstrating how to understand and explain geolocation of RF signals from basic diagrams to the final principle simulation examples and makes recommendations for the future final products of geolocation of RF signals. Starting with RF signals, the book progressively examines various signal bands – such as VLF, LF, MF, HF, VHF, UHF, L, S, C, X, Ku, and, K and the corresponding geolocation requirements per band and per application – to achieve required performance objectives of up to 0° precision. Next follows a step-by-step approach of RF geolocation techniques and concludes with notes on state-of-the-art geolocation designs as well as advanced features found in signal generator instruments. The book also includes the best mathematical techniques employed for geolocation of RF signals at 100 MHz to 18 GHz or even 60 GHz.

The book is designed into two parts taking into consideration the vastness, depth, and resourcefulness of the material. Part I contains Chaps. 1–3 and part II includes Chaps. 4–6.

Part I of the book is intended to engage and immerse the reader with unique, powerful ideas, detailed descriptions and discussions, powerful analysis, important principles and visualization tools, and most of all provide the means to deepen the reader's imagination for future research and development work, applications, and product, and development of future prototypes. Figures make the reader aware of the vastness of opportunities to refine future models and modeling, principle recopies, and analysis tools.

Part II of the book is intended to engage, train, and prepare the reader with powerful principle "recipe secrets" for analyzing, modeling, and simulating GRFS systems. Since this is the first edition of the book, the emphasis here is given in the main principles, algorithm descriptions, best blind signal array processing techniques, recursive algorithms, and adaptive array algorithms.

Chapter 1 is an introduction to the *Geolocation of RF Signals: Principles and Simulations*, hereto are referred to as GRFS, which includes a discussion on GRFS system concept, proper technical definition, and performs the classification of the GRFS systems into outdoor, indoor, air, and underwater GRFS systems, perhaps the most detailed and organized discussion on requirements of GRFS systems, GRFS system main description, a brief discussion on best state-of-the-art GRFS techniques, and finally is concluded with applications of GRFS systems.

Chapter 2 provides a review of the research, investigation, and proposal of the navigation, communications, and geolocation requirements, and capabilities of indoor, urban, suburban, global, and satellite GRFS systems. It has the most unique organization, the most extensive discussion, and the most detailed graphical illustration. It also illustrates what areas and applications are matured, what areas have scare information and what areas need special attention. In this chapter, the reader will become aware of the vastness, depth, complexity, and resourcefulness of this area of research, development, and commercialization of GRFS systems both to the military and civil users.

Chapter 3 builds upon the work already discussed in Chaps. 1 and 2. From this chapter, the reader expects to understand the finer details of RF signals that will connect the information prepared in Chaps. 1 and 2 and also later in the part II in Chaps. 4–6. This chapter includes a great discussion on *RF Signals Main Parameters*, *Best Described RF Signals*, and then discusses several candidates of RF signals for indoor, urban, suburban, global, and satellite GRFS systems.

Chapter 4 starts the part II of the book with the adaptive array algorithms for GRFS systems. A great deal of discussion on this chapter is dedicated to adaptive antenna array employing a blind adaptive algorithm which can be exploited to extract signals with unknown characteristics coming from unknown locations based only on very limited knowledge of the received signal properties. These signals may be RF sources of interference to a desired GPS signal, Mobile phone, wireless network, two-way radio, satellite TV, FM station, etc. and whose locations might be determined once these signals are extracted and illustrate the performance of the blind algorithm by comparing the extracted signals with the original signals for very simple signal designs in 2001 and 2002 and more contemporary signal designs in 2010 and the estimated signal locations with the corresponding actual signal locations up to 0° precision.

Chapter 5 incorporates the discussion of the best recursive linear algorithms for adaptive array processing which enables these algorithms and systems to be implemented in real time or near-real time. There are three main principles discussed in Chap. 5: gain in computation time, i.e., perform a computation faster; reduction of computation memory, i.e., utilize as little software and hardware resources as possible; and improvement in robustness, i.e., maintain stability.

Chapter 6 discusses adaptive array beamforming for interference mitigation for GRFS systems. Dr. Progri reveals the research and development process by demonstrating how to understand, explain, model, and simulate four most recognized adaptive array beamforming processing techniques for interference mitigation for GRFS systems which are: (1) adaptive temporal selective attenuator (ATSA); (2) adaptive spatial selective attenuator (ASSA); (3) adaptive spatial temporal selective attenuator (ASTSA); and (4) an improved adaptive spatial temporal selective attenuator (IASTSA) (or an ASTSA with restored phase); from basic diagrams to be utilized to the principle simulation examples and makes recommendations for the future final products of geolocation of RF signals.

With six chapters and a variety of topics, young or experienced professionals have many tools to analyze, model, simulate very complex RF signal models, build complex and sophisticated real-time digital signal array processing capabilities into existing systems, or propose future passive systems. *Geolocation of RF Signals: Principles and Simulations* should be a very useful tool for the Department of Defense government agencies which are looking for further research and development in the area of GRFS systems. The book is very useful for large corporations which dictate and produce future requirements for GRFS transmitters such as satellites, mobile array transmitters. The book should be an indispensable guide for small research and development companies which rely on government contracts and also on collaboration from large corporations because further discussion on this book is based upon strong and close collaboration between small and large businesses. This book should offer a unique opportunity to Ph.D. students to engage in very complex and sophisticated analysis, modeling, and cutting edge research. The material of this book can be also taught in part or as a whole in small or large conferences such as IEEE RadarCom, IEEE Globecom, ION/IEEE PLANS, and ION GNSS, etc.

This book would have been impossible without the numerous opportunities I had while working with a number of outstanding people whose name and contributions I would like to acknowledge publically.

I would like to thank my high-school math teacher Gergji Papanikolla and Fredi Fundo who prepared me to win in three mathematical, national high-school competitions at "Themistokli Gërmenji" High School in Korça, Albania from 1986 to 1989. During my undergraduate university studies I would like to express immense gratitude to my professors Jorgo Malita, Raimonda Bualoti, and especially to Niko Thomo at the Polytechnic University of Tirana, Tirana, Albania. Professor Jorgo helped me especially with my theoretical mathematical background while I was completing the proofs of the theorems of his books on Calculus I, II, III, and IV. Raimonda was the first person to introduce me to some of the recursive algorithms for solving complex linear system of equations. Professor Niko Thomo was the first to introduce me to famous book on *Mathematical Methods for Physicist* (which is referred in this manuscript) which became a great foundation for my graduate mathematical preparation from 1990 to 1994.

I would like to express in-depth gratitude to Professors Alex Emanuel, Kevin Clements, and Reinhold Ludwig in the Electrical and Computer Engineering (ECE) department at Worcester Polytechnic Institute (WPI) who helped me during my Master's thesis. Professor Ludwig pushed my mathematical–theoretical skills to the limit with his broad, difficult, and extensive homework and projects on

Electromagnetic Theory. Professor Clements enabled me to gain an extensive research and experience on solving complex linear system of equations with applications on power systems and power system analysis and software. Professor Emanuel, as my M.S. Thesis coadvisor, helped me with my writing, organization, and presentation skills just to name a few. Professor Emanuel also helped me with my principle discussion and with the methodology of the project in general, from 1996 to 1997.

My first three and a half years industry experience came from Quantum Corp. and Mayflower Communications Inc. I would like to thank Dr. T. Upadhyay, Dr. W.E. Vander Velde, and other members of the technical staff at Mayflower Communications Inc. who helped me in the area of sample matrix inversion for interference mitigation and with my analysis on mutual coupling and channel errors from 1997 to 2000.

Six and a half years of my Ph.D. studies became without doubt the most solid building block for my knowledge of signals, systems, communications, linear and numerical algebra, and geolocation background. During this time, I became without doubt the Ph.D. student in the ECE department at WPI with most publications. I am indebted to Professor William R. Michalson who supervised my Ph.D. dissertation and introduced me to systems such as GPS, GNSS, Indoor Geolocation Systems, and hardware and real-time embedded software requirements on interference mitigation techniques from 1998 to 2003. Dr. Matthew C. Bromberg became a great resource and we worked together in a few projects which are: in blind adaptive equalization, statistical adaptive array signal processing, and recursive algorithms.

Later on, I would like to thank Dr. Scott Hensley, the Tutorials Chair for IEEE RadarCon09, who enabled me to prepare my first tutorial on Geolocation of RF Signals and give this tutorial at the IEEE RadarCon09. I would like to thank Dr. Paul Rosen, General Chair of the IEEE RadarCon09 who enabled me to work with other technical members of organizing committee of the IEEE RadarCon09 in Pasadena, California in May 2009.

I would also like to thank anonymous reviewers and especially my editor Steven Elliot of Springer who have greatly enabled me to improve the manuscript from 2009 to 2010.

I would also like to thank the following organizations IEEE, ION, ComSoc and AESS, RIN which have enabled to publish some of my early research work in the area of geolocation of RF signals from 1998 to 2010.

Finally, I would like to thank my mom, Lumturi, my dad, Fiqiri, my sister, Ana, Mrs. Elizabeth Demir, and Dr. Peter Demir who have been very supportive while writing this book and throughout my professional career.

To conclude, I would like to thank first my corporation Giftet Inc which I hope will greatly benefit from this publication with future R&D work, proposals, and contracts; and second, the public and the readers who will buy, read, and refer from this book. Their comments and suggestions will be considered the most valuable asset for future editions and further investigations and studies.

Worcester, MA Ilir Progri

Contents

Chapter 1
Introduction to Geolocation of RF Signals

Abstract *Geolocation of RF Signals: Principles and Simulations* offers an overview of the best practices and innovative techniques in the art and science of geolocation over the last 20 years. It covers all research and development aspects including theoretical analysis, RF signals, geolocation techniques, key block diagrams, and practical principle simulation examples in the frequency band from 100 MHz to 18 GHz or even 60 GHz. Dr. Progri reveals the research and development process by demonstrating how to understand and explain GRFS systems from basic diagrams to the final principle simulation examples and make recommendations for the future final products of GRFS systems. Starting with RF signals, the book progressively examines various signal bands – such as VLF, LF, MF, HF, VHF, UHF, L, S, C, X, Ku, and K and the corresponding geolocation requirements per band and per application – to achieve required performance objectives of up to 0° precision. Next follows a step-by-step approach of RF geolocation techniques and the book concludes with notes on state-of-the-art geolocation designs as well as advanced features found in signal generator instruments. The book also includes the best mathematical techniques employed for geolocation of RF signals at 100 MHz–18 GHz or even 60 GHz. The principle simulation examples which are discussed in great detail during the second part of the book in Chaps. 4–6 offer invaluable insights – all-in-one source for the beginner, the experienced, expert analysts, and professionals.

1.1 Geolocation of RF Signals Systems

A geolocation of RF signals system, just like any other system, requires a proper technical definition; its utilization is indispensable in many facets of life; it has a historical background and moment of conception in the past; it has a progression in the present time; it has a certain field of applicability and vision towards the future (see Fig. 1.1) (Chap. 1 of [1]).

The book gathers a unique collection of block diagrams, signal diagrams of power spectrum descriptions, principle receipts, and "principle secrets" usually treated as other hard-to-find information. All the best known geolocation

I. Progri, *Geolocation of RF Signals*,
DOI 10.1007/978-1-4419-7952-0_1,

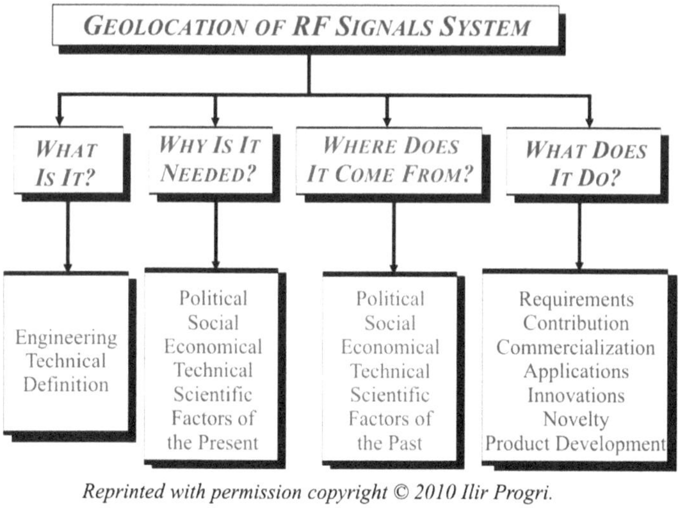

Fig. 1.1 An overview of a geolocation of RF signals system. Reprinted with permission © Ilir Progri

techniques are well illustrated with practical examples awaiting implementation in future military industrial products. Written in a refereed journal style, the book is an invaluable treatment of all aspects of RF geolocation – all-in-one source for the beginner engineer and the more experienced professionals.

The main purpose of this book is to research, investigate, and propose the navigation, communications, and geolocation properties, requirements, and capabilities of several candidate radio frequency (RF) signals in the entire frequency band of 100 MHz–18 GHz in open outdoor, suburban, urban, and indoor environments and also in closed indoor environments. In order to accomplish this effectively, we will (1) introduce the best state-of-the-art geolocation of RF signals techniques in Chap. 1; (2) address the requirements of systems for geolocation of RF signals in Chap. 2; (3) extensively describe RF signals in the desired frequency spectrum of 100 MHz–66 GHz in Chap. 3; (4) describe in great detail blind geolocation of RF signals in Chap. 4; (5) address the computation complexity for successfully achieving the desired objectives in signal processing starting with recursive algorithms such as Cholesky and MGSO in Chap. 5; (6) address the recursive generalized eigen-value solution in Chap. 5; (7) investigate certain applications of geolocation of RF signals systems, which include signal suppression and interference mitigation applications starting with a GPS adaptive temporal selective attenuator in Chap. 6; (8) investigate an adaptive spatial temporal selective attenuator in Chap. 6; (9) investigate an improved adaptive spatial temporal selective attenuator in Chap. 6; and conclude our book with a chapter on summary and conclusions in Chap. 6.

Chapter 1 is organized as follows. We briefly provide the history and background of geolocation of RF signals systems in Sect. 1.2. Next, we discuss the geolocation of RF signals system concept in Sect. 1.3, which includes the domain, the environment, and the methodology. We also illustrate under ideal conditions the generic principle of operations of a geolocation of RF signals system. In Sect. 1.4, we provide a proper technical definition and perform the classification of the geolocation of RF signals systems into outdoor, indoor, and underwater geolocation of RF signals systems. Requirements of geolocation of RF signals systems are assessed in Sect. 1.5. The geolocation of RF signals system's main description is provided in Sect. 1.6. This description includes the RF signal emitters with unknown location and frequency; RF geolocation channel; and the geolocation of RF signals receiving unit which typically can be an array of antennae. In contrast to indoor geolocation or outdoor geolocation systems which are designed to operate with known signals, i.e., the transmitter and receiver are designed to yield maximum signal reception within the domain and the environment, in the case of geolocation of RF signals we have absolutely no idea of the signals we are receiving and also of the distance of the RF sources. Therefore, it is expected that the complexity and engineering methodology in designing geolocation of RF signals systems is several orders of magnitude higher than those of indoor geolocation systems. This is the reason why geolocation of RF signals systems has maximum performance achievable of any system under normal conditions and yield up to 0° angle of arrival (AOA) estimation precision accuracy. The best state-of-the-art geolocation of RF signals techniques (or algorithms) are discussed in Sect. 1.7. Applications of geolocation of RF signals are discussed in Sect. 1.8. A generic geolocation of RF signals system is provided in Sect. 1.9. Chapter 1 is concluded in Sect. 1.10.

1.2 History and Background

For the complete discussion on the history and background on geolocation, the reader should refer to [1] and also Dr. Progri's upcoming book on *Indoor Geolocation Systems: Theory and Applications.*

1.3 Geolocation of RF Signals System Concept

There are three main components that constitute any geolocation system: (1) the *domain*, (2) the *environment*, and (3) the *methodology* as shown in Fig. 1.2 [1].

The *domain* of any geolocation of RF signals system is a well-defined space and time coordinate system which is also defined as the reference system. For example, the World Geodetic System 1984 (or WGS'84) is an Earth-Centered-Earth-Fixed (ECEF) coordinate system [1]. The WGS'84 system is fixed with respect to the Earth, but it is moving with respect to the Sun. Another well-known reference system is an Earth-Centered-Inertial (ECI) and, in general, the Earth is not fixed with respect to this system [1]. A space–time reference system can also be a local system such as a North East Down (NED) frame, which is widely used for local positioning and navigation and also geolocation of RF signals.

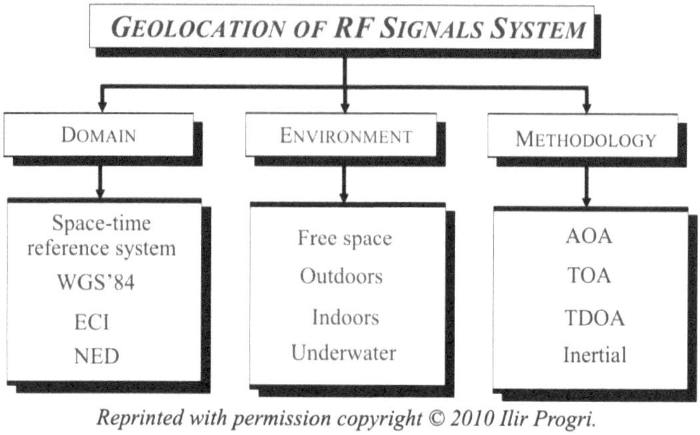

Fig. 1.2 Geolocation of RF signals system's main components: domain, environment, and methodology. Reprinted with permission © Ilir Progri

The *environment* is the set of local or global physical properties of the medium of operation of any geolocation of RF signals system. For example, the free-space physical properties of the medium are different from those in the air, from those on the surface of the Earth, from those inside buildings, and from those underwater and so forth [1].

The *methodology* for achieving geolocation of RF signals is a set or system of techniques, methods, principles, analysis, and rules for regulating, mapping, or interconnecting the appropriate theoretical approach (or the idea or concept or algorithms) to the appropriate geolocation of RF signals systems application. The domain, the environment, and the methodology form *the art and science* of geolocation of RF signals systems.

For example, commonly studied radio techniques for geolocation of RF signals systems are the AOA positioning, the *time of arrival* (TOA) positioning, and the *time difference of arrival* (TDOA) positioning. The TOA and TDOA techniques become applicable methodologies only after the invention of the radio in the beginning of the twentieth century. For example, satellite-based geolocation systems use TOA for localization and Loran C and mobile phone positioning systems employ TDOA for doing the same thing [1].

Now that we know which are the main components of a geolocation of RF signals system, we shall illustrate its concept of operations. For that, we have assumed that the domain is a hypothetical ENU frame, the environment is outdoors, and the methodology is AOA estimation. In this case we have obtained a geolocation of RF signals system, which is shown in Fig. 1.3.

There is one main RF source which is a military airplane and the geolocation of RF signals source consists of a passive linear 16 element array with the elements directions and formation as shown in Fig. 1.3. This passive element array forms a beam with a pattern as shown in Fig. 1.3. This particular passive wideband array is capable of driving a minimum of sevens deep nulls while maintaining gain in a

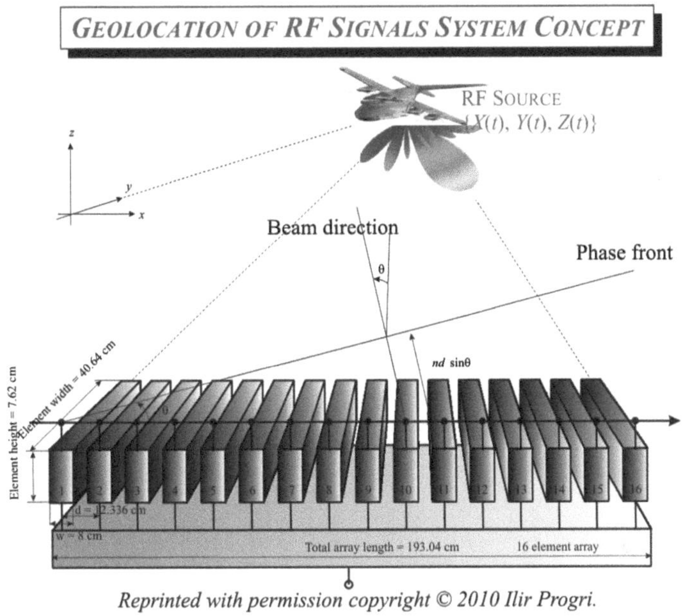

Fig. 1.3 A generic geolocation of RF signals system diagram. Reprinted with permission © Ilir Progri

commanded direction when the antenna outputs are processed via algorithms such as MUSIC as we will explain later in the book.

The important aspect to recognize at this stage is that there is a way of electronically commanding or steering the gain on a particular direction of the array. Therefore, we will define the AOA or the steering angle θ as the angle between the direction of the source and the angle between the direction of the maximum gain of the array or the phase front.

1.4 Technical Definition and Classification

Geolocation of RF signals is defined as the problem of precise localization (or geolocation) of spatially separated sources emitting electromagnetic energy in the form of *radio signals* within a certain frequency bandwidth by observing their received signals at spatially separated sensors (or array elements) of the geolocation of RF signals system (taken and modified from [1]). *Geolocation of RF signals* is of considerable importance occurring in many fields, including radar, sonar, mobile communications, radio astronomy, seismology, unmanned air vehicle (UAV) for intelligence gathering information, emergency and rescue personnel, mining and agriculture, drilling, aviation, ground transportation, naval, etc.

"*Radio* is the practice or science of communicating over a distance by converting localization information into electromagnetic waves and transmitting these directly through space, without connecting wires, to a receiving set, which changes these into signals appropriate for performing localization" [1]. A [*radio*] *geolocation of RF signals* system is a navigation system which continuously receives *radio* signals from spatially separated transmitters and utilizes a *radio* receiver with N array elements to resolve at a minimum the DoA of $N-1$ spatially separated sources (see Figs. 1.3 and 1.4).

Since we are interested in radio geolocation of RF signal systems, the coin term *radio* is understood to mean a *radio system*, a *radio transmitter*, a *radio receiver*, or a *radio channel*, and therefore, it is under-emphasized and only used on special occasions. Based on the application *environment*, a geolocation of RF signals system can be classified into three main categories: *outdoor*, *indoor*, or *underwater geolocation* of RF signals *system* (see Fig. 1.5).

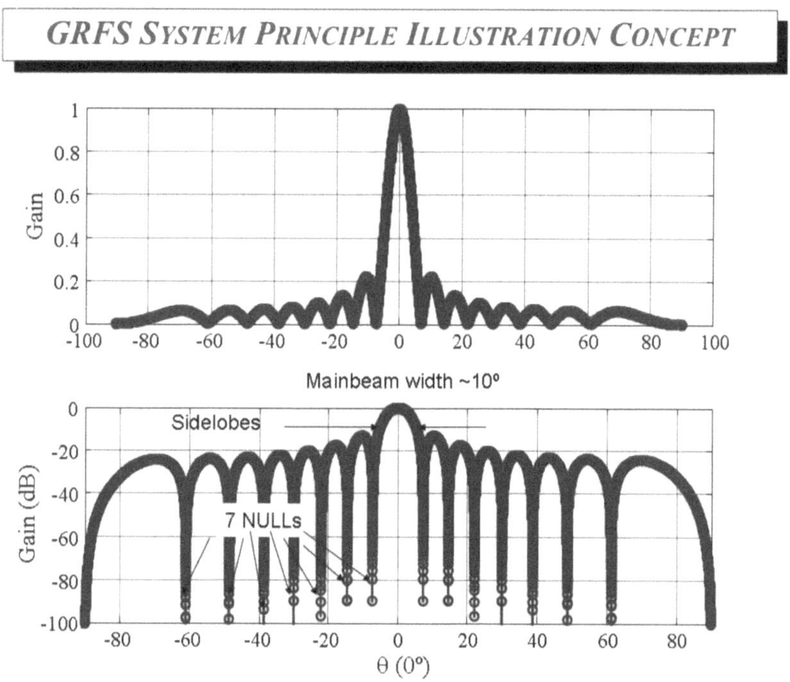

Reprinted with permission copyright © 2010 Ilir Progri.

Fig. 1.4 Normalized aperture (absolute units) and (dB) vs. angle (θ°) of a 16 element linear array showing a mainbeam width of ~10° and 7 deep NULLs on the order of −80 dB. Reprinted with permission © Ilir Progri

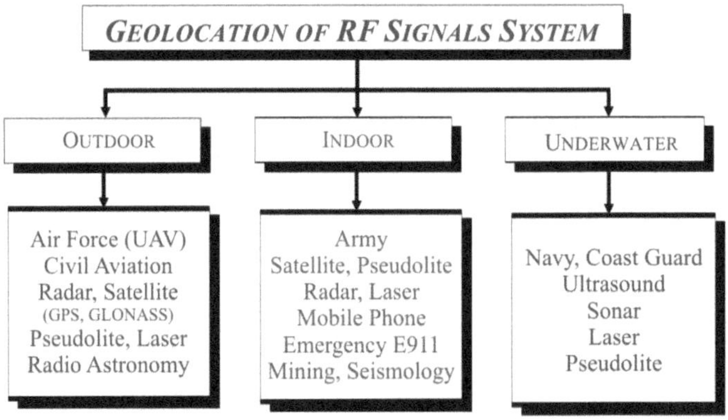

Fig. 1.5 Technical classification of geolocation of RF signals systems. Reprinted with permission © Ilir Progri

1.5 Requirements of Geolocation of RF Signals Systems

The main requirement of GRFS systems is to provide precise real-time direction of arrival and distance of any RF source in the frequency band of 100 MHz–18 GHz or even 60 GHz from the receiving antenna array to within 0° AOA and cm level accuracy for distance measurements assuming that the signal structure of the received signal is largely unknown.

In the companion book "*Indoor Geolocation Systems: Theory and Applications*," Dr. Progri discuss how we can achieve cm level position and cm/s velocity accuracy 99.9% of the time from a pseudolite-based indoor geolocation system from several proposed signal designs (or structures) such as C-CDMA, OFDMA, and MC-CDMA. It is anticipated that the "Geolocation of RF Signals: Principles and Simulations" is a much tougher problem than "*Indoor Geolocation Systems: Theory and Applications*," although both are based on the same principle of operations: angle measurements and distance measurements.

1. *Accuracy*: is the measure of correctness of the estimated position of the RF source from the actual position (or location of the RF source). It can be estimated either based on the direct line-of-sign (LOS) measurements such as distances or as the direction of arrival measurements. The more accurate the direction of arrival measurements, the more precise the accuracy. It is desired that up to 0° error for AOA precision accuracy for identifying RF signals of interests and cm level position or distance measurement.
2. *Adaptation/reconfiguration*: it is the internal capability of these systems to adaptively reconfigure themselves to account for future changes of the signal designs and also of the user needs [2].

3. *Affordability*: it is defined as the amount of investment (in US$ or other currencies for other countries) that is required to execute a proper business plan to produce a prototype or sustain the broadcast of RF signals or sources for a certain period of time. Geolocation of RF signals systems should be affordable with the current commercial of the shelf hardware or should be built using custom hardware.
4. *Area of coverage*: it is defined as the area in which its diameter is equal twice the maximum effective range. It is desired that the area of coverage is from few meters to several kilometers, or even several hundreds of kilometers, or even maybe several thousands of kilometers. This is also a function of the frequency for wireless local area networks and for satellite communications and geolocation. For example, the 60 GHz band is unsuitable for long-range (>2 km) communications; therefore, this part of the spectrum can be dedicated entirely to short-range (<1 km) communications [3]. For a detailed description of the area of coverage, the reader is reminded to look at the following section which is the main description section.
5. *Authenticity/originality/relay-based sources:* it is defined as the internal characteristic of the RF signal or source that distinguished it from other sources or other replicas of itself. There are two kinds of RF sources: (a) authentic (original, genuine, real, actual, veritable) RF sources; and (b) relay-based (fake, duplicate, replica, in-genuine, imaginary, not actual, ghost) RF sources [4]. In this publication we are going to assume that all the signals are authentic or genuine, i.e., coming from directly original RF sources. This is a very important and delicate problem; distinguishing and differentiating signals from relay-based (or replica, or duplicate) RF sources will require a much higher level of sophistication and complexity on the system analysis, concept, and overall design and with substantial amount of work which in many respects might be considered as very sensitive and proprietary information which can only be disclosed to the US military and the Department of Defense (DoD) agencies. (For example in the Movie ("Under Siege 2") the lives of the people in Washington, DC depended on the quick and accurate detection, differentiation, and distinction of "Ghost" Satellites as opposed to the real "Grazer 1" which the DoD wanted to eliminate, which was a weapon space-based satellite system that was designed by the DoD to be invisible by all other nations.) Imagine how difficult it is to detect, differentiate, and geolocate authentic RF sources when we have an environment with many "ghost" signals which we have not designed.
6. *Bandwidth*: is defined as the effective or useful spread of the RF signal in the frequency domain. It is desired that the bandwidth can vary from 10 MHz to even 100 or several hundreds of MHz. For ultrawide band signals or signals with wide bandwidth, it is certainly desired to require very high bandwidth at the RF front-hand and also for signal processing to overcome the problem of poor coverage at higher frequencies [1]. This concept is discussed in great detail in Chap. 3.
7. *Bandwidth utilization*: In contrast to *Indoor Geolocation Systems: Theory and Applications* in which we have control over the transmitted signals, here in

geolocation of RF signals we have no control on the transmitted signals and on the ways signals have been modulated. This requirement will be exploited to the extent that we perform adaptive array processing. More discussion about this requirement is provided in Chaps. 4 and 6.

8. *Challenges/issues/impairments/obstacles*: it is an array of internal, external, economic, technical, and environmental factors that will impede the progress, performance, operation, production, manufacturing, etc. of RF signals and sources. There should be an intelligent enumeration, modeling, analysis, discussion, assessment of challenges, issues, impairments, and obstacles associated with geolocation of RF signals systems.

   The main impairments or errors for geolocation of RF signals in the measurements of location metrics are *multipath fading* and *no line-of-sight* (NLOS) conditions due to shadow fading [5]. There are four major phenomenon influencing the wireless communications channel: (1) *multipath fading*, (2) *multipath delay spread*, (3) *multipath cochannel interference* [6], and (4) the *realistic number of RF sources* in the environment. While there is a lot of discussion in the literature showing how to handle the first three, I have found very little discussion on the last one with exception of my Ph.D. Dissertation [1]. Channel state information (CSI) should also be considered as not available from the transmitter point of view. Perhaps for *Indoor Geolocation Systems: Theory and Applications*, we might consider CSI as being available for certain applications. However, we should mention that the accurate determination of the DOAs should provide enough information on the condition of the CSI. For example, we will discuss how MIMO-based spatial multiplexing technology and coded OFDM turn multipath delay spread into a benefit [7]. Other impairments, which are implementation specific, include *I/Q mismatch, phase noise, carrier frequency offset*, and *sampling frequency offset* [8, 9]. The grand theme of the first edition is to focus on the principles and simulations of geolocation of RF signal systems as depicted in Chaps. 4 and 6. Perhaps the second edition of this book and *Indoor Geolocation Systems: Theory and Applications* will contain more implementation and application specific assessment of these important impairment parameters. Other future effects to be studied and analyzed are *spatial correlation, various AOA distributions, mutual coupling, near-field scatterers* (NFS), and *channel capacity* [10].

9. *Commercialization strategy/market/manufacturing/sales*: is defined as a portion of the business plan that enables the production, manufacturing, sales, ownership of RF signals, and sources. These requirements should be discussed in a standalone product development and marketing as part of the company's business plan. Although we may use certain market data to enable a better understanding of certain technical parameters for the most part, these requirements will not be discussed in this book.

10. *Compatibility*: refers to that property of RF signal design and GRFS systems that ensures coexistence of GRFS systems, principles, signal design, components, and technologies that enables as little changes as possible to ensure

that one GRFS system is capable of performing the same tasks as another GRFS system. Compatibility is considered an important factor throughout, the design, deployment, test, and operation of Galileo; modernization of GPS and GLONASS; GPS-III, GPS IV, QZSS, etc. This is the reason why Chapter 3 has an extensive and detailed discussion on compatibility of RF signals.

11. *Diversity*: is defined as the dissimilarity in ingredients of GRFS systems and sources that produce dissimilar RF signals. Geolocation of RF signals should make good use of different types of diversities (or the dissimilar ingredients) such as array, space, time, frequency, bandwidth, capacity, phase, and polarization diversities to mitigate the effects of fading, delay spread, and cochannel interference [6, 11]. Antenna array diversity and adaptive antenna arrays, which mitigate fading and cochannel interference, have been in commercial use at the wireless base stations for many years. Other important diversities include space–time receivers, space–time coding, and space-division multiplexing (SDM) systems aiming at the wireless broadband access for performance improvement [6]. Other types of diversities that are not discussed or exploited are bandwidth and capacity diversities. Bandwidth diversity may be exploited as a mechanism to combat interference and jamming from the signal's design point of view. To a certain extent, this technique is exploited in GPS in which the ephemeris data are available both at the C/A ~ 2 MHz and P(Y) code ~20 MHz. Capacity diversity, on the other hand, can also be exploited to increase system resistance against cochannel interference and jamming. The simplest way of increasing capacity is by adding more antenna elements. Bandwidth diversity will be discussed more in the *Indoor Geolocation Systems: Theory and Applications* book. When strong interference is present, diversity alone is not enough to combat interference and jamming. For these cases, smart antennas and adaptive antenna arrays should be considered to combat interference and provide accurate geolocation information [6]. Much more discussion on the adaptive processing will be provided in the following chapters.
12. *Expendability*: is the internal quality of GRFS systems that are not worth maintaining because better systems and technologies come up. GRFS systems should become either partially or fully expandable (if they are not partially or fully upgradable) as a result of agreements among the participating parties. Expendability is also related to the compatibility and interoperability.
13. *Flexibility*: it is defined as the ability of GRFS system to respond to various present or future changes in the requirements or specifications [12]. This requirement is achieved if intelligence is built into the architecture of the GRFS systems which will enable them to achieve all other requirements. Reference [12] illustrates how flexibility relates to other requirements such as run-time adaptation and reconfiguration of the WIND-FLEX transceivers. The approach we are aiming is to illustrate how flexibility relates to other system requirements from the principle simulation examples. Furthermore, flexibility of a GRFS system depends on the number of independent parameters that can be controlled and adapted during the system design phase or reconfigured

during the geolocation phase with the aid of advanced implementation techniques facilitated by means of software define radios [13]. Because this is such an important definition in Chaps. 4 and 6, we are going to refresh our material taking into consideration the range of all the independent parameters that constitute the degrees of freedom of the GRFS system under consideration and also focus on a methodology for selecting the optimum set of these parameters to yield the optimum (or highest possible) performance.

14. *Friendliness or user's convenience*: is that internal quality of RF signals and sources that enables its acceptance from various groups or population of users. The description, modeling, analysis, discussions, results, conclusions, and functionality must take into consideration user convenience and must allow for ease of use and friendliness of usage.
15. *Frequency of operation/frequency reuse*: it is defined either as the center or reference frequency of RF signals or source. Chapter 3 has a detailed discussion about this parameter. It is desired that the frequency of operation is from 100 MHz to 18 GHz or even 66 GHz for different IEEE 80211, IEEE 802.15, IEEE 802.16, and IEEE 802.20 standards, etc. At these frequencies, we are going to discuss several signals and systems such as VLF, LF, MF, HF, VHF, UHF, L, S, C, X, Ku, and K and the corresponding geolocation requirements per band and per application – to achieve required performance objectives of up to 0° precision. On the other hand, frequency reuse can increase data capacity without compromising range by means of sectorization and polarization in that the use of orthogonally polarized channels can effectively double the number of available channels [14].
16. *Global convergence*: it is a capability of GRFS systems to identify, sense, process, and converge to a global solution of all RF signals in the environment and also to simultaneously differentiate and compute all RF signals and their respective directions of arrival and distances. It is desired that global convergence is also another important characteristic of all the geolocation of RF signals systems.
17. *Integration*: it is defined as the capability of GRFS systems to be integrated with other systems, components, hardware, software, firmware, etc. GRFS systems should also allow for integration of both navigation and communication services for mutual benefits for both services [2]. Although provision will be given to the accuracy of the geolocation techniques, it is readily implied that geolocation information should be made readily available to communication service payloads.
18. *Interference*: is defined as the unwanted (or unwelcomed) in-band or out-of-band RF signals or their secondary effects. A good understanding of interference is devoted in the book. Interference is classified as: (1) *unintentional interference*; (2) *intentional interference*; and (3) *hostile jamming*. For example, multipath should also be classified and treated as unintentional interference. Sometimes multipath can be useful for determining the direction of arrival of the RF source when no direct line-of-sight signal is available; in other cases, multipath can be detrimental when it tries to deteriorate the line-of-

sight signal. In the later case, multipath acts as unintentional interference to the desired signal path. Another type of unintentional interference comes as a result of the multipath effect of signals coming from other RF sources. This type of interference is known as *multiaccess interference*. Depending on the nature of the RF source, this type of interference can be classified as: (1) *unintentional interference*; (2) *intentional interference*; and (3) *hostile jamming*. Chapters 4 and 6 deal specifically with this qualitative and quantitative analysis, modeling, simulation, and ways to mitigate this type of interference.

19. *Interoperability*: Refers to a property of GRFS systems ability to work together with other telecommunications devices taking into account RF signals spectrum as defined by the International Telecommunications Commission (ITU) Spectrum Monitoring handbook, all possible interfaces, and other factors that impact GRFS systems performance [15, 16]. GRFS systems should be interoperable with all the existing systems because they are considered passive systems. For example, GRFS systems should be interoperable with Mobile phones, GPS receivers, cordless phones, and FM radios, just to name a few telecommunications technologies that are approved by the ITU [15, 16]. Interoperability is further discussed in Chap. 3.
20. *Mobility*: is that characteristic of GRFS systems that takes into consideration relative RF sources and/or users' motion (or movement). For the most part, GRFS systems should work well on stationary cases, i.e., if no Doppler and for zero user velocity or no mobility. In reality, movements of portable stations as well as movement of nodes (such as pseudolites, base stations, mobile stations, etc.) cause Doppler effects as frequency shifts and spectrum broadening of the received signal [3]. Doppler effects at 60 GHz are very severe because they are proportional to the frequency. For example, if a node moves at a speed of 1.5 m/s (walking speed), the Doppler spreads that results at 60 GHz is 1,200 Hz [3]. Since we are here, we should also mention that for this book we are not concerned with the attitude information, i.e., we are only interested in finding the AOA or the direction of arrival of the transmitting antenna array of the transmitting unit. Future research on this subject should consider attitude information.
21. *Modeling*: the representation, often mathematical, of RF, IF, baseband, or other intermediate processes, concepts, or operations of GRFS systems, often implemented by a computer program such as MATLAB, Simulink, etc. Modeling should be seen as the most important requirement for accurate representation of RF signals, the environment, and also the processing that occurs at the receiving unit. The refinement of the existing and the development of the new RF signal models, channel models, and advanced receiver signal processing should be seen as the core of the modeling requirements for this book.
22. *Modularity*: the use of individually distinct functional units, as in assembling or reproducing a GRFS system in hardware, software, firmware, etc. GRFS should be built as a modular system making use of the software-defined radio technology [17]. Modularity is also related to expandability, compatibility, interoperability, and reusability, just to name a few requirements.

23. *Number of antennas/number of sources*: it is the number of identical RF antennas of the GRFS receiving unit which is directly linked to the number of RF sources. It is desired that M be a large number of antennas (or antennae). Very simple adaptive algorithms which are based on M receive antennae should be able to differentiate or discriminate up to $M-1$ RF sources. On the other hand, exploiting the knowledge of training sequence and/or properties of received signals (e.g., constant envelope, finite alphabet, cyclostationarity (see Chap. 4)) allows for increased number of detection and differentiation of RF sources that share the same available recourse (e.g., time, frequency, codes) [18].
24. *Physical dimensions, power, size, and weight*: these requirements are important; however, they are not considered here because this book deals primarily with principles and simulations, just to give an idea about the weight, size, and power of these systems. The size will vary from a very small portable handheld such as a mobile phone device to an array of radio telescopes, which will occupy hundreds of square kilometers. Conversely, the weight will vary from a few grams to hundreds of thousands of tons; and the power will vary from a few mille watts to several MW or several hundred of MW.
25. *Privacy/safety/security*: GRFS systems have their sole purpose to provide a passive, very accurate, very affordable, very reliable, very secure and safe, and so forth solution for geolocation of RF signals of interests within the effective range of operations. If an attack comes as an RF threat such as a deliberate interference or jamming, then GRFS systems should be able to identify those threats. Moreover, through extensive processing, identification, tracing, and tracking, etc., GRFS systems should be able to provide solutions to many concerns on security. Chapter 2 is dedicated to provide an extensive overview to the security, safety, and privacy concerns associated with these amazing technologies.
26. *Processing*: perhaps the most important assumption and requirement for GRFS systems is that processing occurs only at the receiver, i.e., these systems are considered as receiver processing only. Because the signals (i.e., the RF sources that transmit those signals) that we are trying to geolocate are man-made signals or signals transmitted from manmade devices, we are going to discuss those signals in the context of helping us understand the analysis, modeling, processing, design, and implementation of the receiver's unit. All the techniques, all the methods, and all the wireless systems that discuss receiver processing only could potentially be applied to GRFS systems by keeping in mind that here we are interested in obtaining the direction of arrival and AOA and then of the distance for the sources that have transmitted those signals. This is the reason that the majority of the material is devoted to processing or to the array and signal processing to obtain geolocation information for the AOA of the RF sources. We are going to devote very extensive discussions on Chaps. 4 and 6 with the main purpose to refresh the material in those chapters, to add more up to date references for those chapters, to add more up to date modeling of those techniques and systems, to add more up to date principle simulation examples for those chapters, and to better connect the material with the rest of the book and also with the rest of the material currently

reported in the literature. For example, for the *highest sampling rate of wideband applications,* field programmable gate arrays (FPGAs) are still an attractive implementation technology as opposed to traditional DSPs and general purpose processors (GPPs) which might become valuable alternatives for low-sampling rate wideband applications [19].

27. *Reliability*: is an attribute of GRFS systems that should consistently produce the same accurate geolocation results, preferable meeting, or exceeding the other requirements that we have discussed so extensively and thoroughly. Reliability is of two kinds: (a) *coverage reliability* and (b) *link* or *channel reliability*. At this state of the research and for this particular publication, we are going to assume that reliability is not an issue because the focus of this book is on *Geolocation of RF Signals: Principle Simulation* examples and not so much on the effects of the environment and on the nature of the RF transmitters. On the second edition of this publication we would like to consider the reliability for a more realistic prototype or several realistic prototypes of geolocation of RF signals systems.
28. *Scalability*: the ability of GRFS systems to adapt to increased demands, such as to different environments, bandwidth, bit rates, and all possible RF signals of interests in the frequency band of 100 MHz–66 GHz considered simultaneously. With an increase of data and signal density (data rates and area of coverage), a scalable GRFS system architecture is an essential parameter for future implementations [14]. It is suggested that a proper selection of the modulation scheme has a significant impact on the network scalability because modulation can be altered to support higher data rates [14]. For example, OFDM, OFDMA, QPSK, and higher-order (or M-ary) QAM (in IEEE 802.16 standard) have been successfully used in high-bit-long-range scalable applications [14].
29. *Signal/system description*: the intelligent act(s), language(s), interface(s), process(s), drawing, or technique(s) of describing, representing, picturing, etc. GRFS systems to a technical audience. It is desired that GRFS systems contain one main antenna array section, one main RF front-end section, and also one main baseband signal processing section. Details on the description of the RF front-end will be given in the following section. Chapter 2 has a detailed discussion on the description of GRFS systems; Chap. 3 considers RF signals to a reasonably good detail and depth; and Chaps. 4 and 6 are dedicated on the descriptions of the baseband signal processing of geolocation and beamforming sections and techniques.
30. *Synchronization*: is the adjustment of certain internal signal or device's periodicities of two or more electrical or mechanical devices so that the incoming RF signal periods are equal or integral multiples or fractions of generated RF signal. For the most part, geolocation of RF signals systems under consideration should not be affected by synchronization as part of the novel, adaptive, global, and "blind" channel estimation technique. Chapter 4 discusses in great detail our technique and we have demonstrated the effectiveness of our technique without relying on synchronization. For most state-of-the-art and most

legacy systems, synchronization is an issue that is usually preceded by a synchronization (*sync*) slot for timing phase, timing frequency, and frequency offset estimation [8, 9]. In *Indoor Geolocation Systems: Theory and Applications* we are going to discuss this issue because for these systems we are interested in distance estimation and accurate, real geolocation coordinates. However, one of the techniques discussed in Chap. 6 requires proper synchronization to achieve full optimality.

This concludes the requirements of GRFS systems section. We do not rule out other requirements as this material is considered for further expansion and as other recommendations are provided from the technical readership.

1.6 Geolocation of RF Signals Main Description

Figure 1.6 illustrates the geolocation of RF signals systems' main description range and field of operation. As illustrated in the Fig. 1.6, geolocation of RF signals systems' main description contains five different segments: (1) indoor segment of operation; (2) urban segment of operation; (3) suburban segment of operations; (4) global segment of operation; and (5) satellite segment of operations.

The *grand vision* of the geolocation of RF signals' main description is that we are striving for a *ubiquitous Geolocation of RF Signals systems capability* from the

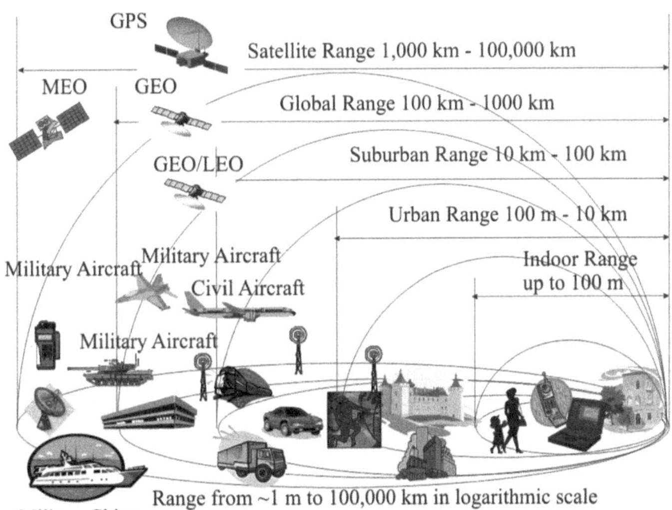

Fig. 1.6 Geolocation of RF signals' main description. Reprinted with permission © 2010 Ilir Progri

indoor segment to the satellite segment in *three major steps*. The *first major step* is the *grand vision* of this edition which is a methodical, academic, journalistic approach of the description, analysis, modeling, simulations, discussions, documentation, and publication of the RF signals of interest, system requirements, geolocation of RF signals techniques, methods, etc. in which the location of the user is assumed known. The *second major step* is the *grand vision* of *Indoor Geolocation Systems: Theory and Applications* in which we are going to design RF signals and systems to provide very accurate location of the user 99.999% of the time. The *third major step* of our *grand vision* will be a new publication perhaps such as *Ubiquitous Geolocation Systems: Theory and Applications* in which we are trying to achieve the requirements of both steps only from the user point of view. This is by all accounts a very ambitious vision and very daunting task; nevertheless, we are taking the very first major step in this grand vision and I am certainly very proud to make a contribution to the navigation and geolocation community. The reader should be able to visualize by now that *ubiquitous geolocation* is a much tougher problem than *ubiquitous communications*, i.e., *ubiquitous communications* is a necessary condition for *ubiquitous geolocation* and *ubiquitous geolocation* is a sufficient condition for *ubiquitous communications*. Another way of expressing this is as follows: while we might be close to visualizing a world coverage network such as 4G networks to achieve worldwide global and ubiquitous communications, we are virtually very far from achieving worldwide coverage ubiquitous geolocation, let alone in terms of GRFS systems.

Coverage is defined as the *effective range* for which geolocation of RF signals systems can reliably *detect*, *discriminate*, and *geolocate* wireless users belonging to a particular type [7]. For example, since cordless phones' effective range is only indoors (~10 m), we would expect that the coverage area for cordless phones is indoors to within their range (~10 m) and so forth. Outside of the coverage area, other signals can be treated as pure background RF noise signals. In Chap. 3 in which we try to make a detailed discussion on the RF signals of interests, by the same token, we also discuss the environment with regard to the coverage area, i.e., a classification of their effective range. Due to the enormous complexity of the signal processing that takes place on the receiver for geolocation of RF signals systems, we are going to defer the wireless channel analysis, modeling, simulation, assessment, and impact until the separate publication of *Indoor Geolocation Systems: Theory and Applications*, which will be published after this book.

One very interesting link between coverage and the frequency of operation is that the geolocation of RF signals systems in the 2–11 GHz band should include all the ranges from indoor to satellite range and therefore will provoke a greater discussion and we are going to devote a great length of the material in the book because it is driven by the need for nonline-of-sight operation [20, 21 (more to be discussed in Chap. 3)].

However, geolocation of RF signals systems in the frequency range of 10–66 GHz should include a line-of-sight necessity in suburban, urban, and indoor areas as these signals are characterized by very high data rates and quite short effective range as the result of rain and foliage attenuation [20, 22].

Chapter 3 contains a detailed description and discussion on the RF signals of interests such as OFDM, OFDMA, etc. For example, IEEE 802.16 specifies two types of OFDM signals: one simple identified as OFDM and the other as OFDMA. OFDM aims at less challenging application, quite short range, and eventually indoors [20, 22]. OFDMA, on the other hand, can be utilized for indoors, urban, and suburban ranges.

Home wireless networking and home geolocation of RF signals are perhaps the most attractive approach for the home due to the elimination of the cost of wire pulling and the challenges of using the existing wires ([23–25] and (Chap. 3)). In order to give a brief summary of the existing and future competing technologies and associated standards and advocacy groups, we have the following:

IEEE 802.11 at 2.4 GHz (in 2002–2008), 5 GHz (today); 10 Mb/s (in 2002–2008), 54 Mb/s (today); QoS support.

HomeRF at 2.4 GHz; ~10 Mb/s; QoS support

Bluetooth at 2.4 GHz; ~1 Mb/s; QoS support

HyperLAN at 2.4 GHz (in 2002–2008); 5 GHz (today); 10 Mb/s (in 2002–2008), 54 Mb/s (today); QoS support

Ultra wideband at 3–6 GHz; 100 Mb/s; QoS support

Bluetooth connects to wireless devices without relying on the line-of-sight communications through walls or other nonmetal objects within a 10-m effective range (or even up to 100 m if the transmitter's power is increased) [26]. For this reason, the most useful Bluetooth applications are a mobile phone in a pocket or a briefcase asking as a modem for a laptop or PDA [26].

For the satellite segment of operation, the global standard for the Universal Mobile Telecommunications System (UMTS) was in the frequency bands of 1,885–2,025 and 2,110–2,200 MHz, with only a subband of 30 MHz reserved to the satellite component [27]. It is suggested that for mobile broadband satellite services, the feeder links should operate in the Ka (20–30 GHz) and extra high frequency (EHF) (40–50 GHz) via low Earth Orbit (LEO at altitude elevation ranging from 1,500 to 6,000 km) and Medium Earth Orbit (MEO at altitude elevation at approximately ~37,500 km) [27]. (I had a hard time understanding the numerical results of this paper [27]. Therefore, I think that other numerical references need to be looked at to come up with more conclusive numerical results.)

A good description of the frequency allocation for evaluation of aeronautical communications for personal and multimedia services is given in [28]. Further details of the spectrum allocation will be given in Chap. 3.

1.7 Best State-of-the-Art Geolocation of RF Signals Techniques

Best state-of-the-art geolocation of RF signals techniques provide an independent measure of the TOA bases techniques because it is difficult to accurately measure AOA, DOA, and RSS [5]. Another geolocation of RF signals technique for the purpose of geolocation is to provide a visualization of the possible RF source instead of an estimation of RF source location coordinates. In these cases, a possible region of mobile location will be determined from both geometric and statistical triangulation algorithms [5].

Other geolocation of RF signals techniques such as *pattern recognition* (also known as *location fingerprinting*) which utilize premeasurement-based location information for quasi-stationary mobile nodes (or stations) work better than traditional methods and Kalman filter-based tracking techniques [5]. The major drawback of these techniques still lies in substantial efforts required in generation and maintenance of signature databases taking into consideration constant changes in the working environment [5].

Direction of arrival estimation through array processing provides sufficient precision to many applications [2]. Many such techniques have been proposed and recommended for cellular networks such as observed timing difference (OTD) and enhanced timing difference (E-OTD) for GSM and observed timing difference of arrival (OTDOA) for UMTS [2]. For short-range cells, pure cell ID methods and higher frequencies should be considered.

MUSIC and ESPIRIT are also used to determine the direction of arrival and to determine the desired signals. However, in a real environment (or under realistic conditions), there are too many DOAs to properly detect, and therefore, these algorithms may not perform as well as the algorithms discussed in chap. 4.

It is already been mentioned that MIMO-based spatial multiplexing technology and coded OFDM turn multipath delay spread into a benefit and also can be used successfully in nonline-of-sign environments [7].

It is known that substantial improvements of the system performance are obtained as a result of the use of *adaptive modulation* by exploiting the margins of the SNR available at any time/location [7].

Smart antennas refer to the use of multiple antenna array elements at the RF front-end section followed by intelligent signal processing and coding at the intermediate frequency or baseband section of geolocation of RF signals systems [7]. In this context, geolocation of RF signals systems can be defined as smart antennas that provide very accurate direction of arrival and distance information on RF sources of interests [7]. Moreover, although smart antennas and adaptive arrays are a category of research that is mature and well-researched and commercialized both by the military and the civilian community with the aim to protect the desired signals of interest, the focus of this book and of this research is mainly on the *Geolocation of RF Signals: Principles and Simulations* taking into consideration those same aspects of the smart antennas and adaptive arrays material and references that are already published in the literature and also of the novel research that I have performed in the last decade or so. I am certainly going to highlight both the similarities and the differences of these techniques in Chaps. 4 and 6.

Best state-of-the-art adaptive algorithms that are discussed in this book are classified as *nonblind* and *blind* [29]. Chapter 4 discusses blind algorithms that do not relay on a training sequence, but use the received signal spatial and temporal characteristics to update a weighted value [29]. Chapter 6 discusses nonblind algorithms which are used in GPS to suppress interference.

Best state-of-the-art adaptive antenna systems can be classified to operate in one of the two distinct modes: *diversity* or *beamforming* [30]. Assuming statistical independence among the signals at the antenna elements, diversity techniques help increase the signal-to-noise ratio and reduce the likelihood of deep fades [30]. On the other hand, if we assume coherence among the signals at the antenna elements, a narrow beam can be created at the direction of the desired user [30]. We should also note that either *diversity* or *beamforming* techniques are either *blind* or *nonblind* and, for the purpose of GRFS, they are extensively discussed in Chaps. 4 and 6 [30].

1.8 Applications of Geolocation of RF Signals

There are many interesting applications for geolocation of RF signals systems.

Other interesting applications are remote sensing, pollution monitoring, meteorological measurement, real-time monitoring of seismic or coastal regions and terrestrial structure, agriculture support, etc. [2].

Other interesting applications consider the geolocation of airships and aerodynamic aircraft which are readily deployed in disaster management areas, or relocated, expanded, or updated with new payloads [2].

Standardization of mobile phone positioning for 3G and 4G systems is certainly one viable commercial application of the geolocation of RF signals systems [29, 31]. 3G or 4G mobile phone positioning systems will be discussed more extensively in the *Indoor Geolocation Systems: Theory and Applications* book.

1.9 A Generic Geolocation of RF Signals System

Examples of implementation from the CMOS design point of view of wireless LAN chipsets are given in [12, 32, 33] (just to name a few). For the most part, most or all of the references provided here [1–3, 5–7, 11, 12, 15, 16, 20–29, 31–134] should be sufficient to enable the reader to visualize what we are about to do here which is to introduce a generic GRFS system.

Regardless of the specific details [1–3, 5–7, 11, 12, 15, 16, 20–29, 31–134], a generic GRFS system contains three main components as shown in Figs. 1.7 and 1.8. These components are RF *transmitter(s)* (or we may also refer to them as RF *source (s)*), RF *channel*, and GRFS *receiver(s)*. The interpretation for the three components is as follows. First, as far as RF transmitters are concerned, someone else has designed them. RF channel is what Mother Nature has provided to us as far as its RF properties or characteristics. The GRFS receiver (or unit) is the only component that we are analyzing, modeling, simulating, designing, and so forth taking into consideration signals coming from RF transmitters and properties of the RF channel.

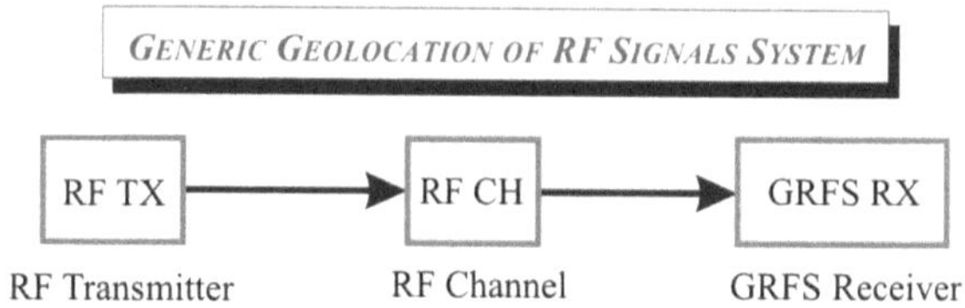

Reprinted with permission copyright © 2010 Ilir Progri.

Fig. 1.7 Components of a generic GRFS system. Reprinted with permission © 2010 Ilir Progri

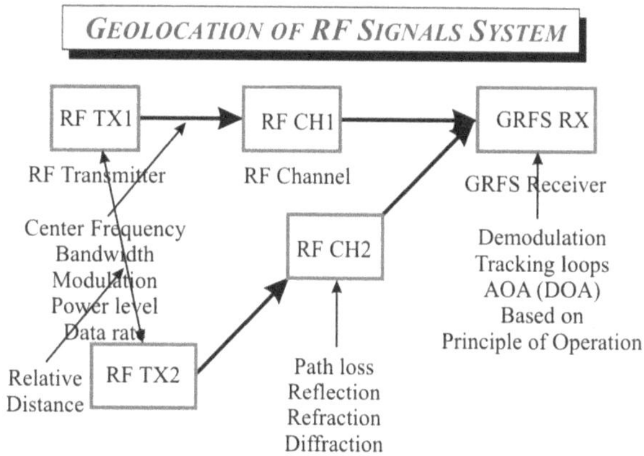

Reprinted with permission copyright © 2010 Ilir Progri.

Fig. 1.8 Components of another generic GRFS system. Reprinted with permission © 2010 Ilir Progri

An RF *transmitter* (or *source*) is a device (or apparatus), which generates electromagnetic waves based on signals that are encoded in a manner that improves their ability to render communication and localization and sends the electromagnetic waves into space by means of an antenna (or antenna array typical these days). Important characteristics of a transmitter include the number of antenna elements, RF center frequency band from 100 MHz to 66 GHz, the RF bandwidth (from 10 MHz to several GHz), the RF-transmitted power level, encoding and modulation type such as, OFDM, OFDMA in UWB, MC-CDMA, W-CDMA in GSM, CDMA in GPS etc., and data rate from 50 B/s (CDMA in GPS) to 100 MB/s (OFDMA in UWB).

A GRFS *receiver* is a device (or apparatus) that captures electromagnetic waves coming from RF transmitters in the environment by means of an antenna array and converts them into electric signals for extracting localization information, such as the AOA or direction of arrival or even distance from these RF transmitters, by means of an adaptive digital signal processing algorithm based on its principle of operation. Electromagnetic waves are received by means of an antenna array and are converted into electric signals, which are: (a) demodulated and utilized for tracking the code and the phase of the transmitted signal or (b) utilized by means of the adaptive signal processing algorithm based on its principle of operation to yield the desired AOA or direction of arrival or even the distance from the [unknown] transmitter.

A RF *channel* is the medium (or the environment) that enables the electromagnetic waves from the transmitting RF antenna array to the GRFS receiving antenna array. The transmitted electromagnetic waves passing through a channel are subject to various channel impairments such as (1) path loss, (2) reflection, (3) refraction, or (4) scattering.

1.9.1 Principles of Operation

The spherical polar coordinate system (r, θ, ϕ) with origin at the center of the antenna array should be used to provide the relationship between the coordinate of the *i*th RF source x_i, y_i, z_i for $i \in \{1\ldots J\}$ where J is the number of sources and its spherical polar coordinates (r_i, θ_i, ϕ_i) (see Fig. 2.7 in [34]). Section 2.5 in [34] provides all the necessary equations for the direct and indirect transformations between (r_i, θ_i, ϕ_i) and (x_i, y_i, z_i for $i \in \{1\ldots J\}$ where J is the number of sources).

There are three important principles that govern the operation of any GRFS system.

The *first* principle of operation of a GRFS receiver consists in the problem of localization of sources radiating energy by observing their signal received at spatially separated sensors (or antenna array elements). Codara [44] provides an excellent overview and summary of several methods for the estimation of the DOA of narrow-band RF sources of *the same central frequency*, located in the far field of an array of sensors. These processing and source estimation methods are parameterized by the variable θ, which is the AOA in the elevation direction (see Fig. 1.3) [44].

The *second* principle of operation of any GRFS consists in its ability to obtain accurate azimuth angle by means of another elevation antenna array which is orthogonal with the azimuth antenna array should be used to yield an accurate estimation of the azimuth angle ϕ.

The *third* principle of operation of any GRFS consists in ability to render distance estimation; four antenna element arrays (two pairs of orthogonal antenna arrays separated from each other at a considerable length) should be used to yield the desired distance from the *i*th source to each of the pairs of the orthogonal antenna arrays. In Chap. 4, we are getting into necessary mathematical details to support our claims here.

Now that we have determined that four antenna array elements (two pairs of orthogonal antenna arrays separated from each other at a considerable length) are required for yield, the (x_i, y_i, z_i for $i \in \{1\ldots J\}$ where J is the number of sources) assuming that $J < M$, where M is the number of array elements. In Chap. 4, we are also going to consider the most general case for sources with different center frequency.

Taking into consideration these very important GRFS principles of operation, we can focus more on the direction of arrival estimation methods, which include spectral estimation; minimum variance distortionless response estimator; linear prediction; maximum entropy; maximum likelihood; various eigenstructure methods; multiple signal classification (MUSIC) algorithms such as Spectral MUSIC, Root-MUSIC, Constrained MUSIC, and Beam Square MUSIC; minimum norm methods; CLOSEST method; the estimation of signal parameters via rotational invariance techniques (ESPRIT) method; and the weighted subspace fitting method that are perhaps the most researched and studied methods in the last 20–30 years [44]. Because some of these estimation methods are computationally very intensive, we are going to consider in Chaps. 4 and 5 an efficient way of performing most of the algorithmic computations. Performance can also increase as a result of hardware and software implementation in software-defined radio [17].

In Chap. 4, we are also going to provide the derivation and a discussion on the Cramér-Rao lower bound on angle accuracy for the maximum likelihood estimation method based on the development in [45], which can be used as a bound on other angle estimation algorithms discussed in [44]. Because we believe that our method discussed in Chap. 4 achieves the smallest possible angle accuracy (up to 0° angle error), the development and Cramér-Rao lower bound on angle accuracy is critical to support our theoretical analysis and principle simulation example results.

1.9.2 RF Transmitter

A GRFS system, which uses AOA positioning methodology based on the principles of operation discussed in the previous subsection, must contain at least two pairs or orthogonal identical M-element antenna arrays to solve for location of $M - 1$ RF sources (or transmitters). Chapter 3 discusses in great detail RF signals of interests; however, based on a generic RF transmitter diagram shown in Fig. 1.9, we provide a brief description of a generic RF transmitter. The reader also may refer to *Indoor Geolocation Systems: Theory and Applications* for more details on RF transmitters that are employed specifically on indoor geolocation.

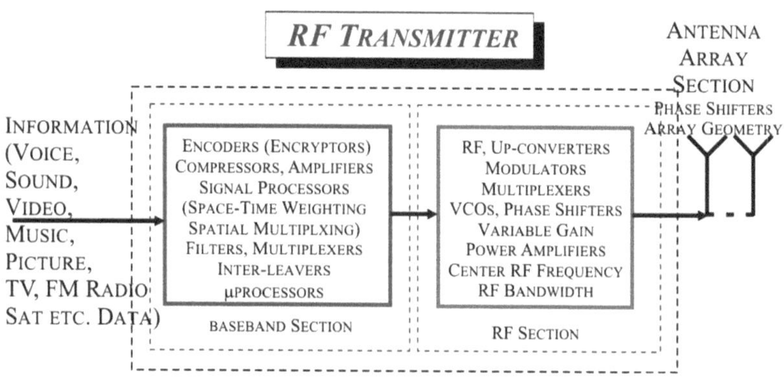

Reprinted with permission copyright © 2010 Ilir Progri.

Fig. 1.9 A block diagram of a generic RF transmitter. Reprinted with permission © 2010 Ilir Progri

A *RF transmitter* is a physical device or system that converts information (voice, sound, video, picture, music, 2D or 3D data (position, navigation, and time (PNT) data), light intensity, motion, temperature variation, pressure variation, weight variation, number of objects, texture, etc.) into electromagnetic waves at certain RF center frequency and bandwidth by means of an array of hardware and software mechanical, electrical and electronic, optical, etc. components, units, subsystems, etc. Regardless of applications, a *RF transmitter* contains three main sections: (a) *baseband section*; (b) *RF section*; and (c) *antenna array section*. The *baseband section* ensures that information (such as voice, sound, video, data in analog domain) is converted, amplified, encoded (or encrypted as a special kind of encoding), filtered, interleaved, signal processed (such as space–time weighting or spatial multiplexing [18]) in baseband signals (in digital domain) by means of dedicated compressors, encoders (or encryption encoders), signal processors, interleavers, amplifiers, filters, convertors, multiplexors, etc. Second, the *RF section* up-converts, multiplexes, and modulates the baseband signal to the desired center frequency by means of RF up-converters, modulators, multiplexers or variable gain power amplifiers, voltage-controlled oscillators, and phase shifters to raise the power of the output signal. Third, the antenna array section broadcasts the signals to the environment as electromagnetic waves by means of an antenna array section which can be further classified as: (a) *resonant*, (b) *nonresonant*, or (c) *UWB antennas* [135]. If beamforming is required at the antenna array section, geometry and phase shifters are utilized in the antenna array section [136, 137]. The *signal structure* describes the process that relates information at the input of the transmitter to the electromagnetic waves transmitted by the transmitted antenna array. The grand theme of Chap. 3 is to discuss, describe, analyze, and assess RF signals or RF signal structures in the frequency band of 100 MHz–66 GHz.

1.9.3 RF Channel

The *RF channel* is the medium (or the environment) between the RF transmitting antenna array and the GRFS receiving antenna array as depicted in Fig. 1.10. The transmitted electromagnetic wave at the RF frequency in the desired frequency band of 100 MHz–66 GHz propagating through the *RF channel* undergoes loss in power and dispersion in direction which is characteristic of the environment as depicted in Fig. 1.6 and provides the following RF channel types: (a) *indoor RF channel*, (b) *urban RF channel*, (c) *suburban RF channel*, (d) *global RF channel*, and (e) *satellite RF channel*.

RF channel modeling is very important for the realistic signal modeling, simulation, design, calibration, and function of GRFS systems. Chapter 2 of this edition discusses realistic application, engagement, and deployment scenarios of GRFS systems, where we made an attempt to discuss and describe all the qualitative properties of RF channel impairments. In *Indoor Geolocation Systems: Theory and Applications*, Dr. Progri dedicates one chapter to provide the most complete details of RF channel characteristics.

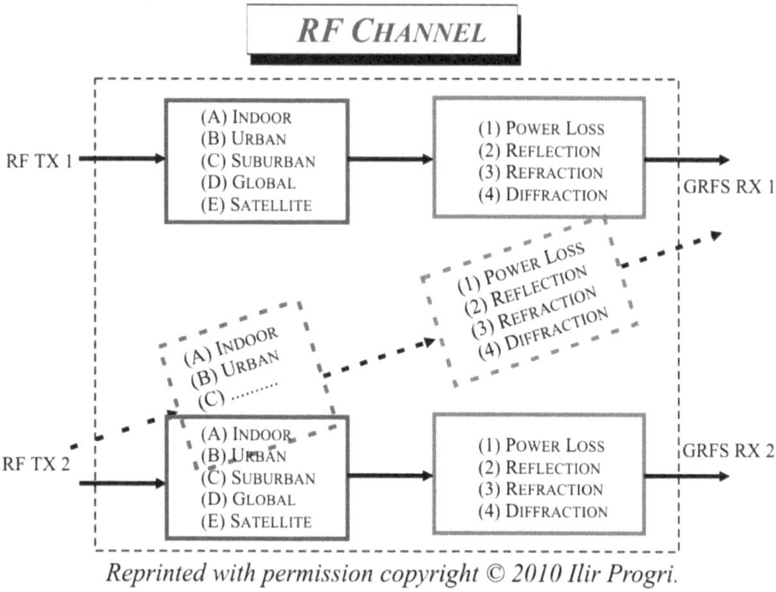

Fig. 1.10 A block diagram of a generic RF channel. Reprinted with permission © 2010 Ilir Progri

1.9.4 GRFS Receiver

A GRFS receiver is a physical device or physical system that converts electromagnetic waves at a certain frequency and bandwidth into information (voice, video, picture, music, and 2D or 3D data (PNT data, light intensity, motion, temperature variation, pressure variation, weight variation, number of objects, texture, etc.)) by means of an *array of hardware* and *software mechanical*, electrical and electronic, optical, etc. components, units, subsystems, etc. Regardless of applications, a *GRFS receiver* contains three main sections: (a) *antenna array* section, (b) *RF front-end* section, and (c) *baseband* section as shown in Fig. 1.11. First, the antenna array section is responsible for receiving the electromagnetic waves and converting them into electric signals at the RF frequency. Second, the *RF front-end section* downconverts, demultiplexes, and demodulates electric signals from the received RF center frequency into baseband digital signal by means of an RF demodulator, demultiplexors, or variable gain power amplifiers, frequency lock loops, and voltage control oscillators, to raise the power of the output signal and RF front-end filter analog to digital converters, etc. The *baseband section* ensures that baseband signals (in digital domain) are converted, filtered, decoded (decrypted), demultiplexed, amplified, deinterleaved, signal processed, phase locked loop, and delay locked loop into information (such as voice, sound, video, picture, data in analog domain) by means of dedicated decoders (decryption decoders), signal processors, amplifiers, filters, downconvertors, deinterleavers, demultiplexors, phase lock loop, delay lock loop, etc. [138–140].

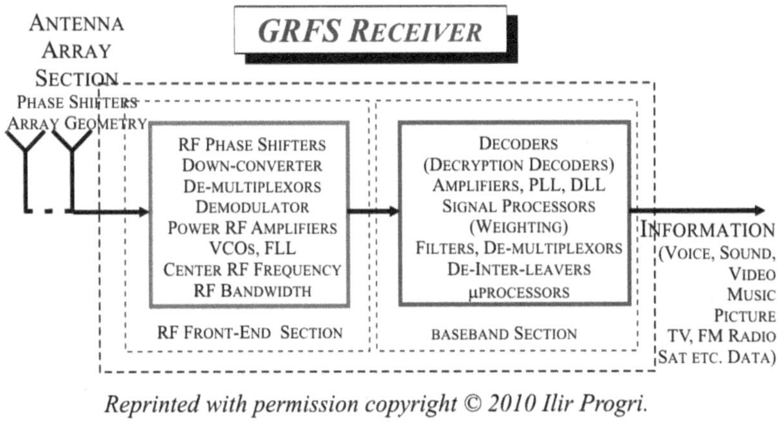

Fig. 1.11 A block diagram of a generic GRFS receiver. Reprinted with permission © 2010 Ilir Progri

The RF-transmitted *signal structure* can be utilized to describe the GRFS receiver structure and processes that relate to electromagnetic waves received by the GRFS receiving antenna array to information at the output of the GRFS receiver. The grand theme of Chaps. 4 and 6 is to discuss, describe, analyze, and assess RF signals or RF signal structures and mathematical processes at the GRFS receiver in the frequency band of 100 MHz–66 GHz [19].

1.10 Conclusions

We conclude here that we have provided a very nice introduction on the geolocation of RF signals systems.

In Abstract, we have given a broad overview of the book. Next, we discussed the overview of the geolocation of RF signals systems in Sect. 1.1. We provided the history and background of geolocation of RF signals systems in Sect. 1.2. Next, we discussed the geolocation of RF signals system concept in Sect. 1.3, which included the domain, the environment, and the methodology. We also illustrated under ideal conditions the generic principle of operations of a geolocation of RF signals system. In Sect. 1.4 we provided a proper technical definition and performed the classification of the geolocation of RF signals systems into outdoor, indoor, and underwater geolocation of RF signals systems. The requirements of geolocation of RF signals systems were assessed in Sect. 1.5, which consisted of: *accuracy, adaptation/reconfiguration, affordability, area of coverage, bandwidth, bandwidth utilization, challenges/issues/impairments/obstacles, commercialization strategy/market/manufacturing/sales, compatibility, diversity, expendability, flexibility, friendliness or user's convenience, frequency of operation, global convergence, integration, interference, interoperability, mobility, modeling, modularity, number of antennas, physical dimensions (power, size, and weight), privacy/safety/security, processing, reliability, scalability, signal/system description, and synchronization.* The geolocation of RF signals system's main description was provided in Sect. 1.6 in five typical environments: (a) indoors, (b) urban, (c) suburban, (d) global, and (e) satellite. These descriptions included the RF signal emitters with unknown location and frequency; RF geolocation channel; and the geolocation of RF signals receiving unit with an array of RF antennae. In contrast to indoor geolocation or outdoor geolocation systems which are designed to operate with known signals, i.e., the transmitter and receiver are designed to yield maximum signal reception within the domain and the environment, in the case of geolocation of RF signals we have absolutely no idea of the signals we are receiving and also of the distance of various RF sources. Therefore, it is expected that the complexity of engineering methodology in designing geolocation of RF signals systems is several orders of magnitude higher than those of indoor geolocation systems. This is the reason why geolocation of RF signals systems have maximum performance achievable of any system under normal conditions and yield up to 0° AOA estimation precision accuracy. The best state-of-the-art geolocation of RF signals techniques (or algorithms) were discussed in Sect. 1.7. Applications of geolocation of RF signals were presented in Sect. 1.8. A generic geolocation of RF signals system was provided in Sect. 1.9, which included three main principles of operations: a generic RF transmitter description; a generic RF channel description; and a generic GRFS receiver description. While this is an outstanding introduction of the book, the reminder of the book is organized as follows:

Chapter 2 discusses in great detail the requirements of GRFS systems. Chapter 3 contains a detailed description of RF signals in the frequency band of 100 MHz–66 GHz. Chap. 4 is probably considered the most important chapter of the book because it will be the best illustration of a typical geolocation of RF signals system starting with the analytical descriptions and interpretation of the principles of operations, with several candidates of RF signals on the environment from FM, wireless LANs, HomeRF, GPS, Bluetooth, GSP, CDMA2000, OFDM, OFDMA, and other standards that we discussed so extensively in Chap. 3; a description of the eigen-value blind adaptive angle estimation method and also of the Cramér-Rao bound lower bound on angle accuracy, several principle simulation examples, summary, and conclusions are included in Chap. 4. Chapter 5 is dedicated to adaptive recursive algorithms to reduce implementation computation complexity of the eigen-value angle blind adaptive angle estimation method. Chapter 6 discusses specific designs on adaptive spatial temporal selective attenuators as GRFS systems to accomplish two things: (a) suppress interference and (b) provide accurate geolocation of interference sources. Chapter 6 is a detailed summary of the most important conclusions of the book. The success of this project depends on the details which is the reason why the rest of the book is so important and quality of the feedback from the technical readership.

References

1. Progri, I., "An assessment of indoor geolocation systems," *Ph.D. Dissertation*, Worcester Polytechnic Institute, May 2003.
2. Avagnina, D., Dovis, F., Ghiglione, A., and Mulassano, P., "Wireless networks based on high-altitude platforms for the provision of integrated navigation/communication services," *IEEE Com. Mag.*, vol. 40, nr. 2, pp. 119–125, Feb. 2002.
3. Smulders, P., "Exploiting the 60 GHz band for local wireless multimedia access: prospects and future directions," *IEEE Com. Mag.*, vol. 40, nr. 1, pp. 140–147, Jan. 2002.
4. Pabst, R., Walke, B.H., Schultz, D.C., Herhold, P., Yanikomeroglu, H., Mukherjee, S., Viswanathan, H., Lott, M., Zirwas, W., Dohler, M., Aghvami, H., Falconer, D.D., and Fettweis, G.P., "Relay-based deployment concepts for wireless and mobile broadband radio," *IEEE Com. Mag.*, vol. 42, nr. 9, pp. 80–89, Sep. 2004.
5. Pahlavan, K., Li, X., and Makela, J.-P., "Indoor geolocation science and technology," *IEEE Com. Mag.*, vol. 40, nr. 2, pp. 112–118, Feb. 2002.
6. Murch, R.D., and Letaief, K.B., "Antenna systems for broadband wireless access," *IEEE Com. Mag.*, vol. 40, nr. 4, pp. 76–83, Apr. 2002.
7. Gesbert, D., Haumonte, L., Bolskei, H., Krishnamoorthy, R., and Arogyaswami, P.J., "Technologies and performance for Non-line-Of-Sight Broadband Wireless Access Networks," *IEEE Com. Mag.*, vol. 40, nr. 4, pp. 86–95, Apr. 2002.
8. Lang, S., Rao, R.M., and Daneshrad, B., "Design and development of a 5.25 GHz software defined wireless OFDM communication platform," *IEEE Com. Mag.*, vol. 42, nr. 6, (*IEEE Radio Com.*) pp. S6–S12, June 2004.
9. Xiang, W., Pratt, T., and Wang, X., "A software radio testbed for two-transmitter two-receiver space-time coding OFDM wireless LAN," *IEEE Com. Mag.*, vol. 42, nr. 6, (*IEEE Radio Com.*) pp. S20–S28, June 2004.
10. Ozdemir, M.K., Avras, E., and Arslan, H., "Dynamics of spatial correlation and implications in MIMO systems," *IEEE Com. Mag.*, vol. 42, nr. 6, (*IEEE Radio Com.*) pp. S14–S19, June 2004.
11. Derryberry, R.T., Gray, S.D., Ionescu, D.M., Mandyam, G., and Raghothaman, B., "Transmit diversity in 3G CDMA systems," *IEEE Com. Mag.*, vol. 40, nr. 4, pp. 68–75, Apr. 2002.
12. Polydoros, A., Rautio, J., Razzano, G., Bogucka, H., Ragazzi, D., Dallas, P.I., Mammela, A., Benedix, M., Lobeira, M., and Agarossi, L., "WIND-FLEX: developing a novel testbed for exploring flexible radio concepts in an indoor environment," *IEEE Com. Mag.*, vol. 41, nr. 7, pp. 116–122, Jul. 2003.
13. Yang, L.-L., and Hanzo, L., "Multicarrier DS-CDMA: a multiple access scheme for ubiquitous broadband wireless communications," *IEEE Com. Mag.*, vol. 41, nr. 10, pp. 116–124, Oct. 2003.
14. Fong, B., Ansari, N., Fong, A.C.M., Hong, G.Y., and Rapajic, P.B., "On the scalability of fixed broadband wireless access network deployment," *IEEE Com. Mag.*, vol. 42, nr. 9, (*IEEE Radio Com.*, vol. 1, nr. 3) pp. S12–S18, Sep. 2004.
15. International Telecommunications Union, *Spectrum Monitoring Handbook*, 2002 Edition. http://www.itu.int/publ/R-HDB-23/en
16. International Telecommunications Union, *Supplement to Handbook on Spectrum Monitoring*, 2008 Edition. http://www.itu.int/publ/R-HDB-53/en
17. Georganopoulos, N., Farnham, T., Burgess, R., Scholer, T., Sessler, J., Warr, P., Platbrood, F., and Souville, B., "Terminal-centric view of software reconfigurable system architecture and enabling components and technologies," *IEEE Com. Mag.*, vol. 42, nr. 5, pp. 100–110, May. 2004.
18. Alexiou, A., and Haardt, M., "Smart antenna technologies for future wireless systems: trends and challenges," *IEEE Com. Mag.*, vol. 42, nr. 9, pp. 90–97, Sep. 2004.
19. Pucker, L., "Has FPGA technology peaked in wideband wireless applications?," *IEEE Com. Mag.*, vol. 42, nr. 9, (*IEEE Radio Com.*, vol. 1, nr. 3) pp. S4–S5, Sep. 2004.

20. Eklund, C., Marks, R.B., Stanwood, K.L., and Wang, S., "IEEE Standard 802.16: a technical overview of the WirelessMAN™ air interface for broadband wireless access," *IEEE Com. Mag.*, vol. 40, nr. 6, pp. 90–96, Jun. 2002.
21. Porcino, D., and Hirt, W., "Ultra-wideband radio technology: potential and challenges ahead," *IEEE Com. Mag.*, vol. 41, nr. 7, pp. 66–74, Jul. 2003.
22. Koffman, I., and Roman, V., "Broadband wireless access solutions based on OFDM access in IEEE 802.16," *IEEE Com. Mag.*, vol. 40, nr. 4, pp. 96–103, Apr. 2002.
23. Teger, S., and Waks, D.J., "End user perspective on home networking," *IEEE Com. Mag.*, vol. 40, nr. 4, pp. 114–119, Apr. 2002.
24. Callaway, E., Gorday, P., Hester, L., Gutierrez, J.A., Naeve, M., Heile, B., and Bahl, V., "Home networking with IEEE 802.15.4: a developing standard for low-rate wireless personal area networks," *IEEE Com. Mag.*, vol. 40, nr. 8, pp. 70–77, Aug. 2002.
25. Nakagawa, M., Zhang, H., and Sato, H., "Ubiquitous homelinks based IEEE 1394 and ultra wideband solutions," *IEEE Com. Mag.*, vol. 41, nr. 4, pp. 74–82, Apr. 2003.
26. Sairam, K.V.S.S.S.S., Gunasekaran, N., and Rama Reddy, S., "Bluetooth in wireless communications," *IEEE Com. Mag.*, vol. 40, nr. 6, pp. 90–96, Jun. 2002.
27. Del Re, E., and Pierucci, L., "Next generation mobile satellite networks," *IEEE Com. Mag.*, vol. 40, nr. 9, pp. 150–159, Sep. 2002.
28. Jahn, A., Holzbock, M., Muller, J., Kebel, R., De Santis, M., Rogoyski, A., Franzrahe, O., Werner, M., and Hu, F., "Evolution of aeronautical communications for personal and multimedia services," *IEEE Com. Mag.*, vol. 41, nr. 7, pp. 36–43, Jul. 2003.
29. Liu, J., Yuan, Y., Xu, L., Wu, R., Dai, Y., Li, Y., Zhang, L., Shi, M., and Du, Y., "Research on smart antenna technology for terminals for the TD-SCDMA systems," *IEEE Com. Mag.*, vol. 41, nr. 6, pp. 116–119, Jun. 2003.
30. Pedersen, K.I., Mogensen, P.E., and Ramiro-Moreno, J., "Applications and performance of downlink beamforming techniques in UMTS," *IEEE Com. Mag.*, vol. 41, nr. 10, pp. 134–143, Oct. 2003.
31. Zhao, Y., "Standardization of mobile phone positioning for 3G systems," *IEEE Com. Mag.*, vol. 40, nr. 7, pp. 108–116, Jul. 2002.
32. Razavi, B., "CMOS transceivers for cellular telephony," *IEEE Com. Mag.*, vol. 41, nr. 8, pp. 144–149, Aug. 2003.
33. Meng, T. H., McFarland, B., Su, D., Thomson, J. "Design and implementation of an all-CMOS 802.11a wireless LAN chipset," *IEEE Com. Mag.*, vol. 41, nr. 8, pp. 160–168, Aug. 2003.
34. Arfken, G.B., and Weber, H.J., *Mathematical Methods for Physicists*, San Diego, CA: Academic Press, 1995.
35. Davies, N., *Europe a History*, New York, NY: Oxford University Press, 1996.
36. Butterfield, H., *The Origins of Modern Science, 1300–1800*, The Free Press, London: NY, 1947.
37. Nebeker, F., "September and October in EE history," *IEEE The Institute*, vol. 25, pp. 9, Oct. 2001.
38. Parkinson, B., Gromov, K., Stansell, T., and Beard, R., "A history of satellite navigation," in *Proc ION-AM*, pp. 17–65, Colorado Springs, CO, June 1995.
39. Parkinson, B.W., Spilker, J.J., Axelrad, P., and Enge, P., *Global Positioning System: Theory and Applications*, vol. 1 and 2, Washington, DC: AIAA, 1996.
40. Parkinson B., and Gilbert, S.W., "NAVSTAR: Global Positioning System – ten years later," *IEEE*, vol. 71, pp. 1177–1186, Oct. 1983.
41. Parkinson, B.W., Spilker, J.J., Jr., Axelrad, P., and Enge, P., *The Global Positioning System-Theory and Applications* (vol. 1, chap. 2, "Overview of GPS operation and design"), Washington, DC: American Institute of Aeronautics and Astronautics, 1996.
42. Misra, P., and Enge, P., *Global Positioning System–Signals Measurements and Performance*, Lincoln, MA: Ganga-Jamuna, 2001.
43. *Webster's New World Dictionary of the American Language*. College Edition, Cleveland, OH: World Publishing, 1957.

44. Codara, L.C., *Smart Antennas*, Boca Raton, FL: CRC, 2004.
45. Manolakis, D.G., Ingle, V.K., Kogon, S.M., *Statistical and Adaptive Signal Processing: Spectral Estimation, Signal Modeling, Adaptive Filtering, and Array Processing*, Boston, MA: MC Graw Hill, 2000.
46. Monmoiner, M., *Privacy and Technologies of Identity a Cross-Disciplinary Conversation* (book chapter 5 "Geolocation and location privacy: the "inside" story on geospatial tracking"), US: Springer, Dec. 2006.
47. Boettcher, P.W., and Shaw, G.A., *Lecture Notes in Computer Science* (book chapter on "Energy-constrained collaborative processing for target detection, tracking, and geolocation"), Berlin: Springer, Jan. 2003.
48. Nerguizian, C., Despins, C., and Affès, S., *Lecture Notes in Computer Science* (book chapter on "Indoor geolocation with received signal strength fingerprinting technique and neural networks"), Berlin: Springer, Jul. 2004.
49. Sahi, P., *The Springer International Series in Engineering and Computer Science:* (book chapter 2 on "Geolocation on cellular networks"), The Netherlands: Springer, Apr. 2006.
50. Harris, M.P., Daunt, F., Newell, M. Phillips, R.A., and Wanless, S., "Wintering areas of adult Atlantic puffins *Fratercula arctica* from a North Sea colony as revealed by geolocation technology," *Journal Marine Biology*, Berlin: Springer, Dec. 2009.
51. Shaffer, S.A., Tremblay,Y., Awkerman, J.A., Henry, R.W., Teo, S.L.H., Anderson, D.J., Croll, D.A., Block, B.A., and Costa, D.P., "Comparison of light- and SST-based geolocation with satellite telemetry in free-ranging albatrosses", Berlin: Springer, *Journal Marine Biology*, vol. 147, nr. 4, August, 2005.
52. Bassiouni, M., and Cui, W., *The Springer International Series in Engineering and Computer Science* (book chapter 7 on "Enhancing terminal coverage and fault recovery in configurable cellular networks using geolocation services"), The Netherlands: Springer, pp. 231–254, Apr. 2006.
53. Ki, Y.M., Kim, J.W., Kim, S.R., and Kim, D.K., *Lecture Notes in Computer Science*, (book chapter on "Modified RWGH and positive noise mitigation schemes for TOA geolocation in indoor multi-hop wireless networks"), The Netherlands: Springer, pp. 231–254, Apr. 2006.
54. Hunter, E., Aldridge, J.N., Metcalfe, J.D., and Arnold, G.P., "Geolocation of free-ranging fish on the European continental shelf as determined from environmental variables", *Journal Marine Biology,* vol. 142, nr. 3, pp. 601–609, Feb., 2004.
55. Thygesen, U.H., and Nielsen, A., *Reviews: Methods and Technologies in Fish Biology and Fisheries* Tagging and Tracking of Marine Animals with Electronic Devices, vol. 9 (book chapter on "Lessons from a prototype geolocation problem"), The Netherlands: Springer, pp. 257–276, Jun. 2009.
56. Liu, X., Ma, H., and Sun, W., *Lecture Notes in Computer Science: Advances in Machine Vision, Image Processing, and Pattern Analysis* (book chapter on "Study on the geolocation algorithm of space-borne SAR Image"), Berlin: Springer, pp. 270–280, Aug. 2006.
57. Green, J.A., Wilson, R.P., Boyd, I.L., Woakes, A.J., Green, C.J., and Butler, P.J., "Tracking macaroni penguins during long foraging trips using 'behavioural geolocation'," Berlin: Springer, *Journal Polar Biology*, vol. 32, nr. 4, pp. 645–653, Apr. 2009.
58. Mielke, C., and Chen, H., *Lecture Notes in Computer Science, Intelligence and Security Informatics: Intelligence and Security Informatics* (book chapter on "Mapping dark web geolocation"), Springer, pp. 97–107, Nov. 2008.
59. Kikiras, P., and Drakoulis, D., "An integrated approach for the estimation of mobile subscriber geolocation," The Netherlands: Springer, *Journal of Wireless Personal Communications,* vol. 30, nr. 2–4, Sep. 2004.
60. Mielke, C., and Chen, H., *Ultra-Wideband, Short-Pulse Electromagnetics 5* (book chapter on "Experimental results from an ultra wideband precision geolocation system"), US: Springer, pp. 97–107, 2002.
61. Luthcke, S.B., Carabajal, C.C., Rowlands, D.D., and Pavlis, D.E., "Improvements in spaceborne laser altimeter data geolocation," The Netherlands: Springer, *Surveys in Geophysics,* vol. 22, nr. 5–6, Sep. 2001.

62. Hunter, E., Metcalfe, J.D., Holford, B.H., and Arnold, G.P., "Geolocation of free-ranging fish on the European continental shelf as determined from environmental variables II. Reconstruction of plaice ground tracks," *Journal Marine Biology,* vol. 144, nr. 4, pp. 787–798, Apr. 2004.
63. Bargshady, N., Nayef A. Alsindi, N.A., and Pahlavan, K., *Lecture Notes in Computer Science: Mobile Entity Localization and Tracking in GPS-less Environments* (book chapter on "Performance of TOA- and RSS-based indoor geolocation for cooperative robotic applications"), Berlin: Springer, pp. 255–266, Sep. 2009.
64. Pahlavan, K., Li, X., Ylianttila, M., Chana, R., and Latva-aho, M., *Lecture Notes in Computer Science: Mobile and Wireless Communications Networks* (book chapter on "An overview of wireless indoor geolocation techniques and systems"), Berlin: Springer, pp. 1–13, Jan. 2000.
65. Gueye, B., Uhlig, S., Ziviani, A., and Fdida, S., *Lecture Notes in Computer Science, Networking 2006. Networking Technologies, Services, and Protocols; Performance of Computer and Communication Networks; Mobile and Wireless Communications Systems* (book chapter on "Leveraging buffering delay estimation for geolocation of internet hosts"), Berlin: Springer, pp. 319–330, Apr. 2006.
66. Gueye, B., Uhlig, S., Ziviani, A., and Fdida, S., *Reviews: Methods and Technologies in Fish Biology and Fisheries, Tagging and Tracking of Marine Animals with Electronic Devices* (book chapter on "Summary report of a workshop on geolocation methods for marine animals"), The Netherlands: Springer, pp. 342–363, Jun. 2009.
67. Palaniswami, M., Sundaram, B., Jayavardhana, R., and Shilton, A., *Informatics in Control, Automation and Robotics II* (book chapter on "Target localization using machine learning"), The Netherlands: Springer, pp. 27–33, Jun. 2007.
68. Rowlands, D.D., Lemoine, F.G., Chinn, D.S., and Luthcke, S.B., "A simulation study of multi-beam altimetry for lunar reconnaissance orbiter and other planetary missions", *Journal of Geodesy,* vol. 83, nr. 8, pp. 709–721, Nov. 2008.
69. Choi, W.-J., and Tekinay, S., "Location-Based Service Provisioning for Next Generation Wireless Networks", The Netherlands: Springer, *International Journal of Wireless Information Networks,* vol. 10, nr. 3, pp. 127–139, Jul. 2003.
70. Thygesen, U.H., Pedersen, M.W., and Madsen, H., *Reviews: Methods and Technologies in Fish Biology and Fisheries, Tagging and Tracking of Marine Animals with Electronic Devices* (book chapter 114 on "Geolocating fish using hidden Markov models and data storage tags"), The Netherlands: Springer, pp. 277–293, Jun. 2009.
71. Lee, S.C., Lee, W.R., and You, K.H., *Communications in Computer and Information Science, Control and Automation* (book chapter on "TDoA based UAV localization using dual-EKF algorithm"), Berlin: Springer, pp. 47–54, 2009.
72. Tanner, S., Stein, C., and Graves, S.J., *Scientific Data Mining and Knowledge Discovery* (book chapter on "On-board data mining"), Berlin: Springer, pp. 345–376, 2009.
73. Jeske, D.R., *Statistics for Industry and Technology, Advances in Mathematical and Statistical Modeling* (book chapter on "Jackknife bias correction of a clock offset estimator"), Boston: Birkhäuser, pp. 245–254, 2008.
74. Arslan, H., and Celebi, H., *Cognitive Radio, Software Defined Radio, and Adaptive Wireless Systems* (book chapter 10 on "Location information management systems for cognitive wireless networks"), The Netherlands: Springer, pp. 291–323, 2007.
75. Drent, R.H., Fox, A.D., and Stahl, J., "Travelling to breed", *Journal of Ornithology,* vol. 147, nr. 2, pp. 185–188, Apr. 2006.
76. Riboni, D., Pareschi, L., and Bettini, C., *Lecture Notes in Computer Science, Privacy in Location-Based Applications* (book chapter 10 on "Privacy in georeferenced context-aware services: a survey"), Berlin: Springer, pp. 151–172, 2009.
77. Xiang-zheng, D., Jin-yan, Z., Ji-yuan, L., and Da-fang, Z., "The global rainforest mapping project JERS-1: a paradigm of international collaboration for monitoring land cover change," *Journal of Geographical Sciences,* vol. 12, nr. 1, pp. 185–188, Jan. 2002.

78. Huang, L., Yamane, H., Matsuura, K., and Sezaki, K., *Advances in Information Security, Secure Localization and Time Synchronization for Wireless Sensor and Ad Hoc Networks* (book chapter on "Location privacy in wireless LAN"), US: Springer, pp. 299–322, 2007.
79. Zhou, R., and Schneider, G., *Lecture Notes of the Institute for Computer Sciences, Social Informatics and Telecommunications Engineering, MobileWireless Middleware, Operating Systems, and Applications* (book chapter on "Location-based botany guide: a prototype of web-based tracking and guiding"), Berlin: Springer, pp. 72–86, 2007.
80. DongChen, E., Shen, Q., Xu, Y., and Chen, G., "High-accuracy topographical information extraction based on fusion of ASTER stereo-data and ICESat/GLAS data in Antarctica", *Science in China Series D: Earth Sciences,* vol. 52, nr. 5 pp. 714–722, May. 2009.
81. Taheri, J., and Zomaya, A.Y., "A combined genetic-neural algorithm for mobility management", *Journal of Mathematical Modeling and Algorithms,* vol. 6, nr. 3 pp. 714–722, Sep. 2007.
82. Yang, J., Zhang, C., Li, X., Huang, Y., Fu, S., and Acevedo, M.F., "Integration of wireless sensor networks in environmental monitoring cyber infrastructure," *Wireless Networks,* vol. 6, nr. 3, pp. 714–722, Sep. 2007.
83. Villafuerte, F.L., and Schiller, J., *Lecture Notes in Computer Science, Ad-hoc, Mobile and Wireless Networks* (book chapter on "An ad-hoc algorithm to estimate distances in wireless sensor networks"), Berlin: Springer, pp. 162–175, 2008.
84. Gardiner, K., Yin, J., and Carswell, J.D., *Lecture Notes in Computer Science, Web and Wireless Geographical Information Systems* (book chapter on "EgoViz – a mobile based spatial interaction system"), Berlin: Springer, pp. 135–152, 2009.
85. Yang, F., Zhang, Z., and Chen, Y., "Maximum-likelihood estimator for coarse carrier frequency offset estimation in OFDM systems," *Journal of Wireless Personal Communications*, vol. 49, nr. 1, pp. 55–66, Apr. 2009.
86. Musick, S., *Foundations and Applications of Sensor Management* (book chapter 11 on "Defense applications"), US: Springer, pp. 257–268, 2008.
87. Frattasi, S., Fitzek, P.H.P., and Prasad, R., *IFIP International Federation for Information Processing, Intelligence in Communication Systems* (book chapter on "A look into the 4G crystal ball"), Boston: Springer, pp. 281–290, 2005.
88. Dil, B., Dulman, S., and Havinga, P., *Lecture Notes in Computer Science, Wireless Sensor Networks* (book chapter on "Range-based localization in mobile sensor networks"), Berlin: Springer, pp. 164–178, 2006.
89. Shen, X., Mark, J.W., and Ye, J., "Mobile location estimation in CDMA cellular networks by using fuzzy logic," *Journal of Wireless Personal Communications*, vol. 22, nr. 1, pp. 57–70, Jul. 2002.
90. Mineno, H., Hida, K., Mizutani, M., Miyauchi, N., Kusunoki, K., Fukuda, A., and Mizuno, T., *Lecture Notes in Computer Science, Knowledge-Based Intelligent Information and Engineering Systems* (book chapter on "Position estimation for goods tracking system using mobile detectors"), Berlin: Springer, pp. 431–437, 2005.
91. Buschmann, C., Hellbrück, H., Fischer, S., Kröller, A., and Fekete, S., *Lecture Notes in Computer Science, Wireless Sensor Networks* (book chapter on "Radio propagation-aware distance estimation based on neighborhood comparison"), Berlin: Springer, pp. 325–340, 2007.
92. Li, J., Li, J., Guo, L., and Wang, P., *Lecture Notes in Computer Science, Advanced Web and Network Technologies, and Applications* (book chapter on "Power-efficient node localization algorithm in wireless sensor networks"), Berlin: Springer, pp. 420–430, 2006.
93. Muthukrishnan, K., and Hazas, M., *Lecture Notes in Computer Science, Location and Context Awareness* (book chapter on "position estimation from UWB pseudorange and angle-of-arrival: a comparison of non-linear regression and Kalman filtering"), Berlin: Springer, pp. 222–239, 2009.
94. Kontkanen, P., Myllymäki, P., Roos, T., Tirri, H., Valtonen, K., and Wettig, H., *Emerging Location Aware Broadband Wireless Ad Hoc Networks* (book chapter on "Probabilistic methods for location estimation in wireless networks"), US: Springer, pp. 173–188, 2005.

95. Doherty, P., and Rudol, P., *Lecture Notes in Computer Science, AI 2007: Advances in Artificial Intelligence* (book chapter on "A uav search and rescue scenario with human body detection and geolocalization"), US: Springer, pp. 1–13, 2007.
96. De la Cruz, A., Laneve, G., Cerra, D., Mielewczyk, M., Garcia, M., Santilli, G., Cadau, E., and Joyanes, G., *Lecture Notes in Geoinformation and Cartography, Geomatics Solutions for Disaster Management* (book chapter on "On the application of nighttime sensors for rapid detection of areas impacted by disasters"), Berlin: Springer, pp. 17–36, 2007.
97. Martínez-de-Dios, J.R., Merino, L., Ollero, A., Ribeiro, L.M., and Viegas, X., *Springer Tracts in Advanced Robotics, Multiple Heterogeneous Unmanned Aerial Vehicles* (book chapter on "Multi-UAV experiments: application to forest fires"), Berlin: Springer, pp. 207–228, 2007.
98. Sun, G., and Guo, W., "A novel indoor geo-location method using MIMO array," *Journal of Electronics (China)*, vol. 23, nr. 6, pp. 810–813, Jan. 2003.
99. Pahlavan, K., Li, X., Ylianttila, M., and Latva-aho, M., *The Springer International Series in Engineering and Computer Science, Wireless Communication Technologies: New Multimedia Systems* (book chapter on "Wireless data communications systems"), US: Springer, pp. 201–214, 2000.
100. Merino, L., Caballero, F., Ferruz, J., Wiklund, J., Forssén, P.-E., and Ollero, A., *Springer Tracts in Advanced Robotics, Multiple Heterogeneous Unmanned Aerial Vehicles* (book chapter on "Multi-UAV cooperative perception techniques"), Berlin: Springer, pp. 67–110, 2007.
101. Sunay, M., *The Springer International Series in Engineering and Computer Science, Next Generation Wireless Networks* (book chapter on "Evaluation of location determination technologies towards satisfying the FCC E-911 Ruling"), The Netherlands: Springer, pp. 157–192, 2002.
102. Yarlykov, M.S., and Yarlykova, S.M., "Signal-detection and signal-processing algorithms for code-division multiple-access satellite mobile communications systems employed simultaneously with satellite radio navigation systems," *Journal of Communications Technology and Electronics*, vol. 51, nr. 8, pp. 874–894, Aug. 2006.
103. Amundson, I., Manish Kushwaha, M., and Koutsoukos, X.D., *Lecture Notes in Computer Science, Mobile Entity Localization and Tracking in GPS-less Environments* (book chapter on "On the feasibility of determining angular separation in mobile wireless sensor networks"), Berlin: Springer, pp. 115–127, 2009.
104. Richton, B., Vannucci, G., and Wilkus, S., *The Springer International Series in Engineering and Computer Science, Next Generation Wireless Networks* (book chapter on "Assisted GPS for wireless phone location – technology and standards"), The Netherlands: Springer, pp. 129–155, 2002.
105. Howard, A., and Tunstel, F., *Frontiers of Geographic Information Technology* (book chapter on "Using geospatial information for autonomous systems control"), Berlin: Springer, pp. 63–84, 2006.
106. Hoeher, P., and Schmeink, K., *Lecture Notes in Electrical Engineering, Multi-Carrier Spread Spectrum 2007* (book chapter on "Joint navigation & communication based on interleave-division multiple access"), The Netherlands: Springer, pp. 97–106, 2007.
107. Howard, A., and Tunstel, F., *Lecture Notes in Computer Science, Algorithmic Aspects of Wireless Sensor Networks* (book chapter on "algorithms for location estimation based on RSSI sampling"), Berlin: Springer, pp. 72–86, 2008.
108. Cianca, E., Sanctis, M.D., Araniti, G., Molinaro, A., Iera, A., Torrisi, M., and Ruggieri, M., *Signals and Communication Technology, Satellite Communications and Navigation Systems* (book chapter on "Integration of navigation and communication for location and context aware RRM"), Berlin: Springer, pp. 25–50, 2008.
109. Song, L., Adve, R., and Hatzinakos, D., *Lecture Notes in Computer Science, Wireless Sensor Networks* (book chapter on "Matrix pencil for positioning in wireless ad hoc sensor network"), Berlin: Springer, pp. 18–27, 2004.

110. Chen, Y., and Rapajic, P., "Human respiration rate estimation using ultra-wideband distributed cognitive radar system," *Institute of Automation, Chinese Academy of Sciences, co-published with Springer-Verlag GmbH, International Journal of Automation and Computing,* vol. 4, nr. 5, pp. 325–333, Oct. 2008.
111. Hämäläinen, M., Saloranta, J., Mäkelä, J.-P., Oppermann, I., and TPatana, T., "Ultra-Wideband Signal Impact on the Performances of IEEE 802.11b and Bluetooth Networks," The Netherlands: Springer, *International Journal of Wireless Information Networks,* vol. 10, nr. 4, pp. 201–210, Oct. 2003.
112. Guvenc, I., Sahinoglu, Z., Orlik, P., and Arslan, H., "Searchback Algorithms for TOA Estimation in Non-coherent Low-rate IR-UWB Systems," *Springer Netherlands, Journal of Wireless Personal Communications,* vol. 48, nr. 4, pp. 585–603, Mar. 2009.
113. Zeger, L.M., and Kobayashi, H., "A Simplified EM Algorithm for Detection of CPM Signals in a Fading Multipath Channel," The Netherlands: Springer, *Journal of Wireless Networks,* vol. 8, nr. 6, pp. 649–658, Mar. 2002.
114. Blake Barber, B.D., Redding, J.D., McLain, T.W., Beard, R.W., and Taylor, C.N., "Vision-based Target Geo-location using a Fixed-wing Miniature Air Vehicle," The Netherlands: Springer, *Journal of Intelligent and Robotic Systems,* vol. 47, nr. 4, pp. 361–382, Dec. 2006.
115. Gezici, S., "A Survey on Wireless Position Estimation," The Netherlands: Springer, *Journal of Wireless Personal Communications,* vol. 44, nr. 3, pp. 263–282, Feb. 2008.
116. Ye, L., Geng,, Z., Xue, L., and Liu, Z., *Lecture Notes in Computer Science,* Computational Science and Its Applications – ICCSA 2007 (book chapter on "A novel real time method of signal strength based indoor localization"), Berlin :Springer, pp. 678–688, 2007.
117. Guoqiang, M., Barş, F., and Anderson, D., *Sensor Networks and Configuration* (book chapter on "Localisation"), Berlin: Springer, pp. 281–315, 2007.
118. Khokhar, S., and Nilsson, A.A., *Lecture Notes in Computer Science*, Wireless Algorithms, Systems, and Applications (book chapter on "Introduction to mobile trajectory based services: a new direction in mobile location based services"), Berlin: Springer, pp. 398–407, 2009.
119. Zàruba, G.V., Huber, M., Kamangar, F.A., and Chlamtac, I., "Indoor location tracking using RSSI readings from a single Wi-Fi access point," The Netherlands: Springer, *Journal of Wireless Networks,* vol. 13, nr. 2, pp. 221–235, Apr. 2007.
120. Gu, Z., and Gunawan, E., "Radiolocation in CDMA Cellular System Based on Joint Angle and Delay Estimation," The Netherlands: Springer, *Journal of Wireless* Personal Communications, vol. 23, nr. 3, pp. 297–309, Dec. 2002.
121. Arslan, H., and Celebi, H., *Cognitive Radio, Software Defined Radio, and Adaptive Wireless Systems* (book chapter on "Software defined radio architectures for cognitive radios"), The Netherlands: Springer, pp. 109–144, 2007.
122. Fontana, R.J., *Ultra-Wideband, Short-Pulse Electromagnetics 5* (book chapter on "Recent applications of ultra wideband radar and communications systems"), US: Springer, pp. 225–234, 2002.
123. Boryssenko, A.O., and Schaubert, D.H., "Electromagnetics-Related Aspects of Signaling and Signal Processing for UWB Short Range Radios," The Netherlands: Springer, *The Journal of VLSI Signal Processing*, vol. 43, nr. 1, pp. 89–104, Apr. 2006.
124. Mazuelas, S., Lago, F.A., Fernandez, P., Bahillo, A., Blas, J., Lorenzo, R.M., and Abril, E.J., "Ranking of TOA Measurements Based on the Estimate of the NLOS Propagation Contribution in a Wireless Location System," The Netherlands: Springer, *The Journal of Wireless Personal Communications*, vol. 53, nr. 1, pp. 35–52, Feb. 2009.
125. Manodham, T., Loyola, L., and Miki, T., "A Novel Wireless Positioning System for Seamless Internet Connectivity based on the WLAN Infrastructure," The Netherlands: Springer, *The Journal of Wireless Personal Communications*, vol. 44, nr. 3, pp. 295–309, Feb. 2008.
126. Frattasi, S., and Monti, M., *Cognitive Wireless Networks*, (book chapter on "Cooperative mobile positioning in 4G wireless networks"), The Netherlands: Springer, pp. 213–233, 2007.

127. Artieda, J., Sebastian, J.M., Campoy, P., Correa, J.F., Mondragón, I.F., Martínez, C., and Olivares, M., "Visual 3-D SLAM from UAVs," The Netherlands: Springer, *Journal of Intelligent and Robotic Systems*, vol. 55, nr. 4–5, pp. 299–321, Aug. 2009.
128. Torrieri, D., *Principles of Spread-Spectrum Communication Systems*, (book chapter on "Code-division multiple access"), US: Springer, pp. 293–386, 2005.
129. Webb, W., "Broadband fixed wireless access as a key component of the future integrated communications environment," *IEEE Com. Mag.*, vol. 39, nr. 9, pp. 115–121, Sep. 2001.
130. Danesh, M., Zuniga, J.-C., Concilio, F., "Fixed low-frequency broadband wireless access radio systems," IEEE Com. Mag., vol. 39, nr. 9, pp. 134–138, Sep. 2001.
131. Izadpanah, H., "A military-wave broadband wireless access technology demonstrator for the next-generation internet network reach extension," *IEEE Com. Mag.*, vol. 39, nr. 9, pp. 140–145, Sep. 2001.
132. Falconeer, D., Ariyavisitakul, S.L., Benyamin-Seeyar, A., and Edison, B., "Frequency domain equalization for single-carrier broadband wireless systems," *IEEE Com. Mag.*, vol. 40, nr. 4, pp. 58–66, Apr. 2002.
133. Sampath, H., Talwar, S., Tellado, J., Erceg, V., and Paulraj, A., "A fourth-generation MIMO broadband wireless system: design, performance, and field trial results," *IEEE Com. Mag.*, vol. 40, nr. 9, pp. 143–149, Sep. 2002.
134. Henry, P.S., and Luo, H., "WiFi: what's next?," *IEEE Com. Mag.*, vol. 40, nr. 12, pp. 66–72, Dec. 2002.
135. Yazdandoost, K.Y., and Kohno, R., "Ultra wideband antenna," *IEEE Com. Mag.*, vol. 42, nr. 6, (*IEEE Radio Com.*) pp. S29–S32, June 2004.
136. Hajirimi, A., Kamijani, A., Natarajan, R., Chunara, R., Guan, X., and Hashemi, H., "Phased array systems in silicon," *IEEE Com. Mag.*, vol. 42, nr. 8, pp. 122–130, Aug. 2004.
137. Smerzi, S.A., Girlando, G., Copani, T., and Palmisano, G., "A Ku-band monolithic receiver for DVB-S applications," *IEEE Com. Mag.*, vol. 42, nr. 8, pp. 132–139, Aug. 2004.
138. Zheng, J., and Lee, M.J., "Will IEEE 802.15.4 make ubiquitous networking a reality?: a discussion on a low power, low bit rate standard," *IEEE Com. Mag.*, vol. 42, nr. 6, pp. 140–146, June 2004.
139. Aklyildiz, I.F., Akan, A.B., Chen, C., Fang, J., and Su, W., "The state of the art in interplanetary internet," *IEEE Com. Mag.*, vol. 42, nr. 7, pp. 108–118, Jul. 2004.
140. Monogioudis, P., Conner, K., Das, D., Gollamudi, S., Lee, J.A.C., Moustakas, A.L., Nagaraj, S., Rao, A.M., Soni, R.A., and Yuan, Y., "Intelligent antenna solutions for UMTS: algorithms and simulation results," *IEEE Com. Mag.*, vol. 42, nr. 10, pp. 28–39, Oct. 2004.

Chapter 2
Requirements for Description of GRFS Systems

2.1 Overview

Requirements for description of Geolocation of RF Signals (GRFS) systems offer an overview of the best practices, classifications, visual interpretation, very important detailed descriptive ingredients, and innovative techniques in the art and science of GRFS over the last 20 years. In Chap. 1 we gave a very brief description of the GRFS systems which consisted only of the description from the local coordinate point of view into: (1) indoors, (2) urban, (3) suburban, (4) global, and (5) satellite. In this chapter, in addition to local coordinate environment description, we are going to add the global coordinate environment description which consists of: (1) water, (2) ground, (3) air, and (4) space descriptions. As we are going to see further in this chapter, there are 39 typical principle system illustration case studies for GRFS systems: (1) four correspond to requirements for description of indoor GRFS systems in Sect. 2.3; (2) eight correspond to requirements for description of urban GRFS systems in Sect. 2.4; (3) nine correspond to requirements for description of suburban GRFS systems in Sect. 2.5; (4) nine correspond to requirements for description of global GRFS systems in Sect. 2.6; and (5) nine correspond to requirements for description of satellite GRFS systems. It covers all research and development aspects including key block diagrams, and practical principle typical descriptions in the frequency band from 100 MHz to 60 GHz (or even 66 GHz). Dr. Progri reveals the research and development process by demonstrating how to understand and explain GRFS' most typical system deployment from basic diagrams to the final principle simulation examples (in Chaps. 4 and 6) and make recommendations for the future final products for research and development of GRFS. Starting with an introduction in Sect. 2.2 where an overview of the requirements for description of GRFS systems is given in both local and global coordinates, the chapter progressively examines various signal bands – such as VLF, LF, MF, HF, VHF, UHF, L, S, C, X, Ku, and, K and the corresponding geolocation requirements per band and per application – to achieve required performance objectives of up to 0° precision. Next follows a step-by-step approach on requirements for description of GRFS techniques and makes suggestions on the best state-of-the-art geolocation designs as well as advanced features found in signal

I. Progri, *Geolocation of RF Signals*,
DOI 10.1007/978-1-4419-7952-0_2, © Springer Science+Business Media, LLC 2011

generator instruments in Chap. 3. Chapter 4 also suggests the best mathematical techniques employed for GRFS at 100 MHz to 18 GHz or even 60 GHz. Some typical principle simulation examples taken from these system description requirements are discussed in a great detail during the second part of the book, which offers invaluable insights, all-in-one source, for the beginner, the experienced, expert analysts, and professionals.

2.2 Introduction

An illustration of requirements for description of GRFS systems are given in Fig. 2.1.

The motivation behind the requirements for description of GRFS systems is that "the use of cryptographic solutions, however, is insufficient to prevent attacks in wireless networks" [56]. Therefore, the identification, differentiation, design, development, deployment, integration, etc., of GRFS systems to identify all the threats coming from RF sources is imperative to the US National Defense Security, to the public safety, to search and rescue operations, to combat mission from around the world, etc.

First, as depicted in Fig. 2.1, we need to come up with a definition and explanation of what a local environment reference is. A local environment reference indicates the local environment range in which a GRFS system is analyzed, deployed, simulated

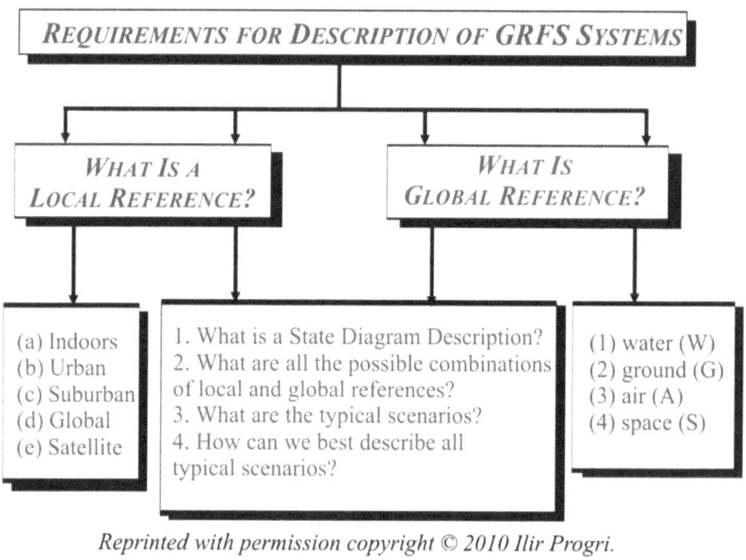

Fig. 2.1 An illustration of requirements of geolocation of RF signals (GRFS) system. Reprinted with permission copyright © 2010 Ilir Progri

etc., which as we have mentioned in Chap. 1 (Fig. 1.6) contains five segments: (1) indoor; (2) urban; (3) suburban; (4) global; (5) satellite.

Second, as depicted in Fig. 2.1, we need to come up with a definition and explanation of what a global reference is. A global reference indicates the global environment range in which a GRFS system is analyzed, deployed, simulated, etc., contains four segments: (1) water (W); (2) ground (G), (3) air (A); (4) space (S).

Third, as depicted in Fig. 2.1, we need to come up with a definition and explanation of what a state diagram is. A state diagram is one in which the global reference is indicated as a state and the local reference represents one aspect of the state diagram, which is indicated with an arc or a state transition path. The visualization and interpretation of the state diagram is obvious in the following sections.

Fourth, what are all the possible combinations of all state transitions in one diagram? The number of combinations of all the state transitions can be determined from the following equation:

$$\begin{aligned} N_C = C_4^1 + C_4^2 + C_4^3 + C_4^4 &= \frac{4}{1} + \frac{4 \times 3}{1 \times 2} + \frac{4 \times 3 \times 2}{1 \times 2 \times 3} + \frac{4 \times 3 \times 2 \times 1}{1 \times 2 \times 3 \times 4} \\ &= 4 + 6 + 4 + 1 = 15. \end{aligned} \quad (2.1.1)$$

So, it appears that there are 15 possible state transition path combinations for each local reference segment. Since we have five local reference segments, there should be up to 75 total number of possible state transition path combinations. We are going to see in much greater detail in the following section that in fact the number of combinations of state transitions paths corresponding to the typical scenarios is only 39.

Fifth, we are going to determine based on the information published in the literature what each typical case study looks like and we are going to make recommendations on what the prospects for future research and development in each case study are.

Sixth, how can we best describe all typical case studies? The main purpose of this chapter and this book is to research, investigate, and make recommendations on navigation, communications, and geolocation properties, requirements, and capabilities of several candidate radio frequency (RF) signals in the entire frequency band of 100 MHz to 66 GHz of all typical state transition case studies of outdoor and indoor environments that we consider next.

This chapter is organized as follows based on the information obtained from [1–138]. First, we are going to research, investigate, and propose the navigation, communications, and geolocation requirements, and capabilities of indoor GRFS systems in Sect. 2.3. Second, we are going to discuss, research, investigate, and make recommendations on the navigation, communications, and geolocation requirements, and capabilities of urban GRFS systems in Sect. 2.4. Third, we are going to discuss, research, investigate, and make recommendations on the navigation, communications, and geolocation requirements, and capabilities of suburban GRFS systems in Sect. 2.5. Fourth, we are going to discuss, research, investigate, and make recommendations on the navigation, communications, and geolocation requirements, and capabilities of global GRFS systems in Sect. 2.6. Fifth, we are going to discuss, research, investigate, and make recommendations on the navigation, communications, and geolocation requirements, and capabilities of satellite GRFS systems in Sect. 2.7. Section 2.8 concludes this chapter.

2.3 Requirements for Description of Indoor GRFS Systems

In this section we are discussing, researching, investigating, and making recommendations on requirements for description of indoor GRFS systems, which have an effective range up to 100 m in any global environment. The state transition path diagram for all indoor GRFS systems is described in Fig. 2.2.

Within 100 m, it is virtually impossible to have any kind of transition path from, let us say, indoor ground environments to indoor water, or air, or space, which is the reason why we have assumed that within 100 m we are either on the ground, in the air, in the water, or in space.

The electronics of GRFS systems working on the ground might be very different from the electronics of GRFS systems working in the air and from those working in space and from those working in the water due to differences in gravity, aerodynamics, dynamics, radiation, temperature, pressure, electric permittivity, magnetic permeability, etc.; however, in this section, as far as we are concerned, the basic principles of GRFS systems remain the same.

This section is organized as follows: first, we describe the requirements for description of indoor ground GRFS systems in Sect. 2.3.1. Second, we discuss the requirements for description of air GRFS systems in Sect. 2.3.2. Third, we consider the requirements for description of space GRFS systems in Sect. 2.3.3. And finally, we consider the requirements for description of water GRFS systems in Sect. 2.3.4.

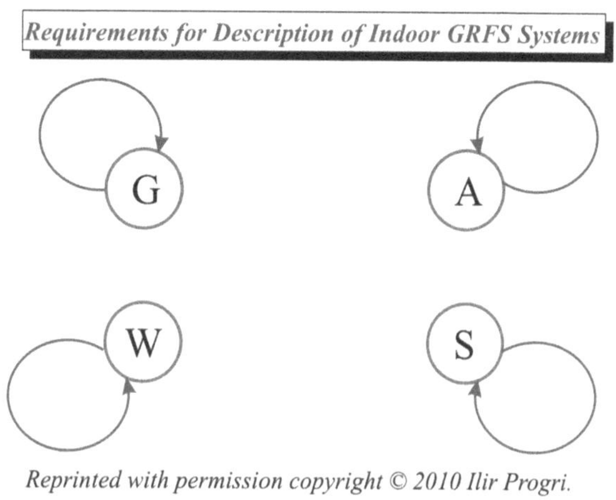

Fig. 2.2 An illustration of the state diagram of requirements for description of indoor GRFS systems. Reprinted with permission copyright © 2010 Ilir Progri

2.3.1 Requirements for Description of Indoor Ground GRFS Systems

For the complete discussion on indoor geolocation systems, the reader should refer to [1] and also to Dr. Progri's upcoming book on *Indoor Geolocation Systems: Theory and Applications*. However, as we have pointed out in Chap. 1, the description of GRFS systems is entirely different from the description of indoor geolocation systems. GRFS systems deal primarily with how to locate RF sources in an indoor environment based on where the user is located; that is, having the user (or GRFS receiver) as the center of the local coordinate system. Indoor geolocation systems deal mainly with locating a user inside based on previously positioned (or known or calculated trajectories of) transmitters (satellites, pseudolites or other positioning sensors).

In the general case, a GRFS system should be able to locate both indoor positioning transmitters and RF sources such as cordless phones, mobile phones, Wi-Fi access points (Worldwide Interoperability for Microwave Access (WiMAX) femtocells [43, 46]), HomeRF, Bluetooth, WLAN, light sensors, medical device sensors [15], wireless personal area network (WPAN) technologies in medical environments to support high efficiency medical care delivery anywhere and anytime [18], WiMedia UWB access point [16], motion sensors, temperature (heat or cold) sensors, smoke (or fire) sensors, seismic sensors, wind sensors, power outage sensor, either fiber, or landline disconnection (or loss of communication) sensor, etc.) A security system with *self-calibrating* and *self-* (or *internal, functional*) *awareness* capability might be very costly today for all homeowners but it might be a necessity for a good number of government facilities, commercial warehouses, retail stores, high class hotels, etc. So, RF sensors will become more and more useful in the future not only to tell us where an object is located but also about the condition (or *internal functional information*) of devices, humans, subsystems, etc. A GRFS system will become a necessary *secondary* (or *diagnostic*) system to locate, monitor, survey, weight, communicate, etc., location, health, status, condition of the primary everyday electric, power, communications, radiation, safety, transportation, etc., systems including humans (Fig. 2.3).

Some practical application examples of indoor ground GRFS systems may include body area networks that will support wireless communications of sensors positioned on a body or other objects [24].

Another example of indoor ground GRFS systems may be a "Human behavior inspired cognitive radio network design" which are supposed to be sensing their operating environment with little or no prior information and learning to adapt their behavior accordingly [36].

Another example of indoor ground GRFS systems is the "Millimeter-wave soldier-to-soldier communications for covert battlefield operations" to enable infantry soldiers of tomorrow – one of the most technologically advanced modern warfare has ever construed by creating the ability to provide information superiority at the

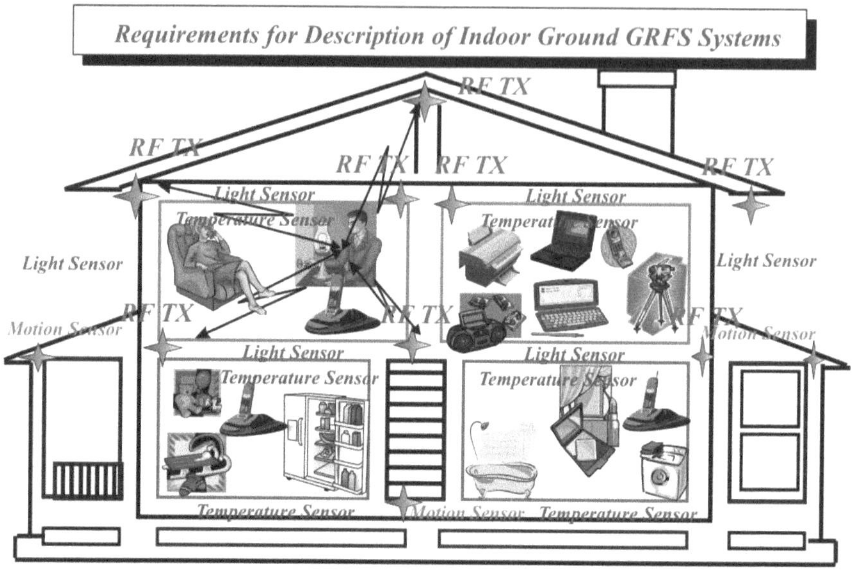

Reprinted with permission copyright © 2010 Ilir Progri.

Fig. 2.3 An illustration of the requirements for description of indoor ground GRFS systems. Reprinted with permission copyright © 2010 Ilir Progri

operational edge of military networks by equipping the dismounted soldier with advanced visual, voice, and data communications [45] or UAVs, and mobile APs [49]. Chapter 4 contains a much greater discussion of this scenario and also provides detailed principle simulation examples on indoor ground GRFS systems.

So, GRFS systems should be able to locate malfunctions or areas of concern in the primary systems and hopefully increase the accuracy repair and the probability of safety of life for any primary system. Moreover, when integrated with indoor geolocation systems, geospatial database, and/or Geographic Information Systems (GIS), and maps, GRFS systems should provide the safest, the shortest, and the best route in case of severe emergency. This requirement is even more critical for the next GRFS systems, which function in the air, in space, and in deep water.

2.3.2 Requirements for Description of Indoor Air GRFS Systems

For the most part, there are many secondary systems in military airplanes, helicopters, or even in commercial airplanes that can quickly and accurately identify faults in the primary systems. However, there are still many improvements that we can make to the existing indoor environments to further enhance the capability of the secondary RF systems. There is a simple explanation why GRFS systems are a much better solution than secondary systems built into the primary system.

Aeronautical communications can be subdivided into two main areas: (1) the safety critical air traffic control (ATC)/air traffic management (ATM) communication which also covers airline communications (AOC, AAC); (2) and the commercial aeronautical passenger communication (APC) [51]. Currently, safety critical communications is mainly based on voice communication using Double-Sideband Amplitude Modulation (DSB-AM) which is over 50-year-old communications technique which uses the available spectrum very inefficiently [51].

There is no surprise that an intruder might corrupt the primary system built on a 50-year-old communications technique; it will be almost impossible for an intruder to corrupt (jam or cause to miss-function) spatially located sensors working independently within an airplane or an indoor environment. Although it will be perhaps the hardest system to design, it will provide for sure the highest level of security, safety, and functionality. (For example, if we were to envision an air system that will read your DNA as you board the plane, that will be something that will make almost impossible for an intruder to board on the plane.) Now, we may not want that level of security on board of every commercial airplane but we will certainly want that level of security for the US Air Force plane that boards the President and the Vice President of the United States of America and maybe other high officials of the Pentagon (or Department of Defense (DoD)) or NATO countries.

Figure 2.4 provides an overview on the requirements for description of indoor air GRFS systems which contains a passenger airplane on the top and also a very small UAV (Shadow-200 UAV $13' \times 10' \times 2.94'$) with three transmitter antennae and four receiver antennae. For example small UAVs can be utilized for a number of case studies. We could use the small UAVs for a number of intelligence gathering missions or a number of other tactical air missions [52]. The Prestigious Defense Science Board of the US DoD performed a study in 2004 that recommended: "UAV and Uninhabited Combat Aerial Vehicle (UCAV) become an integral part of the US force structure, and not an additional asset," and that "UAV and UCAV be allowed unencumbered access to the UN National Airspace System (NAS) outside of restricted areas here in the US and around the world" [55]. A UAV is a low-cost nonpiloted airplane designed to operate in D-cube (Dangerous-Dirty-Dull) situations and although many UAVs exist today; however, with the advent of the commercial UAV's civil applications, the class of mini/macro UAVs is emerging as a valid option in a commercial scenario [52].

Many studies conducted by the Defense Science Board, the office of Science and technology, Government Accountability Office, and the Congressional Research Service Library of Congress have emphasized that soon there will be a significant number of UAVs (600 UAVs were manufactured in the US alone in 2006) operating side-by-side with manned civil aircraft in the Federal Aviation Administration (FAA)'s NAS, in which many UAVs will perform many of the

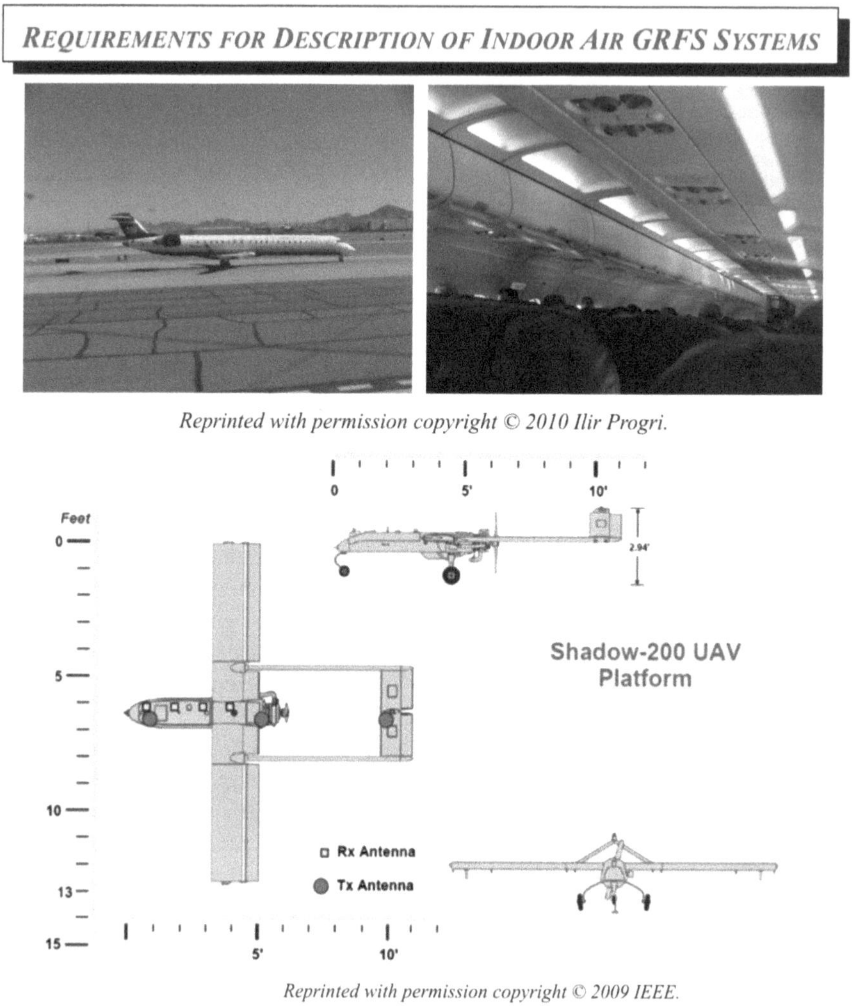

Fig. 2.4 An illustration of the requirements for description of indoor air GRFS systems. Reprinted with permission copyright © 2009 Blum, R.S., Haimovich, A.M., Li, J., *IEEE*; copyright © 2010 Ilir Progri

D-cube civilian missions [55]. Many of the safety certification operations of UAVs in US NAS which includes safety requirements, design, development process, verification, and operational procedures in the planned operational environment are discussed extensively in [55]. There safety requirements are good for the reader to know if a potential client will take the challenge to incorporate the GRFS systems principles of operations in a real-world UAV operational system design.

Further descriptions are provided later in the other air GRFS systems.

2.3.3 *Requirements for Description of Indoor Space GRFS Systems*

Similarly, for the most part there are many secondary systems in space shuttles or space stations that can quickly and accurately identify faults in space shuttle primary systems. However, there are still many improvements that we can make to the existing indoor environments to further enhance the capability of the secondary RF systems. There is a simple explanation why GRFS systems are a much better solution than secondary systems built into the primary system. In space there are different gravity requirements, different space and signal density, and displacement requirements. People and objects may be floating all the time. Therefore, indoor space GRFS systems should be able to locate astronauts, floating objects, faults in lines, or panels that have RF Bluetooth built in (see Fig. 2.5) [107].

Constellation is a human spaceflight program whose goals are gaining experience in operating away from Earth's environment, developing technologies to expand the space frontier, and conducting fundamental science [107].

Constellation was developed through the Exploration Systems Architecture Study, which determined how National Administration Space Agency (NASA) would pursue the goals laid out in the Vision for Space Exploration and the NASA Authorization Act of 2005 [107]. The reader can further understand how NASA's Exploration Systems Architecture Study is further incorporated into this chapter and other GRFS systems are discussed [107] further in this chapter.

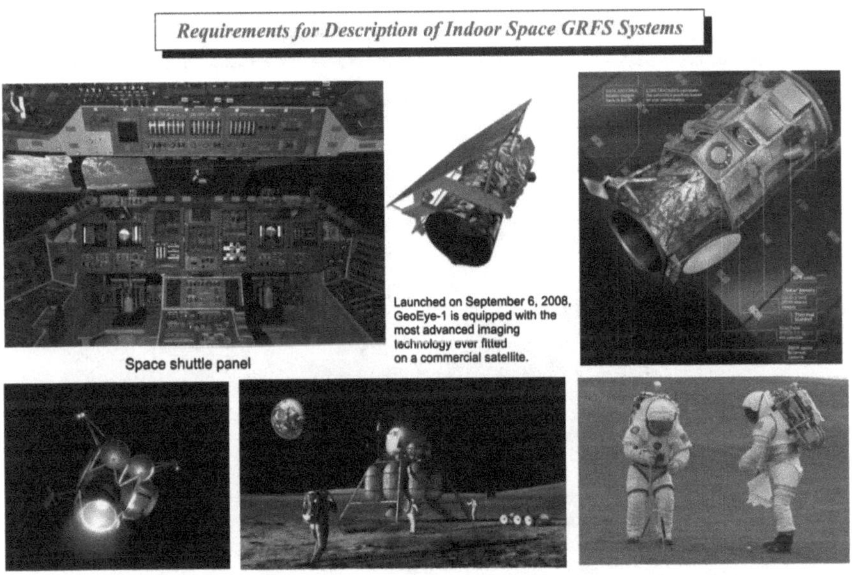

Fig. 2.5 An illustration of the requirements for description of indoor space GRFS systems. Images courtesy of National Administration Space Agency (NASA)

2.3.4 *Requirements for Description of Indoor Water GRFS Systems*

Similarly, for the most part there are many secondary systems in the military and civilian ships, submarines, or space stations that can quickly and accurately identify faults in the military and civilian ships, submarines, or naval vehicles on primary systems. However, there are still many improvements that we can make to the existing indoor environments to further enhance the capability of the secondary RF systems in the military and civilian ships, submarines, or naval vessels.

There is a simple explanation why GRFS systems are a much better solution than secondary systems built into the primary system. In water there are different water pressure requirements, different environment conductivity, permeability, and permittivity requirements. Therefore, indoor water GRFS systems should be able to locate, differentiate, discriminate, and geolocate navy personnel, objects, and faults in lines or panels that have RF Bluetooth built-in as an illustration.

Underwater wireless communications can enable many military applications such as oceanographic data collection, scientific ocean sampling, pollution and environmental monitoring, climate recording, offshore exploration, disaster prevention, assisted navigation, distributed tactical surveillance, and mine reconnaissance [38]. Some of these applications can be supported by underwater acoustic sensor networks (UW-ASNs) which consists of devices with sensing, processing, and communication capabilities that are deployed to perform collaborative monitoring tasks that can be utilized by this Navy destroyer shown in Fig. 2.6 [38].

Fig. 2.6 An illustration of the requirements for description of indoor water GRFS systems. Reprinted with permission copyright © 2009 Griffiths, H., Willis, N., and *IEEE*

2.4 Requirements for Description of Urban GRFS Systems

Urban GRFS systems are defined as GRFS systems in which the urban range of the area of operations is from 100 m up to 10 km on the ground, in the air, in space, or in or on water.

The electronics of GRFS systems working on the ground might be very different from the electronics of GRFS systems working in the air and from those working in space and from those working in the water due to differences in gravity, aerodynamics, dynamics, radiation, temperature, pressure, electric permittivity or magnetic permeability, etc.; however, in this section, as far as we are concerned the basic principles of GRFS systems remain the same.

There are eight urban GRFS systems that we are going to analyze, research, investigate, and make recommendations in this section as depicted in Fig. 2.7. First, we have the requirements for description of urban ground GRFS systems discussed in Sect. 2.4.1. Second, we present the requirements for description of urban air GRFS systems in Sect. 2.4.2. Third, we depict the requirements for description of urban water GRFS systems in Sect. 2.4.3. Fourth, we analyze the requirements for description of urban space GRFS systems in Sect. 2.4.4. Fifth, we discuss requirements for description of the urban ground-to-air (air-to-ground) GRFS systems in Sect. 2.4.5. Sixth, we provide requirements for description of urban ground-to-water (water-to-ground) GRFS systems in Sect. 2.4.6. Seventh, we present the requirements for description of urban air-to-water (water-to-air) GRFS systems in Sect. 2.4.7. Eight and finally, we depict the requirements for description of urban air-to-space (space-to-air) GRFS systems in Sect. 2.4.8.

There are many urban GRFS system that are discussed extensively in this section. One illustration is the Public Safety and Disaster Recovery (PSDR) which extensively relies on Professional Mobile Radio (PMR) communications systems to conduct their daily tactical and emergency operations [12].

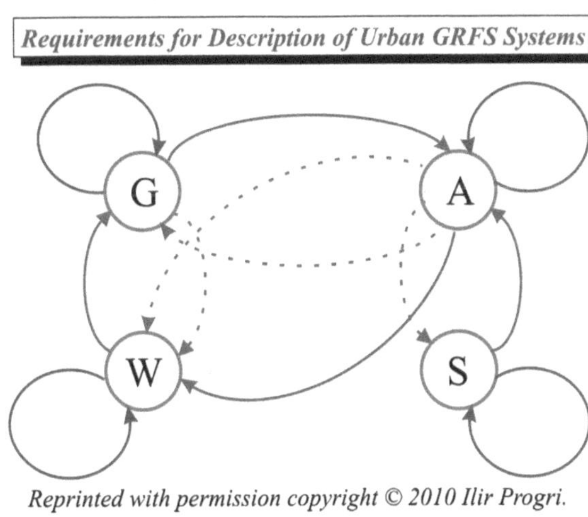

Fig. 2.7 An illustration of the finite state transition diagram of requirements for description of urban GRFS systems. Reprinted with permission copyright © 2010 Ilir Progri

Another example is the wireless broadband (WiBro) system for broadband wireless internet services providing high-speed portable internet access anywhere, anytime at low cost and high data rates [17].

2.4.1 Requirements for Description of Urban Ground GRFS Systems

After indoor ground, urban ground GRFS systems are the most common systems (as depicted in Fig. 2.8) in all or most metropolitan areas, big and small cities, in residential, commercial, or government facilities, in sports arena, university campuses, factories, hotels, etc. In these environments, the multipath distribution is as severe as in indoor ground GRFS systems. Chapter 3 discusses in great detail the kinds of signals that are employed in these environments. The applications range from everyday wireless local area networks, IEEE 802.11, IEEE 802.15, IEEE 802.16, and IEEE 802.20 Wi-Fi, WiMAX [33, 34], 3G WCDMA Mobile Networks [14],

Fig. 2.8 An illustration of the requirements for description of urban ground GRFS systems. Reprinted with permission copyright © 2010 Ilir Progri

FM Radio, Digital TV (DTV), Satellite TV, Low-enforcement networks, and radars to emergency networks such as mobile responder communication networks for public safety [13], the wireless broadband (WiBro) system for broadband wireless internet services [17], and WLAN technologies in medical environments to support high efficiency medical care delivery anywhere and anytime [18]. Another example is the integration of WiMAX with Wi-Fi for optimal pricing and bandwidth sharing using IEEE 802.16e/IEEE 802.11e standards [23, 33, 34] or an evolved cellular system architecture incorporating relay stations [41] or WiMAX femtocells [43].

There are also several proprietary and standard solutions for wireless point-to-point or point-to-multipoint or multipoint-to-point or multipoint-to-multipoint, which include vehicular, hospital, industrial, residential, commercial, etc., application [24, 28] or Universal Telecommunications Mobile Systems (UTMS) case study discussed "On femto deployment architectures and microcell offloading benefits in joint macro-femto deployments" in for cell ranged between 100 m and 10 km were calculated (see urban ground GRFS system from downtown Boston, Fig. 2.8) [46].

Other examples might include wireless relays for broadband access such as fixed, nomadic, or mobile relay stations based on IEEE 802.16e and 802.16j [29]. There are several advantages of wireless relays for broadband access such as: (1) no backhauling required, resulting in lower capital expenditures (CAPEX), and operation expenditures (OPEX); (2) flexibility in locating nodes; (3) within a cell, relays can enlarge the coverage area and increase the capacity of the cell boarders; (4) offer decreased transmit power and interference; (5) fast network rollout, indoor–outdoor service, and macro diversity by way of cooperative relaying [29]. There are also some disadvantages such as increased use of radio resources (in the time domain) and increased number of multiple transceivers in out-of-band relaying (in the frequency domain), additional delays [29].

2.4.2 Requirements for Description of Urban Air GRFS Systems

Urban air GRFS systems have to ensure that in the range of 100 m to 10 km the Air Force aircraft, Navy Aircraft, or Army Helicopter has intelligence about every single RF signal threat in the environment (see Fig. 2.9) [71]. Threats will be interference signals from enemy aircraft radars, enemy aircraft jammers, enemy missiles, etc.

Chapters 4 and 6 provide discussion of these case studies in great detail.

2.4.3 Requirements for Description of Urban Water GRFS Systems

An example of an urban water GRFS system is the design and implementation of a solution for the provision of converged tower and facility management services

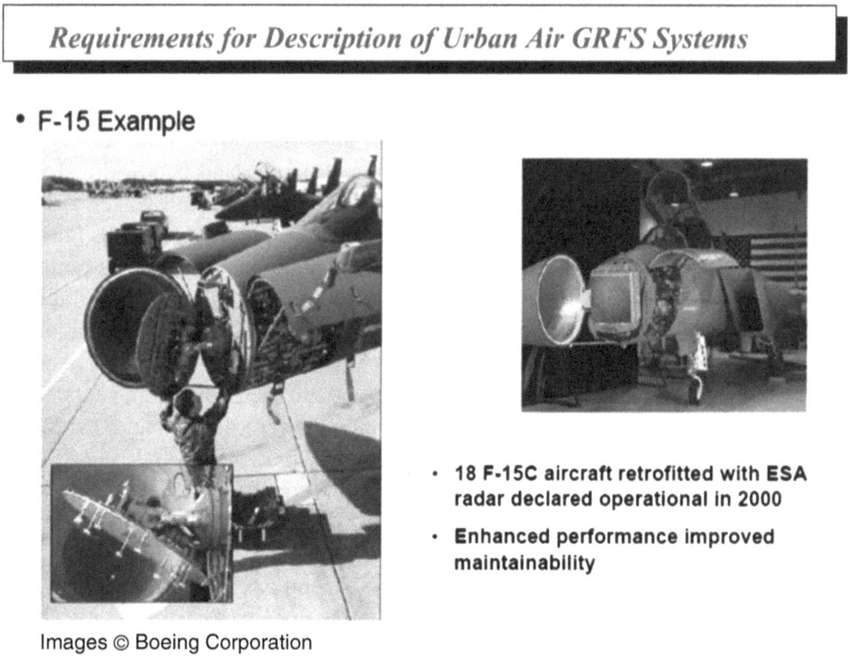

Fig. 2.9 An illustration of the requirements for description of urban air GRFS systems. Reprinted with permission copyright © 2000 Boeing Corp.; copyright © 2009 Williams, J., and *IEEE*

over satellite IP for Greek helicopters [32]. Helicopters generally operate at altitudes of 1,200 m according to rules that apply to instrument flight route and visual flight route (VFR) within the ATC areas, outside of which helicopters are only allowed to operate according to VFR rules [32]. There are also many commercial, recreations, touristic urban water GRFS system that will have the same signal specifications and be able to respond to disaster and safety of life in almost the same way as the urban ground GRFS system in Sect. 2.4.1 in the range of 100 m to 10 km such as the Sydney opera house or many boats sailing in the Sydney's harbor as illustrated in Fig. 2.10.

2.4.4 Requirements for Description of Urban Space GRFS Systems

Urban space GRFS systems constitute the space stations' environment in the 100 m to 10 km as depicted in Fig. 2.11. There is not much going on for urban space GRFS system unless there is a mission to repair the space station or when there is a mission to navigate, rendezvous, and dock a space shuttle in the space station.

Fig. 2.10 An illustration of the requirements for description of urban water GRFS systems. Reprinted with permission copyright © 2009 Brookner, E., and *IEEE*

For example the Lunar outpost, as shown on in Fig. 2.11, will be an inhabited facility on the surface of the Moon which NASA currently proposes to construct over the 5 years between 2019 and 2024. The United States Congress has directed that the U.S. portion, "shall be designated the Neil A. Armstrong Lunar Outpost" [107].

2.4.5 *Requirements for Description of Urban Ground-to-Air (Air-to-Ground) GRFS Systems*

Urban ground-to-air (air-to-ground) GRFS systems are perhaps the most common forms of close combat of infantry ground forces supported by aircraft, helicopter, short-range air missile, etc. This is an environment that certain UAVs might be the most suitable means of close combat intelligence gathering as shown in Fig. 2.12. In the example illustrated in the figure, a UAV identifies a vehicle mounted rocket launcher, other portable RF transmitter, and FM radio stations and communications towers. The two greatest concerns are interference and interoperability.

Requirements for Description of Urban Space GRFS Systems

Fig. 2.11 An illustration of the requirements for description of urban space GRFS systems (SATCOMS IRIDIUM). Images courtesy of NASA

Silent Sentry is a passive sensor that uses emissions from indigenous transmitters as its RF sources. Silent Sentry 3 is tailored for FM radio, but can be extended to other waveforms. Line-of-sight is necessary between receiver/transmitters and aircraft/missiles. A direct reference signal is also necessary at the receiver but line-of-sight is not required. Reflected RF energy is collected at the receiver and compared with the original reference signal to provide detection and tracking information. Track data can be sent to the Silent Sentry track display or to a sensor fusion or command and control system (e.g., via Silent Sentry Byte Stream, OTH-Gold, Link 16, Asterix ...) [70].

Other examples might include "Intelligent sensing and classification in Ad Hoc networks: a case study" such as denial of service (DoS) through intelligent jamming of the most critical packet types of flows in an Ad Hoc network [57].

This is only a very small piece of the network centric warfare (NCW) to enable ground and airborne vehicle-based on-the-move (OTM) and on-the-halt (OTH) network centric connectivity [30]. Urban ground GRFS systems can be used for numerous monitoring, tracking, surveillance, search and rescue operations from the public safety and disaster monitoring of the scales that we have seen in September 11, 2001, in Katrina, etc.

Perhaps the most important example is the improved situational awareness of the military aircraft in both the battlespace and civil airspace which is discussed here in great detail. For the military air communications, navigation, and surveillance

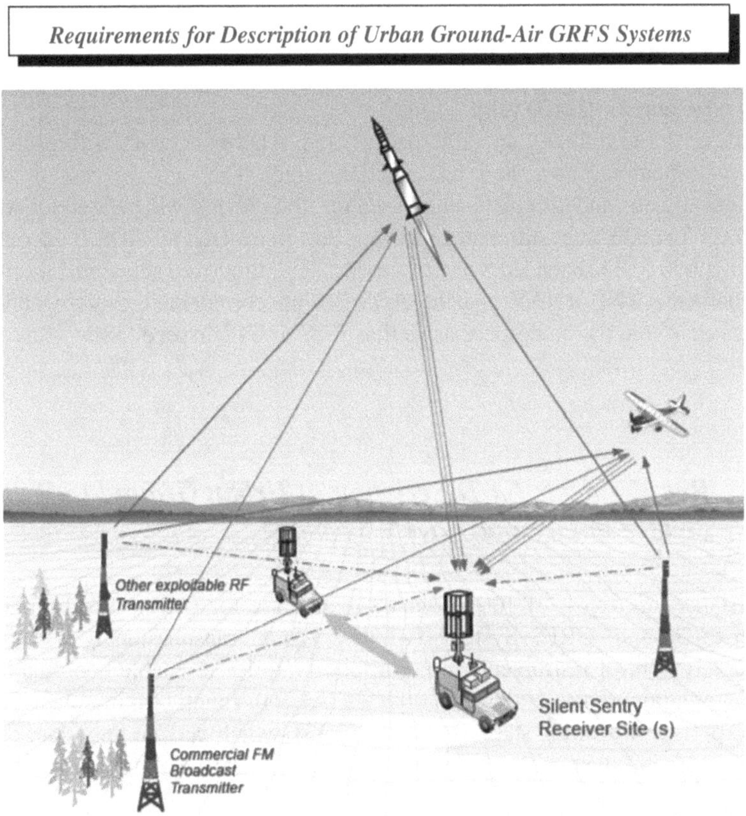

Fig. 2.12 An illustration of the requirements for description of urban ground-to-air (or air-to-ground) GRFS systems. Reprinted with permission copyright © 2009 Griffiths, H., Willis, N., and *IEEE*

(CNS) problems as presented in [58] the interoperability appeared to be a problem. Military aircraft must transition from ground-based navigation aids (VOR/TACAN) to area navigation in performance-based airspace (RNP RNAV) and transition from secondary RADAR surveillance to Automatic Dependent Surveillance Broadcast (ADS-B).

The 1,090 MHz transponder upgrade was proposed to promote safety, facilitate civil interoperability, improve situational awareness, and greatly improve both military and civil air traffic surveillance [58].

With Global Positioning Systems (GPS) installed, the most accurate locating information for each military aircraft will be the self-generated position displayed in that aircraft which was not available to the air traffic controller in 2003. To broadcast this information to the air traffic controller requires installation of a standard data link and a common reference; then, a line-of-site broadcast of the accurate GPS-based *aircraft self-reports* (ADS-B) is possible. These self-reports of aircraft identity WGS-84 geodetic position (Lat-Long-Alt) and velocity vector will be transmitted up to twice a second to the air traffic controller. Aircraft within line-of-sight can use these transmissions to automatically produce a *cockpit display of traffic information* (CDTI) [58].

If all civil and military air traffic participated, ADS-B network will result in an improved situational awareness and aviation safety. The same processing power, modular software, and cockpit displays used for RNP RNAV will be used for ADS-B and CDTI. In 2003, three different data links were considered for ADS-B but only the Mode S 1,090 MHz Extended Squitter was installed. Improved situational awareness from both the RNP RNAV and the ADS-B with considerable military utility is anticipated within the battlespace as well as within civil airspace [58].

2.4.6 Requirements for Description of Urban Ground-to-Water (Water-to-Ground) GRFS Systems

For distances in excess of 100 m, which is the case for Urban Ground-to-Water (Water-to-Ground) GRFS Systems, wireless signal transmission is also crucial to remotely control instruments in ocean observatories to enable coordination of swarms of autonomous underwater vehicles (AUVs), robots, environment (in and around ports), or docking stations (see Fig. 2.13), which play an important role of mobile nodes in future ocean observation networks by virtue of their flexibility and reconfigurability [38].

2.4.7 Requirements for Description of Urban Air-to-Water (Water-to-Air) GRFS Systems

Urban air-to-water (water-to-air) GRFS systems could be very similar to the urban ground-to-air (air-to-ground) GRFS systems but could also be very different from the latter. For urban air-to-water (water-to-air) GRFS systems, the multipath should be less severe than the multipath for ground-to-air (air-to-ground) GRFS systems.

Figure 2.14 illustrates an urban air-to-water (water-to-air) GRFS system in Sydney Harbor (including the Sydney Opera House) that is monitored by the Australian Wegetail Airborne Radar Surveillance System.

Australia has, and is, supporting a significant set of phased array development activities spanning more than 50 years including a wide range of civil and military applications, which ensures a viable and vibrant development of environment across government and industrial laboratories [60]. More discussion is provided in Chaps. 4 and 6.

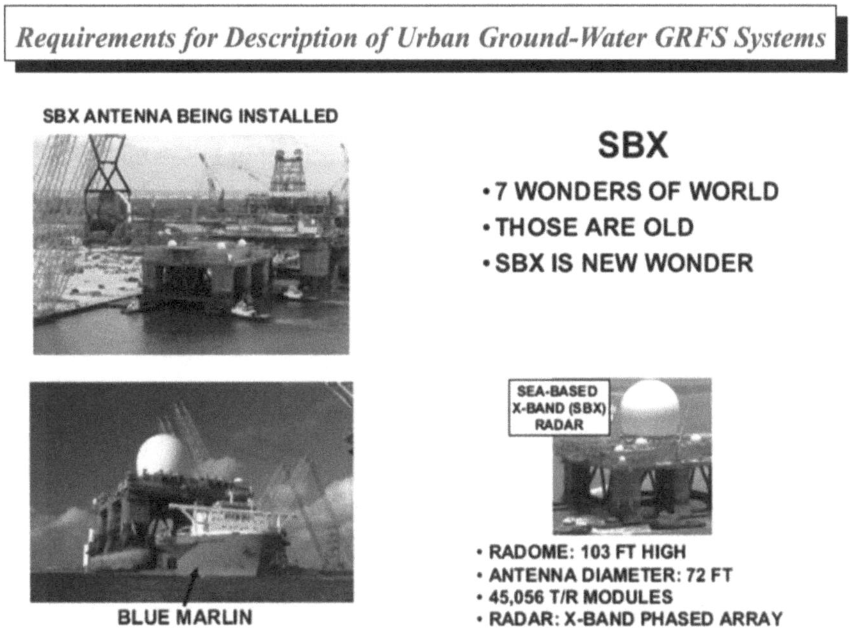

Fig. 2.13 An illustration of the requirements for description of urban ground-to-water (or water-to-ground) GRFS systems. Reprinted with permission copyright © 2009 Brookner, E., and *IEEE*

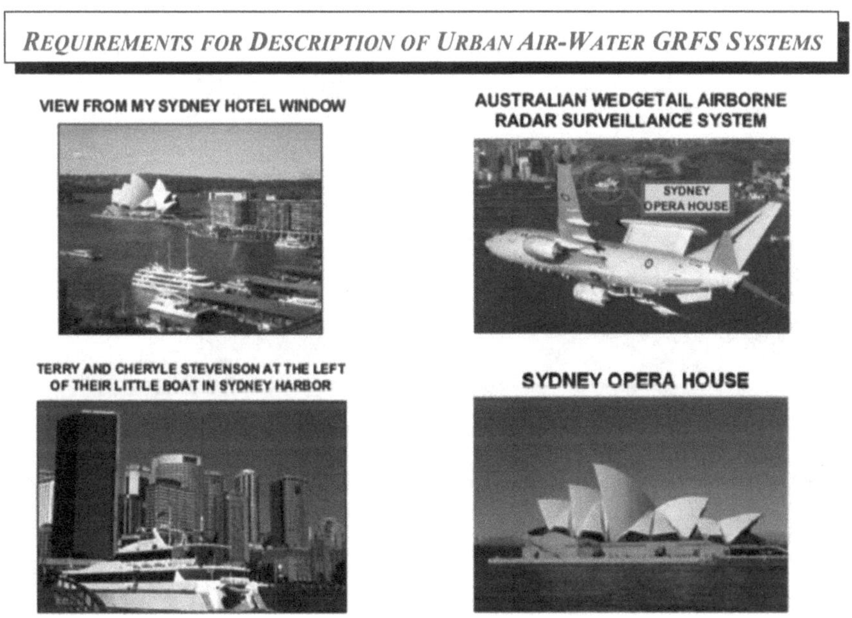

Fig. 2.14 An illustration of the requirements for description of urban air-to-water (or water-to-air) GRFS systems. Reprinted with permission copyright © 2009 Brookner, E., and *IEEE*

2.4.8 Requirements for Description of Urban Air-to-Space (Space-to-Air) GRFS Systems

Urban air-to-space (space-to-air) GRFS systems illustrate the rendezvous and docking of space shuttles to space stations as illustrated in Fig. 2.15. There are, however, many safety requirements for the astronauts that must be taken into consideration. So, from the safety point of view the requirements for description of air-to-space (space-to-air) GRFS systems are at much higher level of complexity and cost than those of the ground-to-air (air-to-ground) GRFS systems even though both environments are very different from each other due to differences in gravity, speed, aerodynamics, range of operation for humans, level of control and coordination of operations, number of people involved, etc.

After the systems are configured, as shown in Fig. 2.15, for lunar flight, the EDS will fire for the 390-s translunar injection (TLI) burn, which will accelerate the spacecraft stack from 28,000 to 40,200 km/h. The TLI burn will be done in the "eyeballs out" fashion, that is, with the astronauts being "pulled" from their seats. After the TLI burn, the EDS is jettisoned [107]. During the 3-day translunar coast, the four-man crew will monitor the Orion's systems, inspect their Altair spacecraft and its support equipment, and, if necessary, change their trajectory to allow the

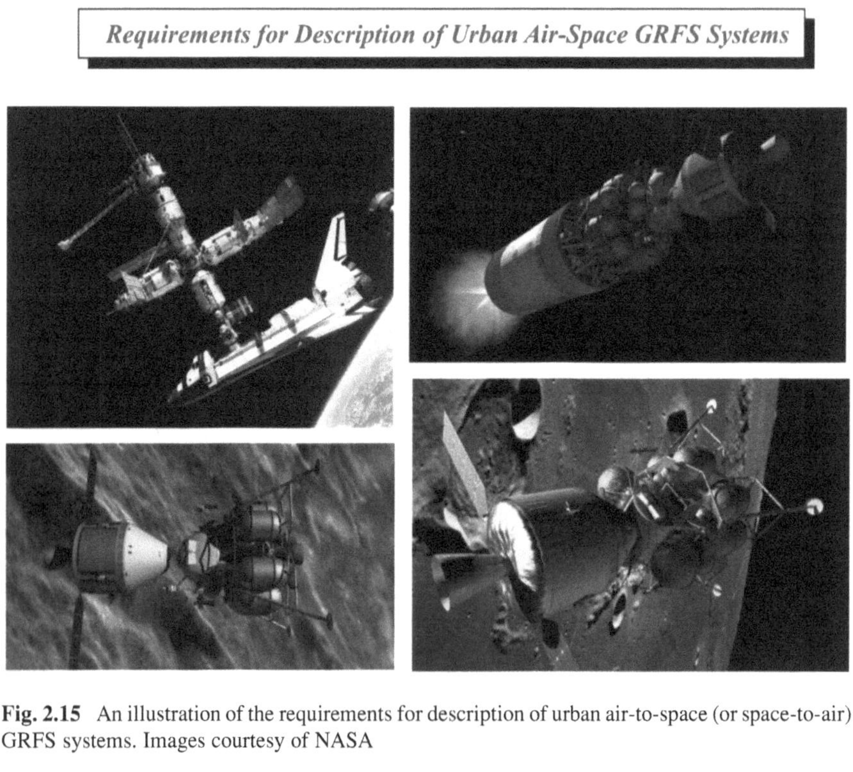

Fig. 2.15 An illustration of the requirements for description of urban air-to-space (or space-to-air) GRFS systems. Images courtesy of NASA

Altair to land in a near-polar landing site suitable for a future lunar base [107]. Three days after TLI, the Orion/Altair combination, approaching the lunar far side, will orient the Altair's engines in the proper direction for the lunar orbit insertion (LOI) burn to begin. Once in orbit, the crew will refine the trajectory and configure the Orion CSM for unmanned flight, upon which all of the crew members will transfer to the Altair, and upon receiving clearance from Mission Control, will undock from the Orion [107].

2.5 Requirements for Description for Suburban GRFS Systems

Suburban GRFS systems are defined as systems in which the suburban range of the area of operations is from 10 km up to 100 km in ground, air, space, or water (see Fig. 2.16). It is hoped that the readers in this section will find out an approach to enable inter- and intracommunity communications, *interoperability* between the commercial, public safety, military, space, etc. communities in a way that was never presented before [30]. An example of a suburban GRFS system includes Wireless Metropolitan Area Networks (WMAN) based on IEEE 802.16 WiMAX technology in the 10–66 GHz frequency spectrum that can achieve maximum transmission range of 50 km [28, 41].

Other standards include IEEE 802.22 for wireless regional area networks (WRANs) serving broadband communications for remote communities, effectively achieved through a cognitive radio (CR) idiom [39]. Another example of suburban GRFS system that we are going to discuss extensively in this section is DARPA's Network Centric Radio System (NCRS), first generation mobile ad hoc network (MANET), designed to enable ground and airborne-vehicle-based OTM and OTH network centric connectivity [30].

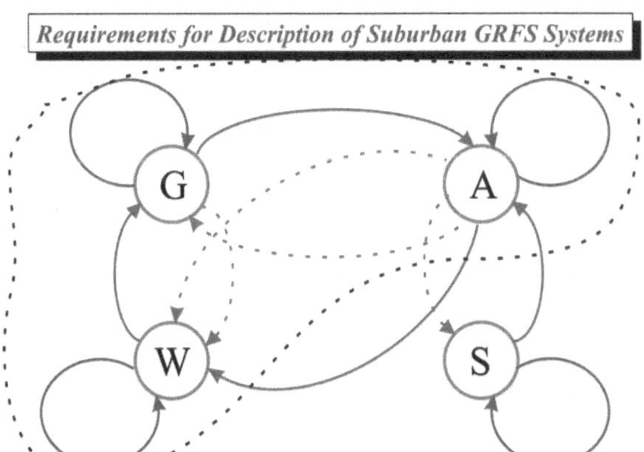

Reprinted with permission copyright © 2010 Ilir Progri.

Fig. 2.16 An illustration of the state diagram of requirements for description of suburban GRFS systems. Reprinted with permission copyright © 2010 Ilir Progri

These and many other systems are only subsystems to the Global Information Grid (GIG) communications systems which is the network fabric with which to build a "system-of-systems" to fulfill the ultimate goal of network-centric warfare [30]. So, as we move forward to the discussion in this section and in the following sections it should become relevant and understandable to the reader that GRFS systems are in fact subsystems to the GIG communications systems that are designed to exploit the signal design of the interoperability of the Joint Tactical Radio Systems (JTRS) among all other signal designs that are part of the network-centric warfare "system of systems."

Taking into considerations the examples presented, we provide an organization of this section which includes the requirements for description of several suburban GRFS systems as depicted in Fig. 2.16. First, we have the requirements for description of suburban ground GRFS systems discussed in Sect. 2.5.1. Second, we present the requirements for description of suburban air GRFS systems in Sect. 2.5.2. Third, we depict the requirements for description of suburban water GRFS systems in Sect. 2.5.3. Fourth, we analyze the requirements for description of suburban space GRFS systems in Sect. 2.5.4. Fifth, we discuss requirements for description of the suburban ground-to-air (air-to-ground) GRFS systems in Sect. 2.5.5. Sixth, we provide the requirements for description of suburban ground-to-water (water-to-ground) GRFS systems in Sect. 2.5.6. Seventh, we present the requirements for description of suburban air-to-water (water-to-air) GRFS systems in Sect. 2.5.7. Eight, we depict the requirements for description of suburban air-to-space (space-to-air) GRFS systems in Sect. 2.5.8. Ninth and finally, we conclude this section with the requirements for description of suburban ground-to-air-to-water (air-to-water-to-ground or water-to-air-to-ground) GRFS systems in Sect. 2.5.9.

2.5.1 Requirements for Description of Suburban Ground GRFS Systems

An example of suburban ground GRFS systems are WRANs which are aimed at providing alternative broadband wireless internet access in rural areas without creating harmful interference to licensed TV broadcasting [19]. For example, a typical 802.22 WRAN base station with radius of the coverage area of 35 km coexisting with a DTV station with radius of the coverage area of 135 km [19], or an effective range for long-term evolution of mobile broadband of 100 km and beyond [40].

Another example of suburban ground GRFS systems are dynamic spectrum access networks (DSANs), also known as NeXt Generation (xG) networks that enable efficient spectrum usage to network users via dynamic spectrum access techniques and heterogeneous network architectures; DARPA aims to dynamically redistribute allocated spectrum based on cognitive radio technologies (see Fig. 2.17) [20].

Another example of a suburban ground GRFS system is WiMAX that is expected to provide high data rate communications in metropolitan area networks

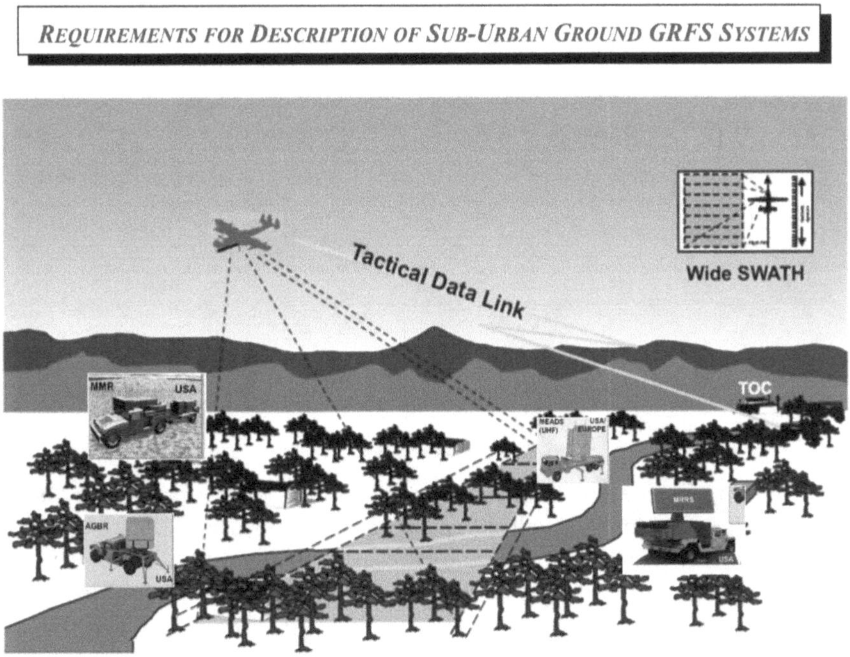

Fig. 2.17 An illustration of the requirements for description of suburban ground GRFS systems. Reprinted with permission copyright © 2009 Davis, M., and *IEEE*

(MANs) based on IEEE 802.16 standard [21, 22]. Details about the signal specifications for WiMAX will be given in Chap. 3.

Other examples include universal mobile telecommunications system (UMTS) multimedia broadcast multicast service (MBMS), digital video broadcasting to handheld (DVB-H), terrestrial digital multimedia broadcasting (T-DMB), Media-FLO, etc. [24]. Suburban ground GRFS systems are perhaps the most crowded systems in terms of technologies such as cognitive radio, dynamic spectrum access, secondary spectrum tracing, and an array of IEEE standards that we have already mentioned. And while we have underlined all the benefits of these technologies, there are also many risks for many stakeholders such as regulators, spectrum right holders, and spectrum operators [25].

2.5.2 Requirements for Description of Suburban Air GRFS Systems

The US Air Force, Navy, and Army (i.e., military) desire, seek, invite proposals, lead programs and projects to design, develop, demonstrate, and commercialize highly interoperable (compatible or noncompatible) radio systems to enable information to be directly exchanged among multiple organizations via NCW [30].

Fig. 2.18 An illustration of the requirements for description of suburban air GRFS systems. Images courtesy of DoD

Figure 2.18 is an example of an airborne network centric suburban air GRFS system. This flight formation will enable the warfighters to take advantage of all the available information within the battlespace in a rapid and flexible manner. An essential capability of the NCRS that was needed since 2003 was *radio interoperability at the tactical level via the network not the radio* [30].

2.5.3 Requirements for Description of Suburban Water GRFS Systems

Requirements for description of suburban water GRFS systems (see Fig. 2.19) is motivated by the recently vested interest in the "Growth of underwater communication technology in the U.S. Navy" [37]. The office of Naval Research has made a significant investment in both theoretical and applied work over the past 20 years and currently funds multiple programs in acoustic communication which includes topics of research increased throughput, better power efficiency, low probability of detection, and compact implementation as depicted in Fig. 2.19.

Figure 2.19 shows an aircraft carrier on the bottom right and Navy aircraft on the top right. Of particular interests for these applications are techniques applied to adaptive arrays space time signal processing [73].

There are multiple existing civil and military systems that provide precision approach and landing for aircraft; a partial list of existing systems includes [81]:

1. Instrument Landing System (ILS) for commercial and limited military
2. Microwave Landing System (MLS) for commercial (Europe) and very limited Military
3. Precision Approach Radar (PAR) for Military
4. Mobile Microwave Landing System (MMLS) for Military
5. Marine Remote Area Approach and Landing System for Military

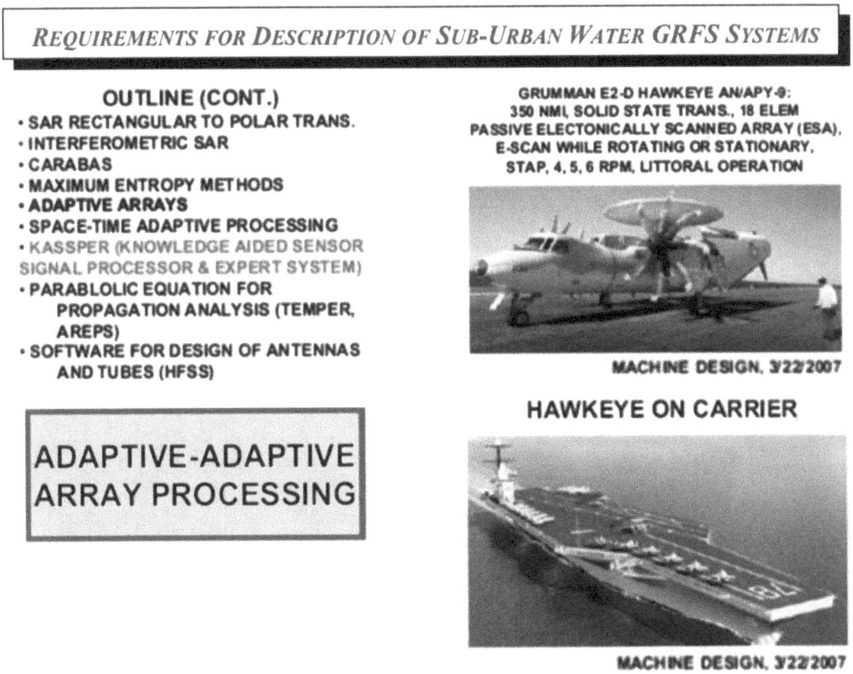

Fig. 2.19 An illustration of the requirements for description of suburban water GRFS systems. Reprinted with permission copyright © 2009 Brookner, E., and *IEEE*

6. The Instrument Carrier Landing System (SPN-46) for Military
7. Joint Precision Approach and Landing System (JPALS) for Navy's Military envisioned meeting the DoD's need for an allweather, antijam, combat-ready, Category II/III aircraft landing system [80–82]

   Navy's JPALS interoperability issues are considered in [82].

2.5.4 *Requirements for Description of Suburban Space GRFS Systems*

Suburban space GRFS systems may include an array of space vehicle in the spherical environment with effective range from 10 km to 100 km (see Fig. 2.20) [71].

There are four different space vehicles shown in Fig. 2.20: (1) SEASAT built in 1978; (2) SIR-A built in 1981; (3) SIR-B build in 1984; and (4) SIR-C built in 1994. They all operate at the frequency 1 GHz.

SEASAT was the first Earth-orbiting satellite designed for remote sensing of the Earth's oceans and had on board the first spaceborne synthetic aperture radar

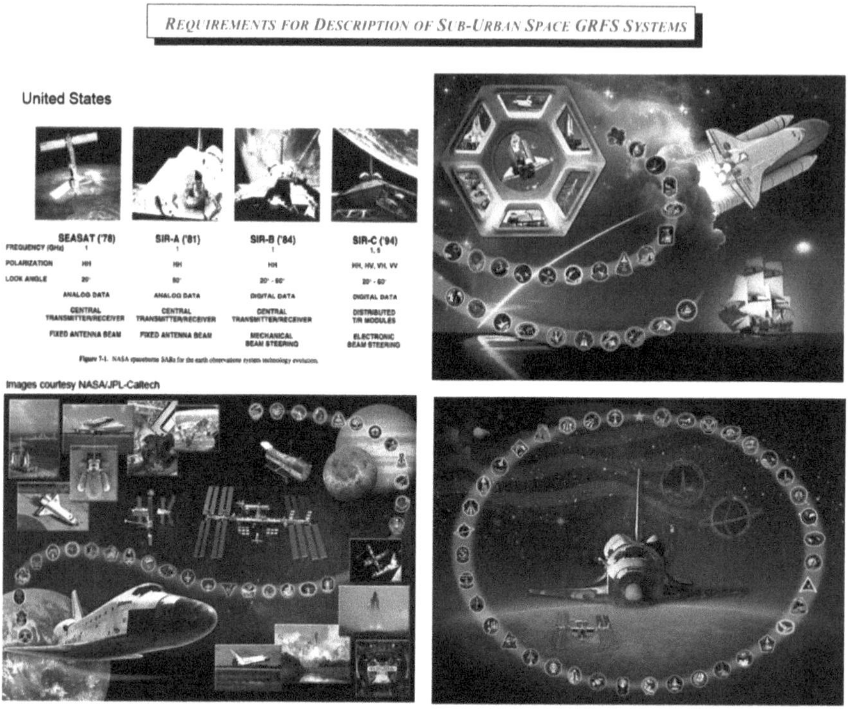

Fig. 2.20 An illustration of the requirements for description of suburban space GRFS systems. Images courtesy of NASA/JPL-Caltech and NASA

(SAR). The mission was designed to demonstrate the feasibility of global satellite monitoring of oceanographic phenomena and to help determine the requirements for an operational ocean remote sensing satellite system. Specific objectives were to collect data on sea-surface winds, sea-surface temperatures, wave heights, internal waves, atmospheric water, sea ice features and ocean topography. SEASAT was managed by NASA's Jet Propulsion Laboratory and was launched on 26 June 1978 into a nearly circular 800 km orbit with an inclination of 108°. SEASAT operated for 105 days until 10 October 1978, when a massive short circuit in the satellite's electrical system ended the mission [74].

SEASAT carried five major instruments designed to return the maximum information from ocean surfaces:

1. Radar altimeter to measure spacecraft height above the ocean surface
2. Microwave scatterometer to measure wind speed and direction
3. Scanning multichannel microwave radiometer to measure sea surface temperature
4. Visible and infrared radiometer to identify cloud, land, and water features
5. SAR L-band, HH polarization, fixed look angle to monitor the global surface wave field and polar sea ice conditions [74].

Many later remote sensing missions owe their legacy to SEASAT. These include imaging radars flown on NASA's Space Shuttle, altimeters on Earth-orbiting satellites such as TOPEX/Poseidon, and scatterometers on NASA Scatterometer (NSCAT), QuikSCAT, and Jason 1. SEASAT was able to detect the wakes of submerged submarines, a discovery not anticipated before launch. The conspiracy theory holds that once this was discovered, the military shut SEASAT down, with a cover story of a power supply short.

Space borne imaging radar missions data (SIR) SIR-A, SIR-B, and SIR-C can be found from [75–77]. SEASAT, SIR-A, and SIR-B SARs operate at 1 GHz as opposed to SIR-C SAR, which operates at 1.5 GHz. SEASAT, SIR-A, and SIR-B SAR signals are only HH polarized as opposed to the SIR-C SAR, which is possibly polarized with all four combinations (HH, HV, VH, and VV). The data in SEASAT and SIR-A is in analog format as opposed to the SIR-B and SIR-C in the digital format. SEASAT, SIR-A, and SIR-B SARs require central transmitter/receiver modules as opposed to the SIR-C SAR which requires distributed T/R modules. Last but not least, SEASAT and SIR-A SARs have fixed antenna beam, SIR-B SAR has mechanical beam steering capability and SIR-C SAR has electronic beam steering [74–77].

With this example we have illustrated what type of satellite radars all future generation of satellites will have.

2.5.5 Requirements for Description of Suburban Ground-to-Air (Air-to-Ground) GRFS Systems

Suburban ground-to-air (air-to-ground) GRFS systems may include many examples of detection, monitoring, tracking and surveillance, etc. as illustrated in Fig. 2.21. Due to the increased range, it might get more difficult to precisely identify, differentiate, and geolocate all the RF emitters in the environment. So, the level of differentiation of a suburban GRFS system is entirely different from the level of differentiation of an urban GRFS system. A suburban ground-to-air GRFS system should be able to locate where the large objects are located as opposed to an urban GRFS system which might be able to even tell how many people are in a certain warfare environment.

There is a wide range of military and civilian applications in which UAVs might be employed successfully such as remote environmental research, pollution assessment and monitoring, fire-fighter management, security, target detection, recognition, and surveillance, etc. [52]. For example in Fig. 2.21, bottom right, a squadron of UAVs can be utilized to monitor QUEEN MARY II Flying Cruise or in Fig. 2.21, top left, UAVs are employed for target detection and recondition in heavy foliage [61].

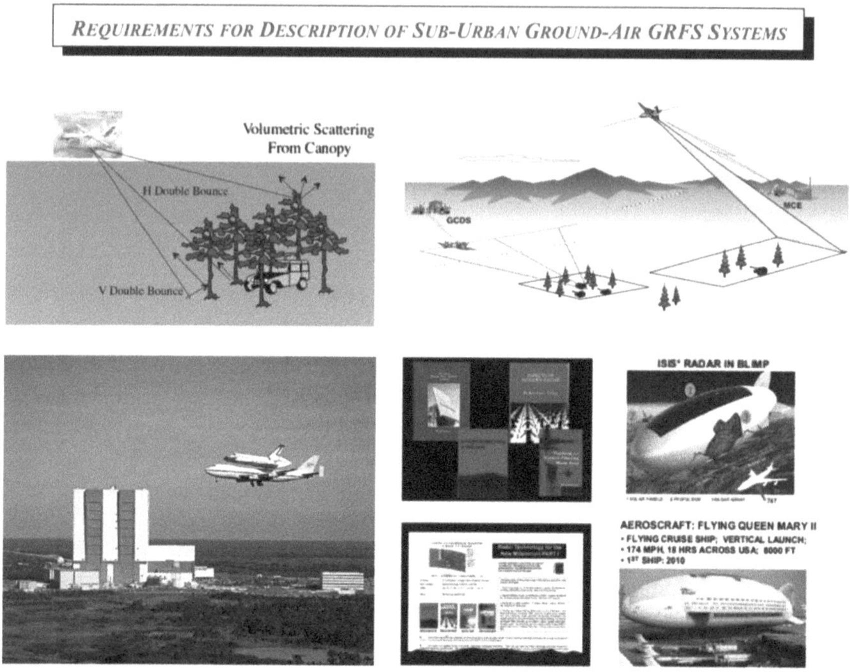

Fig. 2.21 An illustration of the requirements for description of suburban ground-to-air (or air-to-ground) GRFS systems. Reprinted with permission copyright © 2009 Davis, M., Brookner, E., *IEEE*; Image on *bottom left* is courtesy of Wikimedia Foundation, Inc

2.5.6 Requirements for Description of Suburban Ground-to-Water (Water-to-Ground) GRFS Systems

Maritime networks are one of the least studied network configurations and they probably represent the biggest challenge for information and presentation in this chapter [35].

It is hoped that this chapter (and perhaps) this book will expand the readers' perspective on maritime networks and suburban ground-to-water (water-to-ground) GRFS systems (see Fig. 2.22). Maritime networks operate in low-bandwidth environments with varying communications capabilities. Naval at sea (maritime) networks are particularly difficult to manage due to their dynamic, heterogeneous, and low-bandwidth connectivity [35]. Applications in maritime networks must operate differently to take into account mobility (link failures) and scare communications resources (especially bandwidth) [35]. The limited bandwidth connecting each maritime ship is often (node) not sufficient to even support the network traffic generated locally [35]. Maritime networks consist of a network operational center (NOC) (for example one such center is the Naval Base in San Diego (or Perl Harbor on the Pacific) and many other Naval Bases on the Atlantic Ocean) that acts as a land-based relay for all satellite

Fig. 2.22 An illustration of the requirements for description of sub-urban ground-to-water (or water-to-ground) GRFS systems. Top Image reprinted with permission copyright © 2006 Northrop Grumman Shipbuilding[1]. Bottom images courtesy of Wordpress and Wikipedia

communication, a limited number of mobile nodes (ships or maritime land/air units), and the bearers that connect them [35].

2.5.7 *Requirements for Description of Suburban Air-to-Water (Water-to-Air) GRFS Systems*

The Navy is also interested in a link from the acoustical world to the RF line-of-sight or satellite communications which might include gateways of various sorts

[1]Reprinted with permission from Northrop Grumman Shipbuilding only granted for the first edition. Northrop Grumman Shipbuilding retains all rights and copyright ownership of the photo. Any further uses of the photo in future editions will require an additional request for permission.

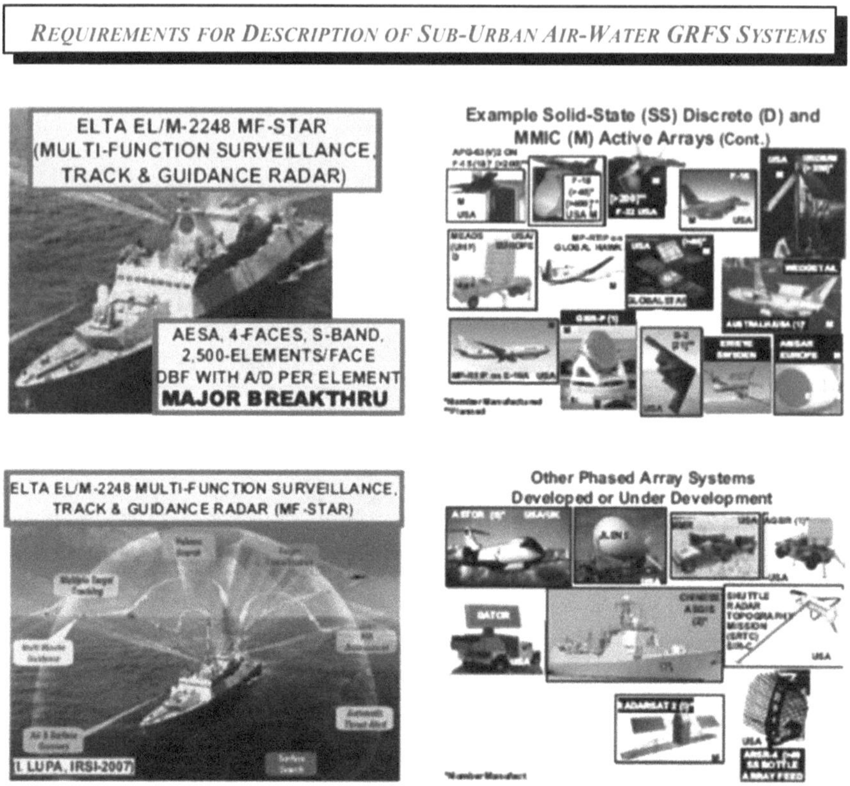

Fig. 2.23 An illustration of the requirements for description of suburban air-to-water (or water-to-air) GRFS systems. Reprinted with permission copyright © 2009 Brookner, E., and *IEEE*

such as small buoys, typically used for coastal applications, and dedicated vehicles, such as solar-powered AUV or gliders used when moorings are not feasible [37]. Several examples of suburban air-to-water (water-to-air) GRFS systems are depicted in Fig. 2.23. These systems should help especially in fighting piracy and ultimately capturing pirate ships [62].

We cannot leave without mentioning: (1) After completing their Lunar Sortie operations, the crew will enter the Altair's ascent stage and lift off from the Moon's surface, powered by a single engine, while using the descent stage as a launchpad (and as a platform for future base construction); (2) upon entering orbit, the Altair docks with the waiting Orion spacecraft, and the crew then transfers themselves and any samples collected on the moon over to the Orion [107].

Other civil applications might include costal boarder monitoring, agriculture and fishery applications, oceanography, communications relays for wide-band applications which can be divided into four large groups: (1) environmental applications; (2) emergency security applications; (3) communications applications; (4) monitoring applications [52].

2.5.8 *Requirements for Description of Suburban Air-to-Space (Space-to-Air) GRFS Systems*

Suburban air-to-space (space-to-air) GRFS systems are depicted in Fig. 2.24. As the space is getting crowded with more and more satellites, space stations, space vehicles, there are more and more opportunities for these systems to become more popular and so, we might see operations for exchange, deployment, communications, etc. from neighboring satellites, space stations and space vehicles, etc. [64]. For example, in Fig. 2.24 we have the integrated symmetrical concentrator (ISC) solar power satellite (SPS) in geosynchronous orbit produced by NASA's Space Electric Rocket Test (SERT) program in 2001 [64, 78] on the left and Tandem X on the right [79].

The concept of deriving terrestrial energy from space-based solar-electric systems using wireless power transfer has captured the imagination of the US government and private stakeholders for over 40 years [78]. Various studies of this concept were conducted during the 1970s, by NASA and the Department of Energy such as

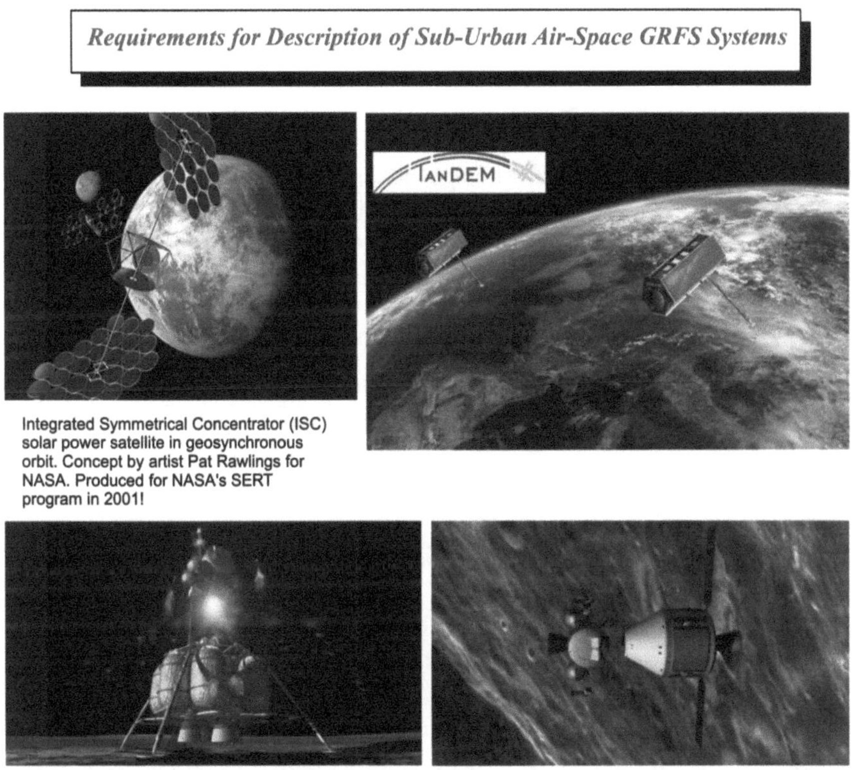

Fig. 2.24 An illustration of the requirements for description of suburban air-to-space (or space-to-air) GRFS systems. *Left* and *bottom images* courtesy of NASA. *Right image* reprinted with permission copyright © 2009 Balmer, R., and *IEEE*

the 1979 Reference SPS System and the 1979 SPS architecture entailed in deploying a series of as many as 60 SPS into geostationary Earth orbit with each system providing power ranging from 5 to 10 GW of continuous energy [78]. This is perhaps one of the applications that most people on earth are not aware of.

On the other hand, Tandem X Satellites, which are radar Satellites positioned for interferometry in a formation flight at distances of only a few hundred meters, the "twins" record data synchronously in the so-called StripMap Mode (3 m ground resolution) and thus acquire the data basis for a global Digital Elevation Model (DEM) of an unprecedented quality, accuracy, and coverage [79]. While a pair of Tandem X Satellites twins is in fact an urban GRFS system, a few pairs of Tandem X Satellites can form a suburban or a global GRFS system.

2.5.9 Requirements for Description of Suburban Ground-to-Air-to-Water (Air-to-Water-to-Ground or Water-to-Air-to-Ground) GRFS Systems

As depicted in Fig. 2.25, which represents a suburban ground-to-air-to-water (air-to-water-to-ground or water-to-air-to-ground) GRFS system, real world applications require fast convergence, robust STAP, and ultrawideband arrays to differentiation between: (1) sidelobe targets; (2) clutter discretes; (3) multiple mainlobe targets in adjacent range cells; (4) range varying nonhomogenous clutter; (5) and not to forget electromagnetic interference. Military vehicles have to operate under rugged terrain conditions, which lead to motion induced antenna pointing errors, such as when antennas are mounted on fast-moving platforms: aircraft and UAVs (see Fig. 2.25) [27].

How critical are the Operational Requirements Document (ORD) (Chaps. 1–3), an Overarching Concept of Operations (Chaps. 1, 2, 4, and 6) and Technical Requirements Document (TRD) (Chaps. 1–3) to enable the design of GRFS systems [50].

2.6 Requirements for Description of Global GRFS Systems

Global GRFS systems are defined as GRFS systems in which the global range of the area of operations is from 100 km up to 1,000 km in any global environment such as ground, air, space, or water.

Taking into consideration the examples presented, we provide an organization of this section which includes the requirements for description of several global GRFS systems as depicted in Fig. 2.26. First, we have the requirements for description of global ground GRFS systems discussed in Sect. 2.6.1. Second, we present the requirements for description of global air GRFS systems in Sect. 2.6.2. Third, we

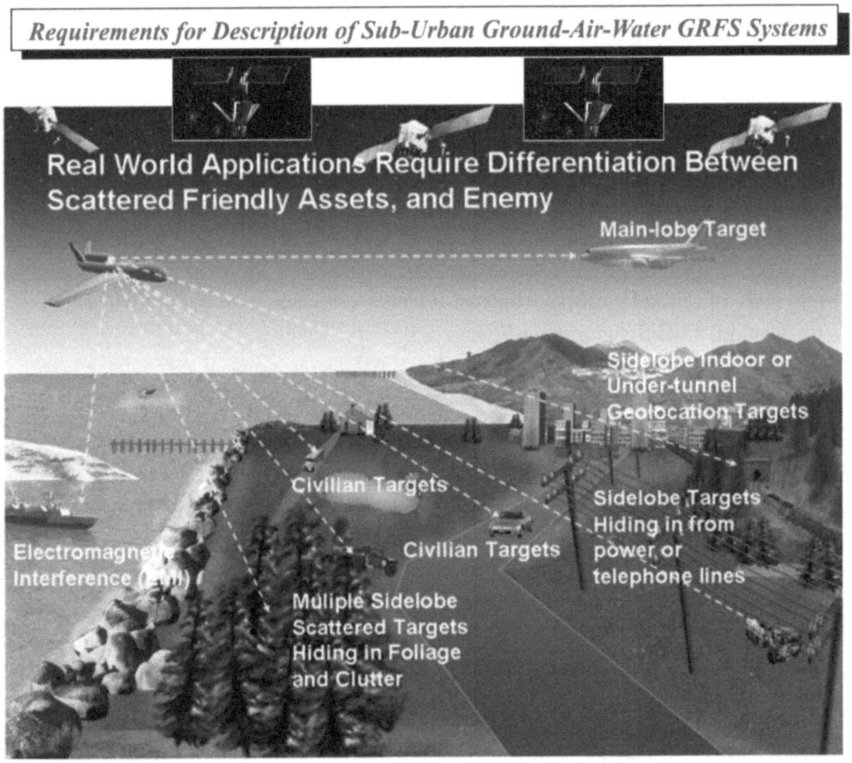

Fig. 2.25 An illustration of the requirements for description of suburban ground-to-air-to-water (or air-to-water-to-ground or water-to-air-to-ground) GRFS systems. Reprinted with permission copyright © 2009 Guerci, J.R., and *IEEE*

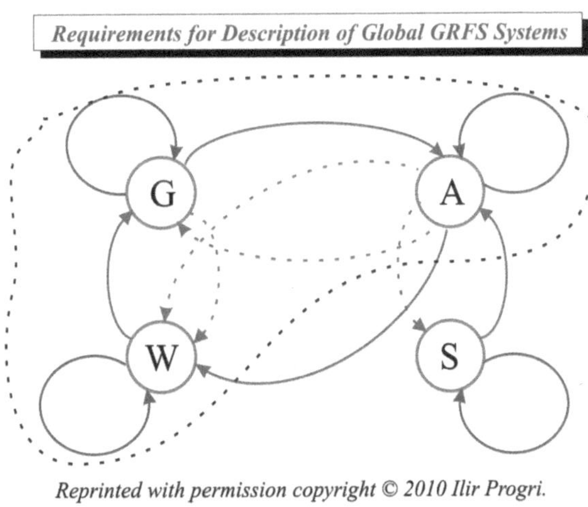

Fig. 2.26 An illustration of the state diagram of requirements for description of global GRFS systems. Reprinted with permission copyright © 2010 Ilir Progri

depict the requirements for description of global water GRFS systems in Sect. 2.6.3. Fourth, we analyze the requirements for description of global space GRFS systems in Sect. 2.6.4. Fifth, we discuss the requirements for description of the global ground-to-air (air-to-ground) GRFS systems in Sect. 2.6.5. Sixth, we provide the requirements for description of global ground-to-water (water-to-ground) GRFS systems in Sect. 2.6.6. Seventh, we present the requirements for description of global air-to-water (water-to-air) GRFS systems in Sect. 2.6.7. Eight, we depict the requirements for description of global air-to-space (space-to-air) GRFS systems in Sect. 2.6.8. Ninth and finally, we conclude this section with the requirements for description of global ground-to-air-to-water (air-to-water-to-ground or water-to-air-to-ground) GRFS systems in Sect. 2.6.9.

As we are going to see in Chap. 3 for the signal design point of view, these systems are made possible only as the result of the existence of the satellite GRFS systems.

2.6.1 Requirements for Description of Global Ground GRFS Systems

An example of a global ground GRFS systems are web services to realize service oriented architecture (SOA) in military communications networks such as shared situation awareness among military units is essential for network-enabled capabilities (NES) operations [26]. In order to enhance interaction within the allied forces there is a focus in NATO on the establishment of a SOA that will focus on rapid reaction, demand more adaptive and efficient solutions for information exchange, and quickly create and dynamically update a relevant picture, which will make military resources available as services [26].

The primary focus of the NATO NEC feasibility study (NNEC-FS) was to develop a NATO concept to adapt, extend, and expand national concepts such as the U.K. NEC and U.S. network-centric warfare (see Fig. 2.27) to the NATO context that will support all communications requirements of the member nations' forces such as communications among people, shared situation awareness, and end-to-end quality of service [26].

2.6.2 Requirements for Description of Global Air GRFS Systems

Global air GRFS systems include global airspace as illustrated in Fig. 2.28. Global air GRFS systems should be able to detect, differentiate, and accurately geolocate each military aircraft or civilian airplane in any kind of situation. A global coverage with acceptable communications [geolocation] performance is still missing today, especially for remote and oceanic areas [51].

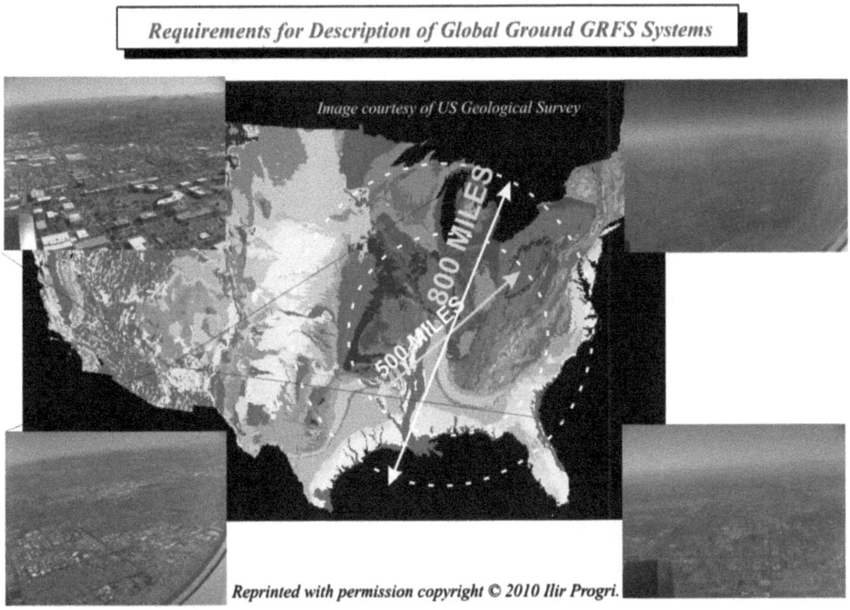

Fig. 2.27 An illustration of the requirements for description of global ground GRFS systems. US map image courtesy of US geological survey. Other four images are copyright © 2010 Ilir Progri

The FAA and Eurocontrol have already identified the upcoming bottlenecks in ATC/ATM communications and have started to develop the "Future Communications Infrastructure" (FCI) under the framework of the International Civil Aviation Organization (ICAO) [51]. Besides the development of new concepts and paradigms, one important part of the FCI is the development of the new aeronautical communications system able to cope with the demands and requirements of future ATC/ATM concepts [51].

2.6.3 Requirements for Description of Global Water GRFS Systems

Global water GRFS systems include global water as illustrated in Fig. 2.29. Global water GRFS systems should be able to detect, differentiate, and accurately geolocate each naval ship or civilian boat in any kind of situation, either in combat engagement or search and rescue operations.

Figure 2.29 illustrates the seismicity of the North Atlantic Ocean from 1975 to 1995 (left), and a more recent maritime modeling and analysis branch photo of the Atlantic Ocean (*right*) [99] an Earth-observing satellite that has provided early detection of ocean storms, including tropical cyclones, and advanced the scientific exploration of global ocean wind patterns, which has also been recognized for

Requirements for Description of Global Air GRFS Systems

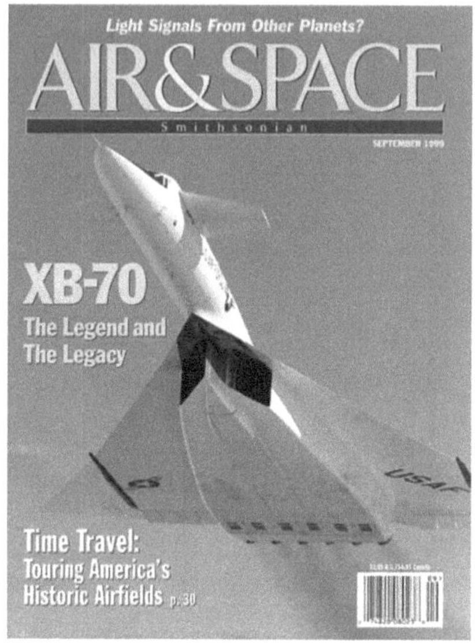

Fig. 2.28 An illustration of the requirements for description of global air GRFS systems. *Left image* Reprinted with permission copyright © 2007 Phil Makanna. *Right image* courtesy of *Air and Space Magazine Smithsonian* 1999

helping scientists better understand our home planet [103]. NASA and the U.S. Department of the Interior Tuesday presented the William T. Pecora Award to NASA's Quick Scatterometer, or QuikScat, mission team [103].

Greater concerns for these systems are "Piracy at sea: Somalia an area of great concern" from the states which have been marked as "weak" or "lawless" [62]. Although these phenomenon have been observed in the Gulf of Aden, near the Arabian Peninsula closer to the Indian Ocean, no one can guarantee that piracy, or smuggling of arms, drugs, human, or kids trafficking does not exists in the Atlantic Ocean.

2.6.4 Requirements for Description of Global Space GRFS Systems

As we have described in suburban space GRFS systems, global GRFS systems as depicted in Fig. 2.30, are satellite-based GRFS systems with global range 100 km–1,000 km. We have a much richer space environment that includes GPS satellite and other satellites which are discussed more extensively in the Satellite Space GRFS Systems. Global space GRFS systems may include surveillance

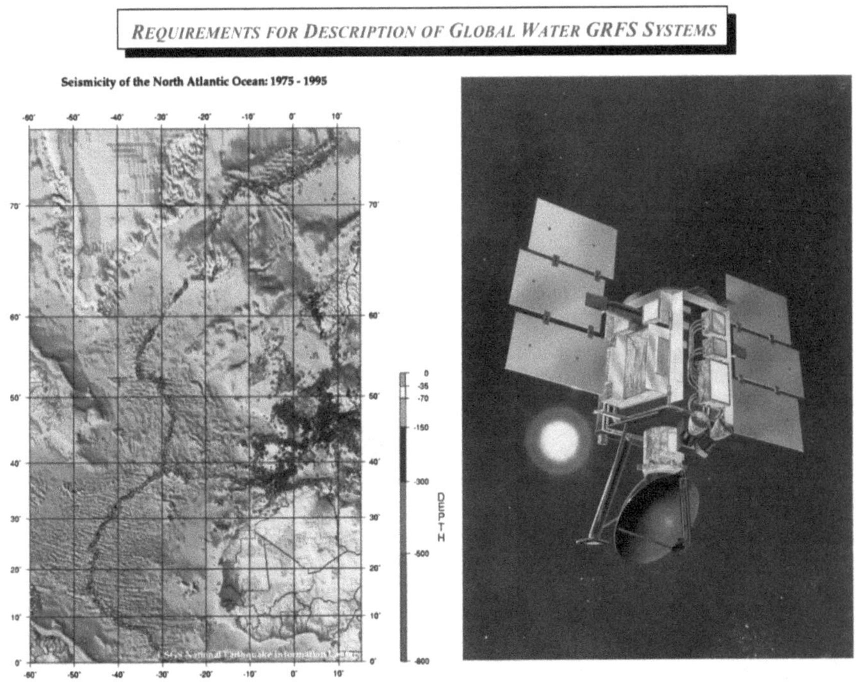

Fig. 2.29 An illustration of the requirements for description of global water GRFS systems. *Left image* U.S. Geological Survey; *Left image* courtesy of NASA

applications as illustrated in Fig. 2.30. In case of GPS, global GRFS systems become a part of the observable satellites from the terrestrial user point of view which could be as much as a third of the total number of satellite in the sky. So if we were to use this observation, then we could also define global space GRFS systems as GRFS systems that include about a third of all space satellites. The reader can also picture that there can only be three mutually exclusive global space GRFS systems.

Huge murals of artwork commemorating three decades of historic explorations and scientific achievements by all five of America's Space Shuttle Orbiters – Columbia, Challenger, Discovery, Atlantis, and Endeavour – now grace the Shuttle Firing Room inside the Launch Control Center (LCC) at NASA's Kennedy Space Center in Florida [116].

2.6.5 Requirements for Description of Global Ground-to-Air (Air-to-Ground) GRFS Systems

An example of a global ground-to-air (air-to-ground) GRFS systems may include a military communications network that consists of a large number of ground-based

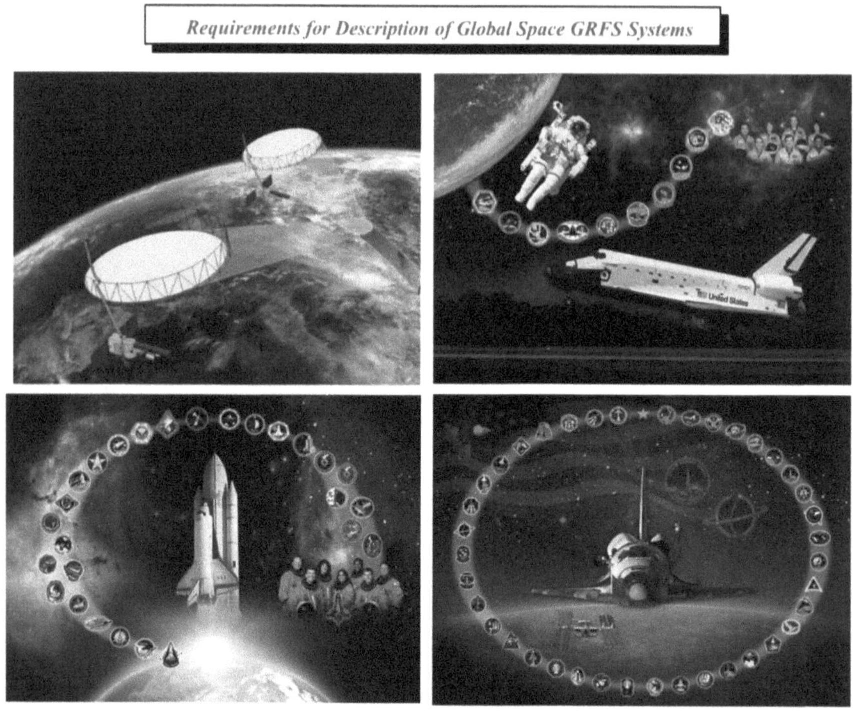

Fig. 2.30 An illustration of the requirements for description of global space GRFS systems (Tandam-L). Images courtesy of NASA

high-mobility vehicles, fast-moving aircrafts, UAVs deployed in intelligence, surveillance and reconnaissance (ISR) (see Fig. 2.31), and several naval vessels [27].

Satellite Communications Systems (SCS) are advantageous when connecting such terminals scattered over large distances; and SCS form by itself a satellite space GRFS system that are discussed more extensively in the following section. The United States Army is currently developing a satellite-based network-centric waveform capable of supporting military applications in highly mobile environments (see Fig. 2.31) [27].

Research to date on tactical wireless communications has focused on increasing bandwidth, improving reliability, and enabling adaptations for focusing on areas such as network coding, dynamic spectrum exploitation, robust routing, protocols, and cross-layer design which should lead to better bandwidth utilization and higher throughput [44]. Some of the most severe issues that these systems face are coming from additional range, interference, mobility, and security which cause severe bandwidth reduction and throughput reduction [44].

The purpose of the ground–air GRFS systems is, perhaps by integration with Blue Force Tracking (BFT) [44] or as part of BFT, to provide warfighters with location information about friendly military forces and also with location of RF interference enemy sources. Illustration details on how this is accomplished in more

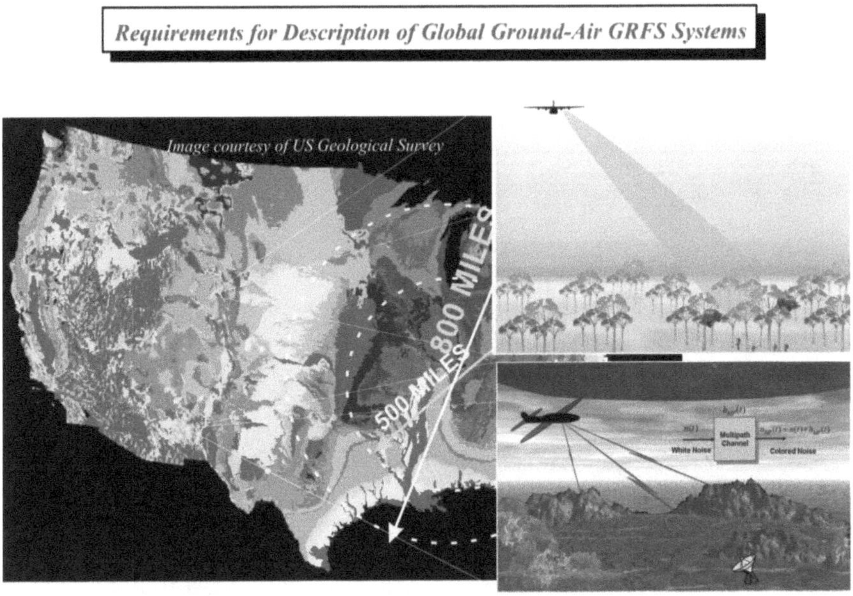

Fig. 2.31 An illustration of the requirements for description of global ground-to-air (air-to-ground) GRFS systems. *Left image* courtesy of US Geological Survey. *Right images* copyright © 2009 Davis, M., Guerci, J.R., and *IEEE*

practical principle simulation examples the reader may obtain further details in Chaps. 4 and 6.

2.6.6 Requirements for Description of Global Ground-to-Water (Water-to-Ground) GRFS Systems

Following the discussion on global ground GRFS systems, ground-to-water will exhibit similar application as global water GRFS systems. Of particular interests are water board patrolling, search and rescue operation by the US Coast Guard, loading and unloading of ships in and around huge ports, monitoring of huge cargo ships, international water patrolling, etc. as shown in Fig. 2.32.

2.6.7 Requirements for Description of Global Air-to-Water (Water-to-Air) GRFS Systems

US Navy has expressed concerns that current passive, phased array antennas are heavy, bulky, and often exhibit poor *aperture efficiency and response linearity* when attempting to design them to cover large RF bandwidths.

Requirements for Description of Global Ground-Water GRFS Systems

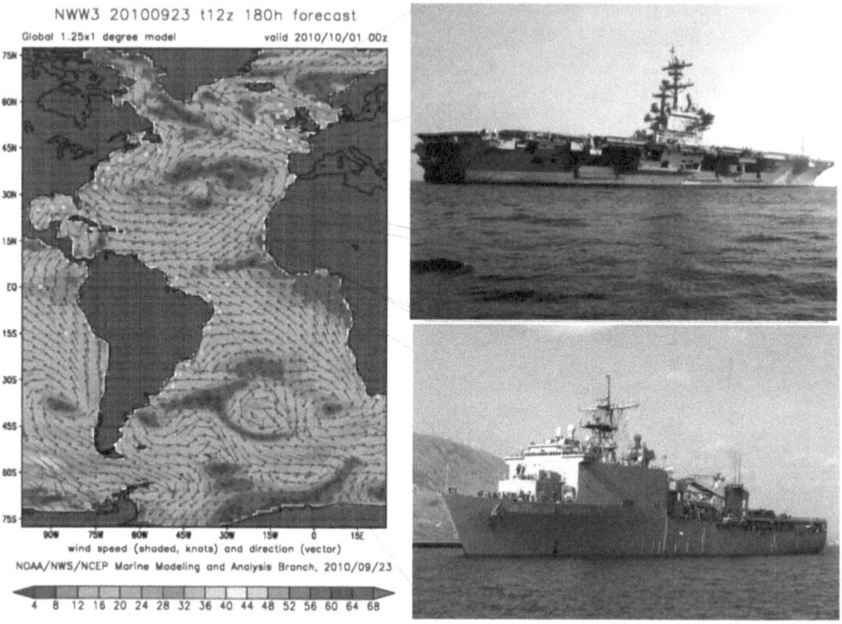

Fig. 2.32 An illustration of the requirements for description of global ground-to-water (water-to-ground) GRFS systems. *Left image* courtesy of US NOAA/NWS. *Right images* courtesy of Wordpress and Wikipedia

This book addresses US Navy's needs for innovative, passive phased array antennas for *drastic improved physical profiles, performance characteristics* to support multiple developmental programs across multiple missions.

Figure 2.33 provides an outstanding illustration of global air-to-water (water-to-air) GRFS systems. As illustrated in Fig. 2.33, a global air-to-water (water-to-air) GRFS system can perform one of the following: (1) horizon search track-white-scan; (2) limited volume search; (3) uplink/downlink; (4) cued acquisition; (5) electronic protection from electronic attach platform; (6) environmental mapping; (7) counter fire; (8) sector search; (9) periscope detection; (10) surface search navigation; (11) target illumination; (12) horizon search track white scan [66].

2.6.8 Requirements for Description of Global Air-to-Space (Space-to-Air) GRFS Systems

Global air-to-space (space-to-air) GRFS systems may include numerous applications such as weather satellites, ozone layer monitoring, ionosphere electronic

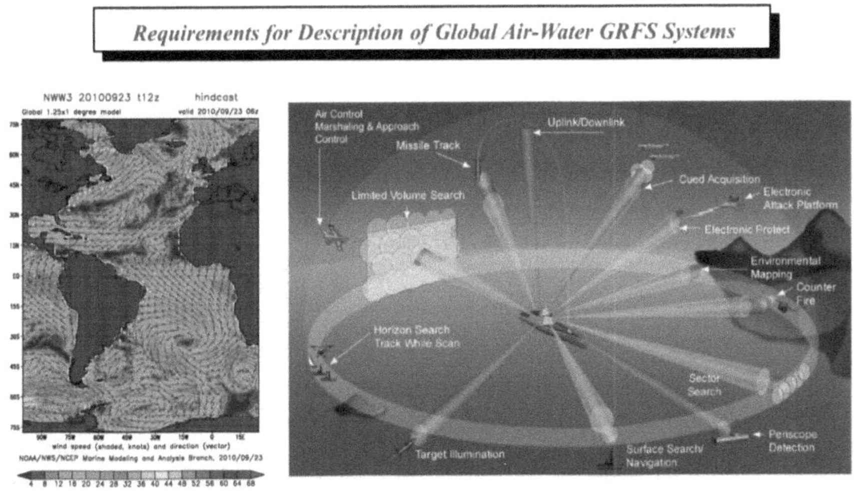

Fig. 2.33 An illustration of the requirements for description of global air-to-water (water-to-air) GRFS systems. *Left image* courtesy of US NOAA/NWS. *Right image* reprinted with permission copyright © 2009 Jeffrey, T., and *IEEE*

content monitoring, etc. Earth observation' satellites are mainly located in low earth orbit (LEO), usually less than 1,000 km from the Earth's surface, and are characterized by the need for downloading huge amounts of data, which are generated by their instruments and are stored onboard during the day [53]. Other applications might include monitoring of health and conditions of other satellites, space stations, space shuttle, etc. as illustrated in Fig. 2.34. It is well accepted that satellites play an established and well-organized role in some "nice" markets such as navigation and localization services, broadcast services, specific observations of Earth observation, and remote sensing [11]. One such system is high-altitude platforms (HAPs) also known as aerial unmanned platforms carrying communications relay payloads and operating in quasistationary positions at altitudes between 15 and 30 km from the surface of the earth [11, 53]. Such systems can be used from telephony and broadband services, navigation systems for providing fleet management and traffic-control services [11, 53]. Other applications might include data-relay satellite systems such as NASA's Tracking and Data Relay Satellite. Other roles of HAPs are support of services such as required navigation performance (RNP) or position navigation, and timing (PNT), to Global Navigation Satellite Systems (GNSSs) [119–138] such as GPS and Galileo, Local Area Augmentation Systems (LAAS), Wide Area Augmentation Systems (WAAS), terrestrial stratospheric Ranging, Integrity, and Monitoring Station (RIMS) network, Local-area Differential GPS (LADGPS), etc., which will be discussed briefly in the *Indoor Geolocation Systems: Theory and Applications* book.

For example, Challenger (Fig. 2.34, top right): This Tribute Display features Challenger, which blazed a trail for other vehicles with the first night landing (STS-8) and also the first landing at Kennedy Space Center (STS-41B).

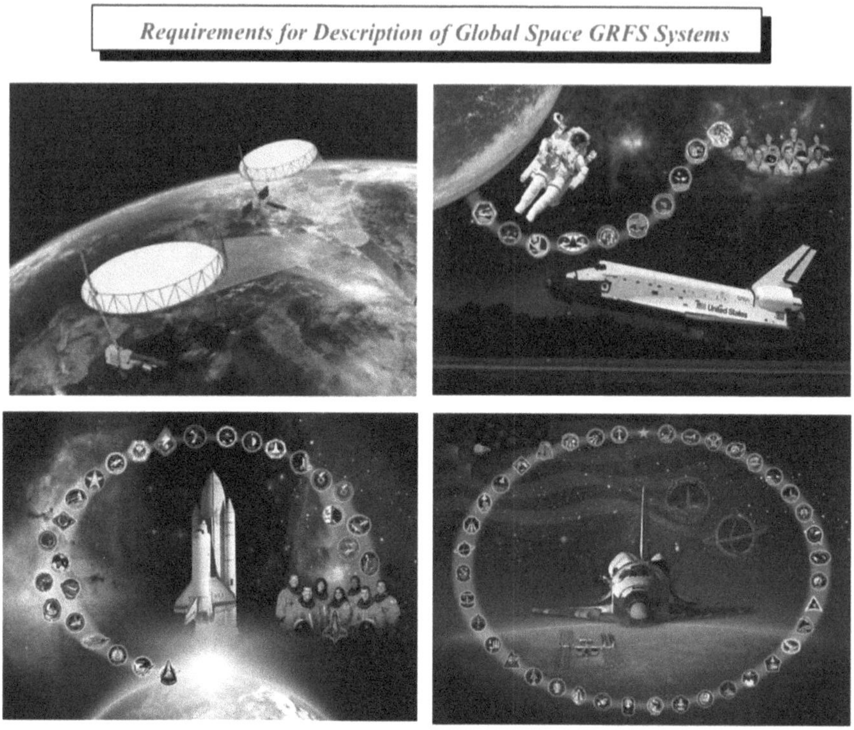

Fig. 2.34 An illustration of the requirements for description of global air-to-space (space-to-air) GRFS systems. Images courtesy of NASA

The spacewalker represents Challenger's role in the first spacewalk during space shuttle mission (STS-6) and the first untethered spacewalk (STS-41B). Crew-designed patches for each of Challenger's missions lead from earth toward our remembrance of the STS-51L crew. Other significant accomplishments include the first night launch with STS-8; the first in-flight capture, repair, and redeployment of an orbiting satellite during STS-41C; the first American woman in space (Sally Ride on STS-7); the first African-American in space (Guion Bluford on STS-8); and the first American woman to walk in space (Kathryn Sullivan during STS-41G). Credit: NASA [116].

2.6.9 Requirements for Description of Global Ground-to-Air-to-Water (Air-to-Water-to-Ground or Water-to-Ground-to-Air) GRFS Systems

A satellite GRFS system can be very handy for detecting insurgents and intruders hiding other foliage [61]. Moreover, if there is a need for rescue operation in heavy rain and high wind, the detection, discrimination, and differentiation range may be affected [61]. A US Navy surveillance aircraft may experience great difficulty detecting humans and vehicles on the ground or even small ships in a harbor if antennas are not above the masking (or hiding) environments which may consists of trees, foliage, larger ships, or heavy wind induced clutter [61].

Giftet Inc proposed that a satellite ground-to-air-to-space-to-water GRFS system will serve the unique purposes of the US Navy for detecting and characterizing manmade objects of any kind as long as these objects have a transmitter which transmits at any of the frequency ranges in 100 MHz to 66 GHz much better than passive sonar systems under any environment, geometry, clutter, signal intensity, density, etc. [48].

As depicted in Fig. 2.35, which represents a satellite ground-to-air-to-space-to-water (all other combinations of four) GRFS system, real world applications require differentiation among many tasks.

Some of the most important tasks include: (1) main lobe targets; (2) side lobe indoor or undertunnel geolocation targets; (3) sidelobe targets hiding in under power or telephone lines; (4) civilian moving targets; (5) multiple side lobe targets hiding in under foliage and clutter; (6) and not to forget electromagnetic interference.

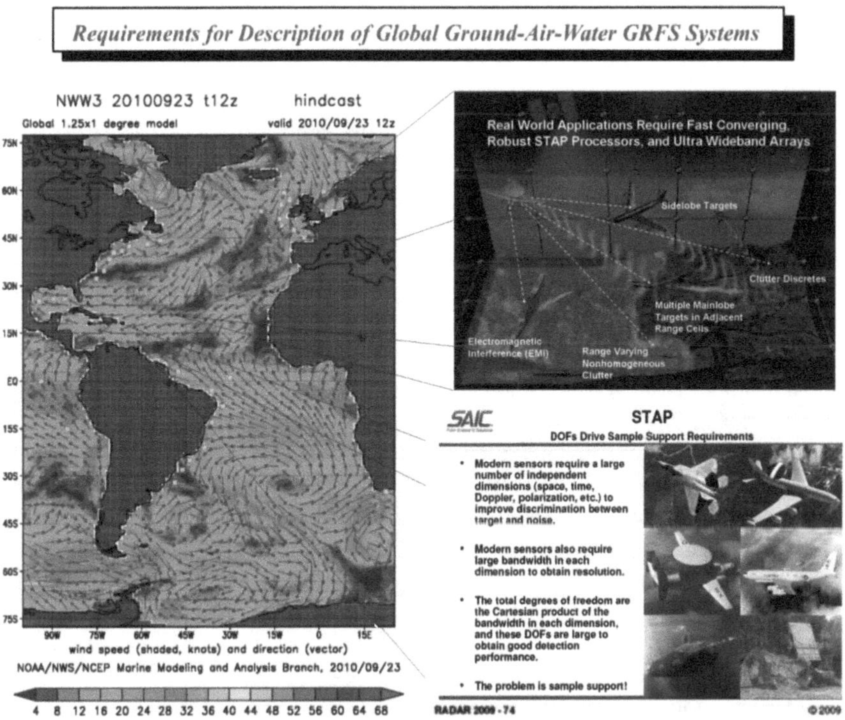

Fig. 2.35 An illustration of the requirements for description of global ground-to-air-to-water (air-to-water-to-ground or water-to-ground-to-air) GRFS systems. *Left image* courtesy of NOAA/NWS. *Right image Top* copyright © 2009 Goldstein, M., Picciolo, M., Griesbach, J., and *IEEE*. *Right image Bottom* copyright © 2009 SAIC

Network-centric operations also referred to as network-enabled capability and network defense, is the cornerstone of modern warfighting, which rely on robust network communications to support timely exchange of information between geographically dispersed entities [44]. Tactical networks which are the basis of the network centric warfighting operations also provide one of the most challenging environments for communications, which included inherently that mobile nodes must communicate by using wireless ad hoc links in hostile radio frequency (RF) environments, creating unreliable networks that have limited bandwidth and variable latency (see Fig. 2.35) [44].

Applications in tactical networks have different, sometimes peculiar requirements; therefore, a one-size-fits-all approach to transport protocol leads to inefficiencies [44] which is the reason why we have the last section of this chapter dedicated to requirements for description of satellite GRFS systems.

2.7 Requirements for Description for Satellite GRFS Systems

Satellite GRFS systems are defined as GRFS systems in which the satellite range of the area of operations is from 1,000 km up to 100,000 km in any global environment such as ground, air, space, or water.

Satellite GRFS systems play an important role for: (1) both military and civilian applications; (2) both for research, development, and commercial needs; (3) both for cutting edge technologies as well as mature and well-established technologies. After the World War II, satellite technologies started to dominate and lead the research and development in aerospace, astronomy, space navigation, radio-cosmology, interplanetary rocket science, radars, celestial navigation, etc. initially by the United States and the former Soviet Union and later by the European Union, Japan, and Australia, and more recently by China and India, and other nations.

There are many advantages of the satellite systems in terms of global coverage, global availability of signals, global means to achieve communications, global time transfer, global location to all the users on the ground, air, water, etc. Moreover, satellite systems give us the much needed information from the universe which is distorted to get from the observatories on earth. Since satellite systems cover almost the entire usable spectrum of frequencies, we are going to restrict our discussions to the desired frequency spectrum of 100 MHz to 66 GHz. Chapter 3 discusses in great detail the signal structure or design or RF signals used in Satellite GRFS Systems. It is without any doubt that one can write an entire book only on Satellite GRFS Systems.

As we have also mentioned previously, it is hoped that this preliminary classification is only going to provide a firsthand overview of the description of the satellite GRFS systems and also an outline for future direction of the research. It is hoped that future editions of this book are going to expand the discussion provided here and include information that will be suggested from reviewers and readers. This is usually the information that is generally not accessible to the author at first hand which is the reason why writing a book for the first time is so important. Without further due let us begin the discussion on the description of requirements of satellite GRFS systems.

At this stage, the reader should be familiar that a GRFS system can be conceptually thought either as a single system in order to address the requirements of a single case study or as a complex system of systems that will be a collection of individual, possible heterogeneous, but functional GRFS systems integrated together to enhance the overall robustness, increase reliability and performance of the overall complex (SoS) system [54]. Although this is a viable option, for the most part we are going to be treating all our case studies as individual and independent GRFS systems and as we gather information for all case studies, we could propose future design that might include concepts of the System of system design and integration in the future editions of the book [54].

Taking into considerations the examples presented, we provide an organization of this section which includes the requirements for description of several satellite GRFS systems as depicted in Fig. 2.36. First, we have the requirements for description of satellite space GRFS systems discussed in Sect. 2.7.1. Second, we present the requirements for description of satellite ground-to-air (air-to-ground) GRFS systems in Sect. 2.7.2. Third, we depict the requirements for description of satellite ground-to-space (space-to-ground) GRFS systems in Sect. 2.7.3. Fourth, we analyze the requirements for description of satellite air-to-water (water-to-air) GRFS systems in Sect. 2.7.4. Fifth, we discuss the requirements for description of satellite air-to-space (space-to-air) GRFS systems in Sect. 2.7.5. Sixth, we provide the requirements for description of satellite ground-to-air-to-water (air-to-ground-to-water or water-to-air-to-ground) GRFS systems in Sect. 2.7.6. Seventh, we present the requirements for description of satellite ground-to-space-to-water (space-to-ground-to-water or water-to-space-to-ground) GRFS systems in Sect. 2.7.7. Eight, we depict the requirements for

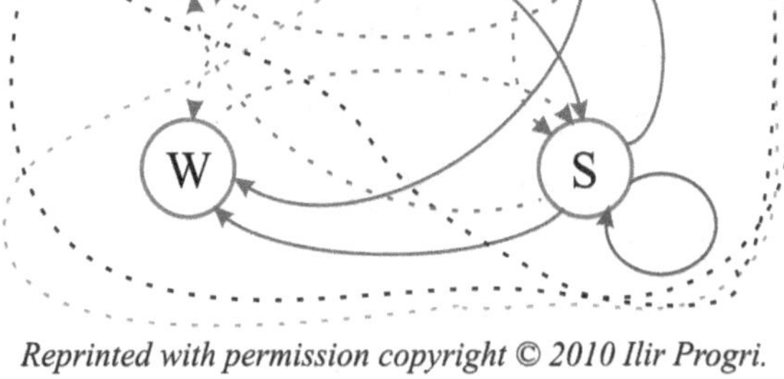

Fig. 2.36 An illustration of the state diagram of requirements for description of suburban GRFS systems. Reprinted with permission copyright © 2010 Ilir Progri

description of satellite air-to-space-to-water (space-to-air-to-water or water-to-space-to-air) GRFS Systems in Sect. 2.7.8. Ninth and finally, we conclude this section with the requirements for description of satellite ground-to-air-to-space-to-water (all other combinations of four) GRFS systems in Sect. 2.7.9.

2.7.1 Requirements for Description of Satellite Space GRFS Systems

Figure 2.37 represents satellite space GRFS systems which are secondary systems to the primary Radar systems shown in the figure because GRFS systems are passive array systems.

There has been a monumental advancement in space exploration from the NASA as depicted in Fig. 2.37 with Jason-1, QuikSCAT, ERBS, ACRIMSAT, Landsat 7,

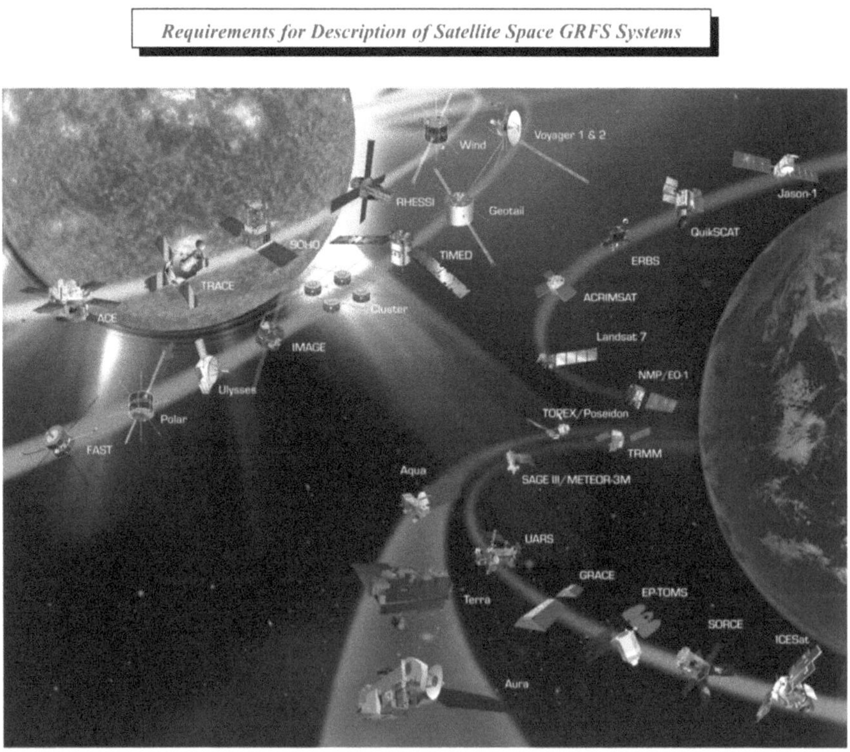

Fig. 2.37 An illustration of the requirements for description of satellite space GRFS systems. Image courtesy of NASA

Table 2.1 Classifications of satellites

Class	Cost	Mass (kg)
Large satellite	$ >100 M	>1,000
Small satellite	$50–100 M	500–1,000
Minisatellite	$5–20 M	100–500
Microsatellite	$2–3 M	10–100
Nanosatellite	$ < M	<10

NMP/EO-1, TRMM, TOREX/Poseidon, Saga III/METEOR 3M, GRACE, EP-TOMS, SORCE, Aura, and Terra [104].

Other space GRFS systems which go beyond the scope of this chapter are voyager 1 and 2, Wind, Geotail, RHSSI, TIMEO, SOHO, Cluster, Image, Trace, Ulysses, Ace, Polar, and Fast. Based on class, cost, and mass satellites can be classified as shown in Table 2.1 [53]:

Recent advances of microelectronics has generated a new species of modern, highly sophisticated, computationally powerful, rapid-response microsatellite (and minisatellites) that have reduced the cost of a single satellite by more than one order of magnitude (see Table 2.1; [53]). These "faster, cheaper, and better" microsatellites now make the implementation of such a disaster network both practicable and affordable as well as offering possibilities for improved weather predictions, real-time monitoring of a stricken area, and nearly immediate restoration of communications services needed for relief efforts [53].

2.7.2 Requirements for Description of Satellite Ground-to-Air (Air-to-Ground) GRFS Systems

An example of a satellite ground-to-air (air-to-ground) GRFS system is shown in Fig. 2.38 which illustrates a space shuttle rocket launch, space shuttle rocket passing through the atmosphere, a space vehicle positioning in orbit, and the return and landing of a space shuttle from its missions [105–108].

These systems are characterized by very high accelerations (or g-s); therefore, during these missions the astronauts' crew is set to be static with respect to the space shuttle rocket during takeoff or landing as shown in Fig. 2.38.

2.7.3 Requirements for Description of Satellite Ground-to-Space (Space-to-Ground) GRFS Systems

An example of requirements for description of satellite ground-to-space (space-to-ground) GRFS system is shown in Fig. 2.39 which depicts most radio telescopes

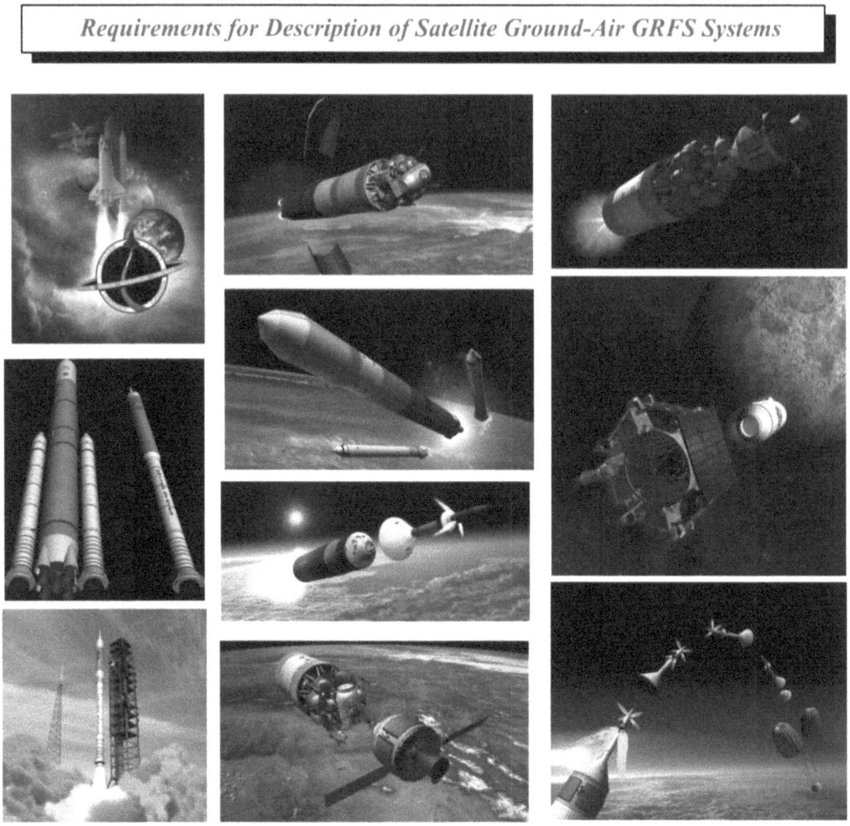

Fig. 2.38 An illustration of the requirements for description of satellite ground-to-air (air-to-ground) GRFS systems. Image on the *left* is courtesy of NASA

as reflectors, such as: (1) Arecibo is 305 m diameter (73,000 m^2) spherical dish (fixed position); (2) Lovell Telescope is the third largest steerable radio telescope in the world; (3) Haystack is 37 m diameter (1,075 m^2) (re-positionable) © MIT; and (4) Proposed Square Kilometer Array (SKA) will be some form of ESA [71].

Another example of a ground-to-space GRFS system includes a description of S-WiMAX: adaptation of IEEE 802.16e for mobile satellite services [42]. It is desirable that Satellite adaptation of WiMAX have baseband affinity with the WiMAX physical (PHY) and medium access control (MAC) layers primarily a power and size efficient dual-mode satellite/terrestrial application-specific integrated circuit (ASIC) and drives a contemporary handheld mobile device [42].

Another example is the digital video broadcast-return channel satellite (DVB-RCS) which includes aeronautical, maritime, and land [59]. This case study will be discussed further in Chaps. 3 and 4.

Fig. 2.39 An illustration of the requirements for description of satellite ground-to-air (air-to-ground) GRFS systems. Reprinted with permission copyright © 2009 Williams, J., and *IEEE*

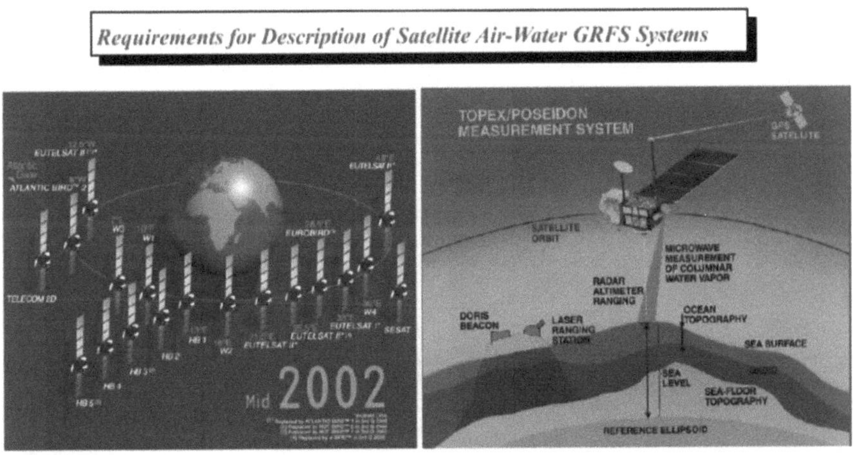

Fig. 2.40 An illustration of the requirements for description of satellite air-to-water (water-to-air) GRFS systems. Reprinted with permission copyright © 2009 Griffiths, H., Willis, N. and *IEEE*; copyright © 2009 Zyl, J.V., and *IEEE*

2.7.4 *Requirements for Description of Satellite Air-to-Water (Water-to-Air) GRFS Systems*

Satellite Air-to-Water (Water-to-Air) GRFS systems are used for a number of remote sensing oceanographic studies as depicted in Fig. 2.40 (*right*).

The oceanographic satellite is equipped with laser altimeter ranging instrument, microwave measurement of columnar water vapor instrument; and is also able to detect a number of laser ranging stations [67, 68].

The TOPEX/POSEIDON Project was a joint US and French mission to develop and operate an Earth-orbiting satellite with sensors capable of making accurate measurements of sea level by means of the NASA radar altimeter (NRA), a fifth-generation US altimeter that provides the primary measurement for the TOPEX/POSEIDON Project altimetric mission [67].

Contrast this with the left of Fig. 2.40 where we have a number of geostationary European satellites that allow for long integration time that are used for satellite DBS-TV monitoring from 2002 until the present days with beams shaped to provide coverage over land [70].

2.7.5 Requirements for Description of Satellite Air-to-Space (Space-to-Air) GRFS Systems

Satellite air-to-space (space-to-air) GRFS systems may include the space shuttle rocket ascension and dissension as illustrated in Fig. 2.41. After jettisoning the Altair to allow it to crash into the lunar far side, the crew, using the onboard engine performs the Trans Earth Injection (TEI) burn for the return trip to Earth. After a

Fig. 2.41 An illustration of the requirements for description of satellite air-to-space (space-to-air) GRFS systems. Images courtesy of NASA

2½ day coast, the crew jettisons the service module (allowing it to burn up in the atmosphere) and then reenters the Earth's atmosphere using a reentry trajectory designed to both slow the vehicle from its speed of 40,200–480 km/h and allow for a West Coast landing [107].

The Orion spacecraft is able to dock with the International Space Station. The six-man crew, the largest number that can fly on an Orion spacecraft, then enters the station and performs its activities for the duration of their flight, usually lasting 6 months, but can be shortened to 4 or lengthened to 8. Once completed, the crew reenters the Orion, which has been kept attached to the station as an emergency "lifeboat," seal off the hatches between it and the ISS, and then undock from the station [107].

2.7.6 *Requirements for Description of Satellite Ground-to-Air-to-Water (Air-to-Ground-to-Water or Water-to-Air-to-Ground) GRFS Systems*

Satellite ground-to-air-to-water (air-to-ground-to-water or water-to-air-to-ground) GRFS systems may include space shuttle rocket launch during takeoff, space shuttle rocket ascension into space, and oceanographic water monitoring as illustrated in Fig. 2.42.

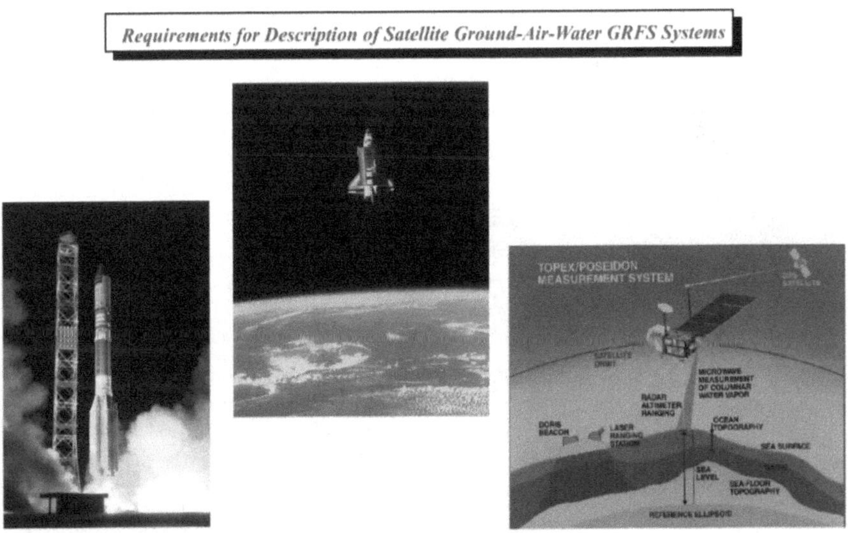

Fig. 2.42 An illustration of the requirements for description of satellite ground-to-air-to-water (air-to-ground-to-water or water-to-air-to-ground) GRFS systems. Image on the *left* is courtesy of 2004–2009 Orbitcast Media LLC. Image in the *center* is copyright © 2009 OrbitalHub. Image on the *right* is copyright © 2009 Zyl, J.V., and *IEEE*

2.7.7 Requirements for Description of Satellite Ground-to-Space-to-Water (Space-to-Ground-to-Water or Water-to-Space-to-Ground) GRFS Systems

Satellite ground-to-space-to-water (space-to-ground-to-water or water-to-space-to-ground) GRFS systems may include space shuttle rocket launch during takeoff, satellite orbiting into space, and oceanographic water monitoring as illustrated in Fig. 2.43.

Another example of a ground-to-space-to-water GRFS system may include a microwave ranging radiometer and aperture synthesis (MIRAS) that was developed by EDAS CASA Espacio with major subcomponents built by companies in Spain and 17 European countries overall [63].

The MIRAS instrument employs 69 individual antenna elements and receivers and two-dimensional aperture synthesis in order to achieve the needed horizontal spatial resolution of the 1.4 GHz brightness temperature measurements [63].

2.7.8 Requirements for Description of Satellite Air-to-Space-to-Water (Space-to-Air-to-Water or Water-to-Space-to-Air) GRFS Systems

The Soil Moisture and Ocean Salinity (SMOS) mission, also known as ESA's Water Mission, is the second one of the European Space Agency's Earth Explorer series launched on 2 November 2009 into a LEO at ~760 km altitude [63].

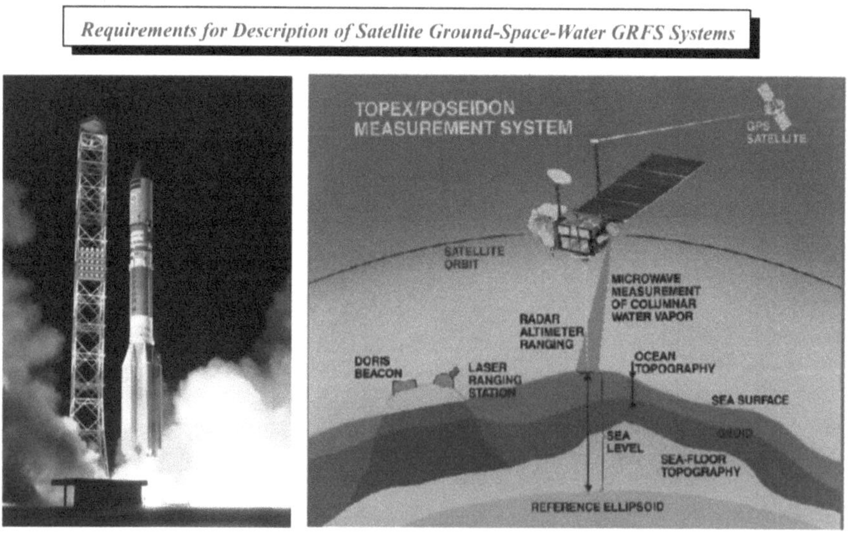

Fig. 2.43 An illustration of the requirements for description of satellite ground-to-space-to-water (space-to-ground-to-water or water-to-space-to-ground) GRFS systems. Image on the *left* is courtesy of 2004–2009 Orbitcast Media LLC. Image on the *right* is copyright © 2009 Zyl, J.V., and *IEEE*

The L-band measurements provide sensitivity to changes in soil moisture and sea surface salinity, but are relatively insensitive to variations in atmospheric effects in water vapor and vegetation cover. These measurements were initially important because they measured and tracked water for agriculture and monitoring desertification which is recently strengthened due to applications for improving weather forecasting and climatology studies [63] or the Katrina Hurricane, or the HUGE BP Oil Spill. The International Space Station or the Hubble Telescope may also be used to gather useful oceanographic information as depicted in Fig. 2.44.

2.7.9 Requirements for Description of Satellite Ground-to-Air-to-Space-to-Water (All Other Combinations of Four) GRFS Systems

Satellite systems are the most effective ways to provide mobile MBMS; its association with hybrid satellite-terrestrial networks (HSTN) enables the formation of cooperative systems by seamlessly combining the most powerful aspects of each network [31]. Satellite system can provide the best and most effective coverage for low-density populations in global and satellite environments, while the terrestrial

Fig. 2.44 An illustration of the requirements for description of satellite air-to-space-to-water (space-to-air-to-water or water-to-space-to-air) GRFS systems. Images courtesy of NASA

Requirements for Description of Satellite Ground-Air-Space-Water GRFS Systems

Fig. 2.45 An illustration of the requirements for description of satellite ground-to-air-to-space-to-water (all other combinations of four) GRFS systems. Images courtesy of NASA

network can provide the highest bandwidth and lowest cost coverage for high-density populations in indoors, urban, and suburban environments [31].

In the end, satellite ground-to-air-to-space-to-water (all other combinations of four) GRFS systems have the largest coverage possible for all geospatial, geointelligence, georeference, etc., applications (see in Fig. 2.45, NASA's exploration roadmap [118]).

2.8 Conclusions

We now conclude this chapter. This is probably the most exciting chapter and the bedrock of the entire book and there are numerous reasons why this is such an exciting chapter.

This chapter has a brand new and original organization which illustrates very vividly the local reference environments (indoor, urban, suburban, global, and satellite) and global reference environment (ground, air, space, and water), a wealth of technical requirements on description and discussions on each GRFS system motivated by extensive world-class literature publications from the *IEEE Communications Magazine*, *IEEE Systems Magazine*, etc.

From this point onward, this chapter will help tremendously the reader to understand the scope, the issues, the interests on each subsystem from the government, commercial, application, usability, etc. point of view, areas that have mature technologies, areas that are lacking in new and innovative research, and system development and deployment.

The most exciting news is not only for the benefits of this book or the research on GRFS Systems per se but also on the need to research and develop many sensors and sensory systems in the context of "systems of systems" that will serve to support many primary systems that are already deployed and will illustrate the need for more sophisticated system integration concept networks and systems in order to meet the requirements of the GRFS systems as proposed in Chap. 1.

The other good news in the context of this chapter is that we have already prepared the ground work for a detailed discussion on RF signals in Chap. 3. This chapter has already provided the template on how Chap. 3 organization should look like and we have already had a great discussion on how the environment looks like and also what the IEEE standards are involved.

In Chaps. 4 and 6 we are going to refer back to this chapter: (1) when discussing case studies; (2) when analyzing principle simulation examples; (3) when assessing deployment scenarios; (4) when presenting new ideas and innovative technologies; (5) when building databases for geospatial solutions and maps; (6) when setting parameterization values for selecting values of different system parameters, etc.

References

1. Progri, I., "An assessment of indoor geolocation systems," *Ph.D. Dissertation*, Worcester Polytechnic Institute, May 2003.
2. Smulders, P., "Exploiting the 60 GHz band for local wireless multimedia access: prospects and future directions," *IEEE Commun. Mag.*, vol. 40, nr. 1, pp. 140–147, Jan. 2002.
3. Porcino, D., and Hirt, W., "Ultra-wideband radio technology: potential and challenges ahead," *IEEE Commun. Mag.*, vol. 41, nr. 7, pp. 66–74, Jul. 2003.
4. Le-Ngoc, T., Leung, V., Takats, P., and Garland, P., "Interactive multimedia satellite access communications," *IEEE Commun. Mag.*, vol. 41, nr. 7, pp. 78–85, Jul. 2003.
5. Bruno, R., Conti, M., and Gregori, E., "Mesh networks: commodity multihop Ad Hoc networks," *IEEE Commun. Mag.*, vol. 43, nr. 3, pp. 123–131, Mar. 2005.
6. Iacovoni, G., Pierotti, D., Parisi, D., and Isopi, M., "Cell capacity of LMDS systems in typical traffic scenarios," *IEEE Commun. Mag.*, vol. 43, nr. 4, pp. 122–129, Apr. 2005.
7. Molinaro, A., Rango, F.D., and Tropea, M., "A scalable framework for end-to-end QoS assurance in IP-oriented terrestrial-GEO satellite networks," *IEEE Commun. Mag.*, vol. 43, nr. 4, pp. 130–137, Apr. 2005.
8. Alkyildiz, I.F., Mohanty, S., and Xie, J., "A ubiquitous mobile communication architecture for next-generation heterogeneous wireless systems," *IEEE Commun. Mag.*, vol. 43, nr. 6, (*IEEE Radio Commun.*) pp. S29–S36, Apr. 2005.

9. Alkyildiz, I.F., and Wang, X., "A survey on wireless mesh networks," *IEEE Commun. Mag.*, vol. 43, nr. 9 (*IEEE Radio Commun.*, vol. 2, nr. 3), pp. S23–S30, Sep. 2005.
10. Ryu, S., Park, S.K., Oh, D., Sihn, G., Han, K., and Hwang, S., "Research activities on next-generation mobile communications and services in Korea," *IEEE Commun. Mag.*, vol. 43, nr. 9, pp. 122–131, Sep. 2005.
11. Cianca, E., Prasad, R., Sanctis, M.D., Luise, A.D., Antonini, M., Teotino, D., and Ruggieri, M., "Integrated satellite-HAP systems," *IEEE Commun. Mag.*, vol. 43, nr. 12 (*IEEE Radio Commun.*, vol. 2, nr. 4), pp. S33–S39, Dec. 2005.
12. Salkintzis, A.K., "Evolving public safety communication systems by integrating WLAN and TETRA networks," *IEEE Commun. Mag.*, vol. 44, nr. 1, pp. 38–46, Jan. 2006.
13. Balachandran, K., Budka, K.C., Chu, T.P., Doumi, T.L., and Kang, J.H., "Mobile responder communication networks for public safety," *IEEE Commun. Mag.*, vol. 44, nr. 1, pp. 56–64, Jan. 2006.
14. Parkvall, S., Englund, E., Lundevall, M., and Torsner, J., "Evolving 3G mobile systems: broadband and broadcast services in WCDMA," *IEEE Commun. Mag.*, vol. 44, nr. 2, pp. 68–74, Feb. 2006.
15. Garawi, S., Istepanian, R.S.H., and Abu-Rgheff, A., "3G wireless communications for mobile robotic tele-ultrasonography systems," *IEEE Commun. Mag.*, vol. 44, nr. 4, pp. 91–96, Apr. 2006.
16. Pavon, J.d.P., Shankar, S.N., Gaddam, V., Challapali, K., and Chou, C.T., "The MBOA-WiMedia specification for ultra wideband distributed networks," *IEEE Commun. Mag.*, vol. 44, nr. 6, pp. 128–134, Jun. 2006.
17. Lee, S.-Q, Park, N., Cho, Ch., Hyongwoo, L., and Ryu, S., "The wireless broadband (WiBro) system for broadband wireless internet services," *IEEE Commun. Mag.*, vol. 44, nr. 7, pp. 106–112, Jul. 2006.
18. Soomre, A., and Cavalacnti, D., "Opportunities and challenges in using WPAN and WLAN technologies in medical environments," *IEEE Commun. Mag.*, vol. 45, nr. 2, pp. 114–122, Feb. 2007.
19. Hu, W., Willkomm, D., Gross, J., Vlantis, G., Gerla, M., and Wolisz, A., "Dynamic frequency communities for efficient IEEE 802.22 operation," *IEEE Commun. Mag.*, vol. 45, nr. 5, pp. 80–87, May 2007.
20. Ji, Z., and Liu, R.J., "Dynamic spectrum sharing: a game theoretical overview," *IEEE Commun. Mag.*, vol. 45, nr. 5, pp. 88–94, May 2007.
21. Lu, K., Qian, Y., and Chen, H.-H., "A secure and service-oriented network control framework for WiMAX networks," *IEEE Commun. Mag.*, vol. 45, nr. 5, pp. 124–130, May 2007.
22. Ni, Q., Vinel, A., Xiao, Y., Turlikov, A., and Jiang, T., "Investigation of bandwidth request mechanisms under point-to-multipoint mode of WiMAX networks," *IEEE Commun. Mag.*, vol. 45, nr. 5, pp. 132–138, May 2007.
23. Niyato, D., and Hossain, E., "Integration of WiMAX and WiFi: optimal pricing for bandwidth sharing," *IEEE Commun. Mag.*, vol. 45, nr. 5, pp. 140–146, May 2007.
24. Stuckmann, P., and Zimmermann, R., "Toward ubiquitous and unlimited-capacity communications networks, European research in framework programme 7," *IEEE Commun. Mag.*, vol. 45, nr. 5, pp. 148–157, May 2007.
25. Chapin, J.M., and Lehr, W.H., "Time-limited leases in radio systems," *IEEE Commun. Mag.*, vol. 45, nr. 6, pp. 76–82, Jun. 2007.
26. Lunch, K., Eggen, A., Hardzic, D., Hafsoe, T., and Johnsen, F.T., "Using web service oriented architecture in military communication networks," *IEEE Commun. Mag.*, vol. 45, nr. 10, pp. 47–53, Oct. 2007.
27. Weerackody, V., and Gonzales, L., "Mobile small aperture satellite terminals for military communications," *IEEE Commun. Mag.*, vol. 45, nr. 10, pp. 70–75, Oct. 2007.
28. Li, B., Yang, Q., Low, C.P., and Gwee, C.L., "A survey on mobile WiMAX," *IEEE Commun. Mag.*, vol. 45, nr. 12, pp. 70–75, Dec. 2007.

29. Soldani, D., and Dixit, S., "Wireless relays for broadband access," *IEEE Commun. Mag.*, vol. 46, nr. 3, pp. 58–66, Mar. 2008.
30. Stotts, L., Seidel, S., Krout, T., and Kolodzy, P., "MANET gateways: radio interoperability via the internet, not the radio," *IEEE Commun. Mag.*, vol. 46, nr. 6, pp. 51–59, Jun. 2008.
31. Kim, S., Kim, H.W., Kang, K., and Ahn, D.S., "Performance enhancement in future mobile satellite broadcasting services," *IEEE Commun. Mag.*, vol. 46, nr. 7, pp. 118–124, Jul. 2008.
32. Apostolacos, S., Manousos, M., Meliones, A., Kavadas, D., Lykakis, G., Manousarides, A., Kardaris, M., and Simeakis, K., "Design and implementation of a solution for the provisioning of converged remote tower and facility management services over Satellite IP Greek heliports," *IEEE Commun. Mag.*, vol. 46, nr. 8, pp. 48–56, Aug. 2008.
33. Etemad, K., "Overview of mobile WiMAX technology and evolution," *IEEE Commun. Mag.*, vol. 46, nr. 10, pp. 31–40, Oct. 2008.
34. Wang, F., Chosh, A., Sankaran, C., Fleming, P.J., Hsieh, F., and Benes, S.J., "Mobile WiMAX systems: performance and evolution," *IEEE Commun. Mag.*, vol. 46, nr. 10, pp. 41–49, Oct. 2008.
35. Kidston, D., and Kunz, T., "Challenges and opportunities in managing maritime networks," *IEEE Commun. Mag.*, vol. 46, nr. 10, pp. 162–168, Oct. 2008.
36. Xing, Y., and Chandramouli, R., "Human behavior inspired cognitive radio network design," *IEEE Commun. Mag.*, vol. 46, nr. 12, pp. 122–127, Dec. 2008.
37. Headrick, R., and Freitag, L., "Growth of underwater communication technology in the U.S. Navy," *IEEE Commun. Mag.*, vol. 47, nr. 1, pp. 80-82, Jan. 2009.
38. Pompili, D., and Akyildiz, I.F., "Overview of networking protocols for underwater wireless communications," *IEEE Commun. Mag.*, vol. 47, nr. 1, pp. 97–102, Jan. 2009.
39. Buljore, S., Harada, H., Filin, S., Houze, P., Tsagkaris, K., Holland, O., Nolte, K., Farnham, T., and Ivanov, V., "Architecture and enablers for optimized radio resource usage in heterogeneous wireless access network: the IEEE 1900.4 working group," *IEEE Commun. Mag.*, vol. 47, nr. 1, pp. 122–129, Jan. 2009.
40. Astely, D., Dahlman, E., Furuskar, A., Jading, Y., Lindstrom, M., and Parkvall, S., "LTE: the evolution of mobile broadband," *IEEE Commun. Mag.*, vol. 47, nr. 4, pp. 44–51, Apr. 2009.
41. Sydir, J., and Taori, R., "An evolved cellular system architecture incorporating relay stations," *IEEE Commun. Mag.*, vol. 47, nr. 6, pp. 115–121, Jun. 2009.
42. Ansari, A., Dutta, S., and Tseytlin, M., "S-WiMAX: adaptation of IEEE 802.16e for mobile satellite services," *IEEE Commun. Mag.*, vol. 47, nr. 6, pp. 150–155, Jun. 2009.
43. Kim, R.Y., Kwak, J.S., and Etemad, K., "WiMAX femtocell: requirements, challenges, and solutions," *IEEE Commun. Mag.*, vol. 47, nr. 9, pp. 84–91, Sep. 2009.
44. Suri, S., Benvegnu, E., Tortonesi, M., Stefanelli, C., Kovach, J., and Hanna, J., "Communications middleware for tactical environments: observations, experiences, and lessons learned," *IEEE Commun. Mag.*, vol. 47, nr. 10, pp. 56–63, Oct. 2009.
45. Cotton, S.L., Scanlon, W.G., and Madahar, B.K., "Millimeter-wave soldier-to-soldier communications for covert battlefield operations," *IEEE Commun. Mag.*, vol. 47, nr. 10, pp. 72–81, Oct. 2009.
46. Calin, D., Claussen, H., and Uzunlioglu, H., "On femto deployment architectures and macrocell offloading benefits in joint macro-femto deployments," *IEEE Commun. Mag.*, vol. 48, nr. 1, pp. 26–32, Jan. 2010.
47. George, S.M., Zhou, W., Chenji, H., Won, M., Lee, Y.O., Pazarloglou, A., Stoleru, R., and Barooah, P., "DistressNet: a wireless Ad Hoc and sensor network architecture for situation management in disaster response," *IEEE Commun. Mag.*, vol. 48, nr. 3, pp. 128–136, Mar. 2010.
48. Shar, P., and Li, X.R., "Passive sonar fusion for submarine C^2 systems," *IEEE A&E Syst. Mag.*, vol. 15, nr. 3, pp. 29–34, Mar. 2000.
49. Sadler, B.M., "Fundamentals of energy-constrained sensor network systems," *IEEE A&E Syst. Mag.*, vol. 20, nr. 8, part. 2, pp. 17–35, Aug. 2005.

50. Collens, J.R., Jr., "Theater battle management core system," *IEEE A&E Syst. Mag.*, vol. 22, nr. 3, pp. 15–20, Mar. 2007.
51. Schnell, M., and Scalise, S., "NEWSKY – concept for NEtWorking the SKY for civil aeronautical communications," *IEEE A&E Syst. Mag.*, vol. 22, nr. 5, pp. 25–29, May 2007.
52. Pastor, E., Lopez, J., and Royo, P., "UAV payload and mission control hardware/software architecture," *IEEE A&E Syst. Mag.*, vol. 22, nr. 6, pp. 3–8, Jun. 2007.
53. Pace, P., and Aloi, G., "Disaster monitoring and mitigation using aerospace technologies and integrated telecommunication networks," *IEEE A&E Syst. Mag.*, vol. 23, nr. 4, pp. 3–9, Apr. 2008.
54. Jamshidi, M., "System of systems engineering – new challenges for the 21st century," *IEEE A&E Syst. Mag.*, vol. 23, nr. 5, pp. 3–19, May 2008.
55. Loh, R., Bian, Y., and Roe, T., "UAVs in civil airspace: safety requirements," *IEEE A&E Syst. Mag.*, vol. 24, nr. 1, pp. 5–17, Jan. 2009.
56. Sampigethaya, K., Poovendran, R., Bushnell, L., Li, M., Robinson, R., and Lintelman, S., "Secure wireless collection and distribution of commercial airplane health data," *IEEE A&E Syst. Mag.*, vol. 24, nr. 7, pp. 14–20, Jul. 2009.
57. Dempsey, T., Sahin, G., Morton, Y.T., and Hopper, C.M., "Intelligent sensing and classification in Ad Hoc networks: a case study," *IEEE A&E Syst. Mag.*, vol. 24, nr. 8, pp. 23–30, Aug. 2009.
58. Frain, S., and Van Sickle, G., "CNS/ATM for tactical military aircraft," in *Proc. 16th Inter. Tech. Mtg. Sat. Div. ION (ION GPS/GNSS 2003)*, Portland, OR, pp. 841–848, Sep. 2003.
59. Tropea, M., Rango, F.D., and Marano, S., "Current issues and future trends: DVB-RCS satellite systems," *IEEE A&E Syst. Mag.*, vol. 24, nr. 9, pp. 15–22, Sep. 2009.
60. Croser, I.T., "Phased array technology in Australia," *IEEE A&E Syst. Mag.*, vol. 24, nr. 12, pp. 24–28, Dec. 2009.
61. Davis, M., "Foliage penetration radar," Tutorial 2, *IEEE RadarCon 2009*, Pasadena, CA, 4 May 2009.
62. Baniela, S.I., "Piracy at sea: Somalia an area of great concern," *J. Navig.*, vol. 63, nr. 2, pp. 191–206, Apr. 2010.
63. Anon. "Soil moisture and ocean salinity (SMOS) mission underway," *IEEE GRSS Newsl.*, p. 9, Dec. 2009.
64. Balmer, R., "SAR interferometry," Tutorial 1, *IEEE RadarCon 2009*, Pasadena, CA, 4 May 2009.
65. Guerci, J.R., "Introduction to STAP and knowledge-aided processing," Tutorial 3, *IEEE RadarCon 2009*, Pasadena, CA, 4 May 2009.
66. Jeffrey, T., "Radar tradeoffs, performance driven specifications and "back-of-the-envelope" performance," Tutorial 4, *IEEE RadarCon 2009*, Pasadena, CA, 4 May 2009.
67. Zieger, A. R., et al., "NASA Radar Altimeter for the TOPEX/POSEIDON Project," *Proc. IEEE*, vol. 79, nr. 6, pp., 810–826, June 1991.
68. Zyl, J.v., "Radar remote sensing for Earth and planetary sciences," Tutorial 6, *IEEE RadarCon 2009*, Pasadena, CA, 4 May 2009.
69. Cloude, Sh., "Radar polarimetry: from science to systems," Tutorial 7, *IEEE RadarCon 2009*, Pasadena, CA, 4 May 2009.
70. Griffiths, H., and Willis, N., "Advances in ballistic radar," Tutorial 8, *IEEE RadarCon 2009*, Pasadena, CA, 4 May 2009.
71. Williams, J., "Electronic scanned array design," Tutorial 12, *IEEE RadarCon 2009*, Pasadena, CA, 4 May 2009.
72. Blum, R.S., Haimovich, A.M., and Li, J., "MIMO radar: a sensing emerging technology," Tutorial 13, *IEEE RadarCon 2009*, Pasadena, CA, 4 May 2009.
73. Brookner, E., "Phased-arrays and radars: basics, past accomplishments, amazing breakthroughs and future trends," Tutorial 14, *IEEE RadarCon 2009*, Pasadena, CA, 4 May 2009.
74. Anon., SEASAT, From Wikipedia, the free encyclopedia, Oct. 31, 2010, http://en.wikipedia.org/wiki/SEASAT.
75. Anon., SIR-A 1982, NASA JPL, http://southport.jpl.nasa.gov/scienceapps/sira.html.

76. Anon., SIR-B 1985, NASA JPL, http://southport.jpl.nasa.gov/scienceapps/sirb.html.
77. Anon., Spaceborne Image Radar (SIR)-C, NASA JPL, http://southport.jpl.nasa.gov/sir-c/.
78. Belvin, F.W., Dorsey, J.T., and Watson, J.J. "Solar power satellite development: advances in modularity and mechanical systems," *Online J. of Space Comm. an Inter. Elect. J.*, issue 16, fall 2009.
79. Astrium GeoInformation Services, an EDAS company, InfoTerra, 2010, http://www.infoterra.de/tandem-x-satellite.
80. International Standards and Recommended Practices Aeronautical Telecommunications, International Civil Aviation Organization (ICAO) Annex 10, Volume I Radio Navigation Aids, July 1996.
81. Boseley, K., and Waid, J., "Shipboard-relative Global Positioning System (SRGPS) demonstration system design," in *Proc. 17th Inter. Tech. Mtg. Sat. Div. ION (ION GNSS 2004)*, Long Beach, CA, pp. 344–353, Sep. s2004.
82. Brown, A., O'Keefe, P., Jovancevic, A., and Ganguly, S., "JPALS performance model using a flexible simulation framework," in *Proc. 18th Inter. Tech. Mtg. Sat. Div. ION (ION GNSS 2005)*, Long Beach, CA, pp. 2923–2929, Sep. 2005.
83. Skidmore, T.A., Cohenour, C., van Graas, F., DiBenedetto, M., and Peterson, B., "JPALS tactical system performance evaluation using a controlled reception pattern antenna," in *Proc. 19th Inter. Tech. Mtg. Sat. Div. ION (ION GNSS 2006)*, Fort Worth, TX, pp. 738–753, Sep. 2006.
84. Goldstein, M., Picciolo, M., and Griesbach, J., "Advanced STAP," Tutorial 16, *IEEE RadarCon 2009*, Pasadena, CA, 4 May 2009.
85. Space Shuttle Atlantis Control Panel, http://media.photobucket.com/image/space%20shuttle%20panel/Mike_F_620/SpaceShuttleAtlantisControlPanel2.jpg.
86. Google Maps Now in HD Thanks to GeoEye-1 Satellite, http://www.dbtechno.com/space/2008/10/10/google-maps-now-in-hd-thanks-to-geoeye-1-satellite/.
87. GeoEye GeoEye-1, http://www.popsci.com/bown/2008/product/geoeye-geoeye-1.
88. Atlantis Meets Mir, March 23, 2008, http://www.nasa.gov/multimedia/imagegallery/image_feature_189.html.
89. Flight formation from US DoD website, http://www.defense.gov/transformation/images/photos/2006-06/Hi-Res/060618-N-4166B-132.jpg.
90. Four F-117 Nighthawks fly in formation, http://www.air-attack.com/images/single/429/Four-F-117-Nighthawks-fly-in-formation.html, http://files.air-attack.com/MIL/f117/f117_formation-20070424.jpg.
91. Shuttle Carrier Aircraft, November 30, 2006, Wikimedia Foundation, Inc., http://en.wikivisual.com/index.php/Shuttle_Carrier_Aircraft.
92. PCU George H. W. Bush (CVN-77) move from Dry Dock 12 to OB1, October 9, 2006. Northrop Grumman Newport News photo by Ricky Thompson, http://www.navsource.org/archives/02/027733.jpg, main page http://www.navsource.org/archives/02/77.htm.
93. USS George H.W. Bush (CVN 77) successfully fired two Evolved NATO Sea Sparrow missiles and two Rolling Airframe Missiles (RAM) for the first time June 25, 2010, http://warweaponry.files.wordpress.com/2010/06/missile-launch.jpg.
94. Navy ship containing 500 marines. Courtesy of Wikimedia, http://upload.wikimedia.org/wikipedia/commons/c/c5/US_Navy_050901-N-0780F-003_The_dock_landing_ship_USS_Ashland_(LSD_48)_and_embarked_elements_of_26th_Marine_Expeditionary_Unit_(MEU),_arrive_in_Souda_Harbor.jpg.
95. Integrated Symmetrical Concentrator (ISC) solar power satellite concept, SpaceRef Interactive Inc., 2001, http://www.spaceref.com/tools/imgcathp.html?cid=41.
96. US Department of Interior, US Geological Survey Map, Monday August 22, 2005, http://pubs.usgs.gov/dds/dds11/kbview2x.html.
97. Air and Space Smithsonian, Smithsonian, March, 2007, http://www.rotors.org/news/20070218.htm.
98. Air and Space Smithsonian, Smithsonian, September, 1999, http://www.coseti.org/images/airspace.jpg.

99. World seismicity maps, regions, north Atlantic Ocean, U.S. Department of the Interior | U.S. Geological Survey, March, 2010, http://earthquake.usgs.gov/earthquakes/world/seismicity/n_atlantic.php.
100. Image courtesy of US National Oceanic and Atmospheric Administration (NOAA), National Weather Service (NWS), http://polar.ncep.noaa.gov/waves/latest_run/nww3_at.anim.3.gif.
101. Image courtesy of US National Oceanic and Atmospheric Administration (NOAA), National Weather Service (NWS), http://polar.ncep.noaa.gov/waves/latest_run/nww3_at.h006h.3.gif.
102. Image courtesy of US National Oceanic and Atmospheric Administration (NOAA), National Weather Service (NWS), http://polar.ncep.noaa.gov/waves/latest_run/nww3_at.h000h.3.gif.
103. NASA's Quikscat Ocean-Observing Satellite Mission Honored, NASA, http://www.nasa.gov/mission_pages/hurricanes/archives/2008/jpl_quikscat.html, http://www.nasa.gov/images/content/291999main_quikscat-large.jpg.
104. Harvard University, atmospheric chemistry, modeling group, current research, August 28, 2009, http://acmg.seas.harvard.edu/curresh.html, http://acmg.seas.harvard.edu/img/nasa_satellites.png.
105. New geostationary satellite called the Sirius FM-5, 2004–2009 Orbitcast Media LLC, http://www.orbitcast.com/archives/sirius-fm-5-satellite-is-up-and-running—can-you-tell-the-difference.html.
106. GOCE Harnesses Ion Propulsion to Capture First 'Gravity Map' of Earth, Popular Science, October 1, 2009, http://www.popsci.com/military-aviation-amp-space/article/2009-10/goce-harnesses-ion-propulsion-capture-first-accurate-gravity-map-earth, http://www.popsci.com/files/imagecache/article_image_large/articles/28_H1.jpg.
107. When the dream become a reality: constellation Program, images courtesy of NASA, http://www.squidoo.com/human_spaceflight_constellation_program.
108. OrbitalHub, The place where space exploration, science, and engineering meet, MIT Open Courseware Aircraft Systems Engineering Lectures, OrbitalHub, 2008–2009, http://orbitalhub.com/?tag=space-shuttle, http://orbitalhub.com/wp-content/uploads/2008/12/mit-space-shuttle-sonic-boom-low.jpg.
109. Space Shuttle Challenger, The U.S. National Archives and Records Administration, 2010, http://www.archives.gov/press/press-kits/picturing-the-century-photos/space-shuttle-challenger.jpg.
110. The nation: we have yet to see the biggest costs of the BP Spill, August, 2010, http://rajpatel.org/wp-content/uploads/2010/08/NASA-Sees-spill-on-May-24.jpg.
111. Evolution of the fuel cell, or hydrogen cell, originally invented for space exploration will eventually become the source of energy in electric vehicles, http://www.tpmonline.com/images%20tpm/shuttle.jpg, 2009, YPMonLine.com, http://www.tpmonline.com/articles_on_total_productive_maintenance/innovations/fuelcells.htm.
112. Fish eye view of Earth and Space Shuttle Atlantis, The Chamorro Bible, 2002–2006, http://chamorrobible.org/images/photos/gpw-20051129-NASA-GPN-2000-001039-fish-eye-view-of-Earth-and-Space-Shuttle-Atlantis-STS-71-19950629.jpg.
113. Columbia, Google, 2010, http://lh5.ggpht.com/__Le_vMi7wPE/RcapPWz1TLI/AAAAAAAAAt8/npz8pJE9Dk0/s160/Columbia.sts-1.01.jpg.
114. Image courtesy of NASA, http://smartregion.org/2009/11/, http://smartregion.org/wp-content/uploads/2009/11/Pat-Rawlings-from-spaceflight.nasa.gov_lunaractivities.jpg.
115. Mars Rover, Image Courtesy of NASA, http://realitypod.com/wp-content/uploads/2009/09/Mars_rover.png.
116. Images courtesy of NASA, http://www.spaceref.com/news/viewnews.html?id=1427, http://images.spaceref.com/news/2010/Endeavour_Tribute_NASA.jpg.
117. Image courtesy of NASA, Kepler observatory locates 700 planets in just six weeks, http://www.thetechherald.com/media/images/201030/Kepler_3.jpg, http://www.thetechherald.com/article.php/201030/5940/Kepler-observatory-locates-700-planets-in-just-six-weeks.
118. NASA exploration roadmap, Image courtesy of NASA, http://images.spaceref.com/news/2009/comm.fig.1.jpg.

119. Strodl, G., Naddeo, G., Samson, J., Dieleman, P., Ferraguto, M., Hardt, H-J., and von der, Gottifredi, F., "System verification: approach, methods, and tools for Galileo," in *Proc. 16th Inter. Tech. Mtg. Sat. Div. ION (ION GPS/GNSS 2003)*, Portland, OR, pp. 2445–2456, Sep. 2003.
120. Maeda, H., "The start of a new era in Global Space – based positioning, navigation, and timing systems QZSS overview and interoperability," in *Proc. 18th Inter. Tech. Mtg. Sat. Div. ION (ION GNSS 2005)*, Long Beach, CA, pp. 0-0, Sep. 2005 (Plenary Presentation).
121. Shaw, M.E., "The start of a new era in Global space-based positioning, navigation, and timing systems," in *Proc. 18th Inter. Tech. Mtg. Sat. Div. ION (ION GNSS 2005)*, Long Beach, CA, pp. 0-0, Sep. 2005 (Plenary Presentation).
122. Ruiz, L., "GALILEO Overall Programme Status," in *Proc. 18th Inter. Tech. Mtg. Sat. Div. ION (ION GNSS 2005)*, Long Beach, CA, pp. 0-0, Sep. 2005 (Plenary Presentation).
123. Klimov, V., Persev, V., Revnivykh, S., Mitrikas, V., Kossenko, V., Zvonar, V., Dvorkin, V., Andruschak, E., and Tyulyakov, A., "GLONASS Status, Performance and Perspectives," in *Proc. 18th Inter. Tech. Mtg. Sat. Div. ION (ION GNSS 2005)*, Long Beach, CA, pp. 0-0, Sep. 2005 (Plenary Presentation).
124. Ballenger, A., "GPS status update ION plenary," in *Proc. 18th Inter. Tech. Mtg. Sat. Div. ION (ION GNSS 2005)*, Long Beach, CA, pp. 0-0, Sep. 2005 (Plenary Presentation).
125. Ripple, M.U., and Vidal, A., "New U.S. GPS policy from a European perspective," in *Proc. 18th Inter. Tech. Mtg. Sat. Div. ION (ION GNSS 2005)*, Long Beach, CA, pp. 407–413, Sep. 2005.
126. Last, D., "We live in interesting times," in *Proc. 19th Inter. Tech. Mtg. Sat. Div. ION (ION GNSS 2006)*, Fort Worth, TX, pp. 18–29, Sep. 2006.
127. Parkinson, B., "Three critical issues for GPtS," in *Proc. 19th Inter. Tech. Mtg. Sat. Div. ION (ION GNSS 2006)*, Fort Worth, TX, pp. 50–58, Sep. 2006.
128. Takac, F., and Walford, J., "Leica system 1200 – high performance GNSS technology for RTK applications," in *Proc. 19th Inter. Tech. Mtg. Sat. Div. ION (ION GNSS 2006)*, Fort Worth, TX, pp. 217–225, Sep. 2006.
129. Journo, S., Poumailloux, J., Piéplu, J-M., and Gouni, P., "Producing the Galileo services from the ground mission segment (GMS)," in *Proc. 19th Inter. Tech. Mtg. Sat. Div. ION (ION GNSS 2006)*, Fort Worth, TX, pp. 504–519, Sep. 2006.
130. Alcantarilla, I., Porras, D., Tajdine, A., Zarraoa, N., and Lévy, J.C., "The benefits of multi-constellation GNSS augmentations," in *Proc. 19th Inter. Tech. Mtg. Sat. Div. ION (ION GNSS 2006)*, Fort Worth, TX, pp. 930–938, Sep. 2006.
131. Kogure, S., Sawabe, M., and Kishimoto, M., "Status of QZSS Navigation System in Japan," in *Proc. 19th Inter. Tech. Mtg. Sat. Div. ION (ION GNSS 2006)*, Fort Worth, TX, pp. 2092–2102, Sep. 2006.
132. Wiley, B., Craig, D., Manning, D., Novak, J., Taylor, R., and Weingarth, L., "NGA's Role in GPS," in *Proc. 19th Inter. Tech. Mtg. Sat. Div. ION (ION GNSS 2006)*, Fort Worth, TX, pp. 2111–2119, Sep. 2006.
133. Bauer, F.H., Moreau, M.C., Dahle-Melsaether, M.E., Petrofski, W.P., Stanton, B.J., Thomason, S., Harris, G.A, Sena, R.P., and Temple, L.P., III, "The GPS space service volume," in *Proc. 19th Inter. Tech. Mtg. Sat. Div. ION (ION GNSS 2006)*, Fort Worth, TX, pp. 2503–2514, Sep. 2006.
134. Lestarquit, L., Harr, J., Grelier, T., Peragin, E., Wilhelm, N., Mehlen, C., Peyrotte, C., "Autonomous formation flying RF sensor development for the PRISMA mission," in *Proc. 19th Inter. Tech. Mtg. Sat. Div. ION (ION GNSS 2006)*, Fort Worth, TX, pp. 2571–2578, Sep. 2006.
135. Hein, G.W., "Towards a GNSS System of Systems," in *Proc. Inter. Tech. Mtg. Sat. Div. ION* (ION GNSS 2007), Fort Worth Convention Center, Fort Worth, TX, pp. 22–30, Sep. 2007 (Plenary Presentation).
136. Helm, A., Montenbruck, O., Ashjaee, J., Yudanov, S., Beyerle, G., Stosius, R., and Rothacher, M., "GORS – a GNSS occultation, reflectometry and scatterometry space

receiver," in *Proc. Inter. Tech. Mtg. Sat. Div. ION* (ION GNSS 2007), Fort Worth, TX, pp. 2011–2021, Sep. 2007.

137. Dow, J.M., Neilan, R.E., and Rizos, C., "The international GNSS Service (IGS): preparations for the coming decade," in *Proc. Inter. Tech. Mtg. Sat. Div. ION* (ION GNSS 2007), Fort Worth, TX, pp. 2136–2144, Sep. 2007.
138. Lestarquit, L., Grelier, T., Harr, J., Peragin, E., Issler, J.-L., Wilhelm, N., Mehlen, C., Garcia, A., and Gerner, J.L., "A spaceborne formation flying RF system in S-band reusing the GPS standards," in *Proc. Inter. Tech. Mtg. Sat. Div. ION* (ION GNSS 2007), Fort Worth, TX, pp. 2435–2441, Sep. 2007.

Chapter 3
RF Signals

3.1 Overview

RF signals offers the best overview of the best practices and innovative techniques in the art and science of RF signals design, signal structure, signal interpretation (which includes propagation, signal density, and absorption models) over the last 20 years in the literature of RF signals [1–144]. This is by all means not a small task considering the fact that the International Telecommunications Union (ITU) Spectrum Monitoring handbook is suggested as a more appropriate reference and structure for discussion of geolocation signals and because the concepts and outdoor principles are well covered in ITU Spectrum Monitoring handbook [11, 12]. It covers all insight aspects including theoretical analysis, RF signals, signal techniques, key block diagrams, and practical principle signal interpretations in the frequency band from 100 MHz to 66 GHz. Dr. Progri reveals the research and development process by demonstrating how to understand and explain a good number of RF signals such as those used in wireless networks, mobile phones (or cellular networks), indoor geolocation systems, amplitude modulation (AM) and frequency modulation (FM) radio, two-way radio, satellite radio, TV broadcasting, satellite TV broadcasting, digital video broadcasting (DVB), Global Navigation Satellite Systems (GNSS), etc. from basic diagrams to be utilized to the principle simulation examples and make recommendations for the future final products of geolocation of RF signals (GRFS) [1–144]. We are going to discuss all these signals here except indoor geolocation systems signals, which are discussed in a separate publication on *Indoor Geolocation Systems: Theory and Applications*, which will be published soon.

Starting with RF signals, the chapter progressively examines various signal bands – such as VLF, LF, MF, HF, VHF, UHF, L, S, C, X, K_u, and, K and makes recommendations for the corresponding geolocation requirements per band and per application – to achieve required performance objectives of up to 0° precision. Next follows a step-by-step approach of RF signal designs and structure interpretations from indoor to satellite environment and concludes with recommendations on state-of-the-art geolocation designs as well as advanced features found in signal generator instruments to be discussed in Chaps. 4 and 6. Chapter 4 includes the best

I. Progri, *Geolocation of RF Signals*,
DOI 10.1007/978-1-4419-7952-0_3, © Springer Science+Business Media, LLC 2011

mathematical techniques employed for GRFS from 100 MHz to 66 GHz. The principle simulation examples, discussed in a great detail in Chap. 4 during the second part of the book, utilize a great deal of signal design knowledge accumulated in this chapter. This chapter offers invaluable insights on RF signals that are not found in any other RF signal textbook or manual, and is an all-in-one source for the beginner, the experienced, expert analysts, and professionals. This chapter also concludes Part I of the book and opens up for the Part II of the book which is on best mathematical techniques and methods on *Geolocation of RF Signals*: *Principles and Simulations*.

3.2 Introduction of RF Signals

RF signals are the raw inputs (or essential elements) with which GRFS systems work. In Chap. 1 we discussed how the RF signals were generated. In Chap. 2 we discussed the requirements for description of GRFS systems. Here is Chap. 3 where we are going to present RF signals. The emphasis here is on how RF signals are classified, presented, understood, reported in the literature, and how this is connected with the rest of the material in the book. Figure 3.1 is especially important because it provides a complete big picture on RF signals.

Based on the initial description in Fig. 3.1, one should arrive at the conclusion that an accurate and complete description of RF signals has a wide range of complexity,

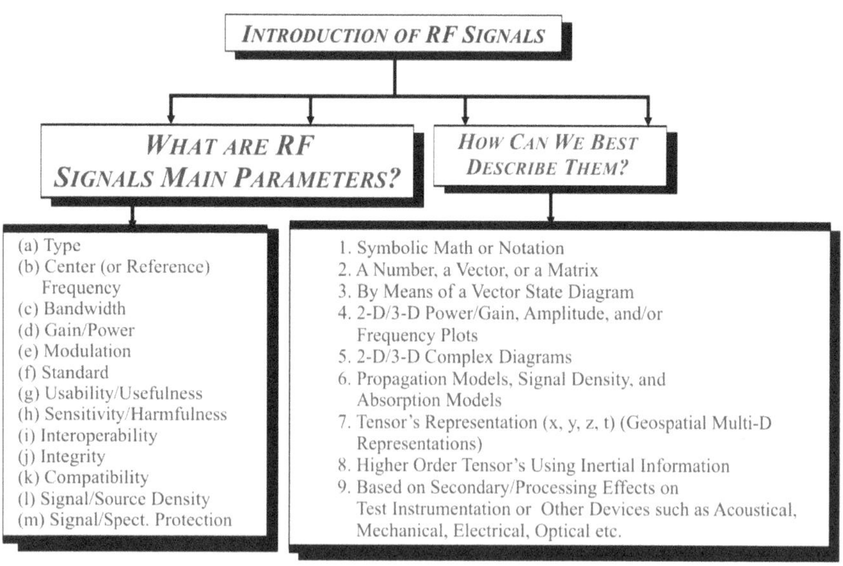

Fig. 3.1 An overview on the introduction of RF signals. Reprinted with permission, © 2010 Ilir Progri

sophistication, representation, and labor involved in it; i.e., it can be something very simple to something extremely complex, laborious, and sophisticated. The main idea is not to frustrate the reader with extremely fancy and highly sophisticated formulas, diagram, and expressions that only a small group of readers might enjoy; but to enable an area for discussion, for research, for implementation, for interpretation, etc. So, taking into account the needs of the readership, the level of depth and representation will be sufficient enough to enable the reader to understand and accurately follow the material and at the same time enable me to discuss RF signals at a level that is acceptable to the graduate students in any accredited US graduate engineering institution.

The remainder of this section is organized as follows: First, we are going to discuss *What Are the RF Signals Main Parameters*? in Sect. 3.2.1. Second, we are going to develop *How We Can Best Describe RF Signals*? in Sect. 3.2.2.

3.2.1 What Are the RF Signals Main Parameters?

As depicted in Fig. 3.1 the RF signals main parameters are: (a) type; (b) center (or reference) frequency; (c) bandwidth; (d) gain/power; (e) modulation; (f) standard; (g) usability/usefulness; (h) sensitivity/harmfulness; (i) interoperability; (j) integrity; (k) compatibility; (l) signal/source density; (m) signal/source protection.

Type: is the first parameter and perhaps the most important identifier of the RF signals that forms a unique class such as: (1) narrow band vs. wide band; (2) AM signals vs. FM signals; (3) frequency division duplex (FDD) vs. time division duplex (TDD); (4) GPS L1, L2, L5 (or code division multiple access [CDMA]) vs. GLONAS signals; (5) power spectrum density (PSD) symmetric vs. PSD A-Symmetric; (6) periodic or A-periodic; (7) stationarity conditions apply vs. nonstationarity; (8) RF signals which require line-of-sign (LOS) reception vs. RF signals that do not require line-of-sign reception (or nonline-of-sign [NLOS]), etc. The LOS and NLOS type signals will be discussed more extensively in the *Indoor Geolocation Systems*: *Theory and Applications* book. As more types of RF signals get generated, the number of types in the class of the RF signals increases with time. One might ask as to how these FDD and TDD technologies relate to communication systems, propagation distances, and the performance conditions of the parameters to follow. It is best to suggest to the ITU community to provide a library of all possible types of the entire RF signals library (at least those that are unclassified) to their latest edition of the ITU Spectrum Monitoring handbook. The second approach is the one followed in this chapter and this book. To a certain extent our discussion is focused on how these FDD and TDD technologies relate to communication systems, propagation distances, and the performance conditions of the parameters to follow.

Center (*or reference*) *frequency*: is the second most important parameter of the RF signals. If RF signals under consideration have symmetry about the center frequency then we can define center frequency as the centerline of the frequency

PSD of the RF signal in the frequency domain otherwise the "center" frequency as in the case of single-side band, chirp pulses, ATSC new digital video channel, or even complex signal over the air need to be deconvolved or demodulated to find the pilot frequency to determine the center frequency, is defined as the *reference frequency* from which the lower cutoff and the higher frequency cutoff of the signal bandwidth are offset or even measured. I.e., the center frequency will either serve as the center of the band for symmetric RF signals or the reference to the band for asymmetric signals. Although RF signals have a wide frequency spectrum, for the purpose of this book and for the most part for the purpose of the research and development, analysis and simulations, product development and deployment, we are going to consider RF signals whose center frequency is in the range of 100 MHz to 66 GHz.

Bandwidth: is defined as the effective or useful spread of the RF signal in the frequency domain. The easiest way to visualize this is that if we were to take the fast Fourier transform (FFT) or if we were to look at the frequency spectrum of the RF signal then we can see that bandwidth is defined as the difference from the higher cutoff frequency to the lower cutoff frequency. It is important to recognize the channel filter shape and the allowable amount of distortion within the band and outside the band which are classified as in-band emissions and out-of-band emissions [15–17]. Every signal generator manufacturer, ex. Agilent Technologies, has well-defined standards for these levels of emissions as defined by the Federal Communications Commission (FCC) and ITU [17]. Chapter 4 contains many useful illustrations of these parameters.

Gain/power: some RF signals can be described with the help of their gain level such as amplitude voltage (or current) or with the help of their power level such as apparent, active, or reactive (VA, W, or VAr). Power representation tends to be more common than gain and for the most part the acceptable units in the literature are dBW or dBm. We are going to define these units later.

Modulation: is the communication technique that enables the variation of one of the other signal parameters such as amplitude, phase, frequency, or type in accordance with a rule or a property [18]. Every communications textbook has great discussions on most common modulations such as AM (or quadrate amplitude modulation (QAM), FM, phase modulation (PM), angle modulation (AM), digital modulations such as binary phase shift keying (BPSK), quadry phase shift keying (QPSK), etc. (example [18]). Again, it is suggested that perhaps the *ITU's Spectrum Monitoring Handbook* (as an appendix) should provide a library of all possible *modulations* of the entire RF signals (at least those that are unclassified) [11, 12].

Standard: is a category for several RF signals such as IEEE 802.11, IEEE 802.15, IEEE 802.16, IEEE 802.20, etc. Some other standards can be the ITU and NATO standards, European Standards, or Japanese Standards, or Chinese Standards, or Korean Standard, etc. To the best of our [engineering community] knowledge, as RF signals are reported in the literature, I am going to provide this information throughout the chapter and book. If the standard field is empty then it can mean the following: (1) The standard exists and I have not been able to find it; (2) The standard could be classified information which means that it is only an unclassified version of the signal that it is considered; (3) The standard is in the making and it is not approved or released at the date of publication. Moreover, it is worth mentioning that industry standards utilize the "*unlicensed frequency bands*" which other "*licensed*" frequency *band* tend to be driven by the ITU and member nations to assure broad adoptions and interoperability with one exception being the WiMax IEEE 802.16 [11, 12].

Usability/usefulness: this is also an important parameter of RF signals. This is a very large subject because the usability and usefulness of RF signals touches every facet of our lives. As far we are concerned, Chap. 2 is the best guide for the usability/usefulness in terms of applications which is also the way that we have organized Chap. 2 and also this chapter. Without too much reiteration, usability/usefulness of the RF signals is defined as the effective useful and detectable range of the RF signals which as described in Chap. 2 ranges from 100 m to 100,000 km.

Sensitivity/harmfulness: this is perhaps the grayest area of all the parameters of RF signals because RF signals have only been around for less than a couple of 100 years as opposed to humans who have been around much longer. Therefore, many interoperability issues related to sensitivity/harmfulness of the RF signals are still under research, investigation, publication, etc. As far as we are concerned, sensitivity is defined as the appropriate level at which an RF signal can be detected and used for processing; i.e., the effective dynamic range. Harmfulness is defined as the parameter in which RF signal(s) power exceeds a certain acceptable useful (or usable) level of an RF signal. Harmfulness is also related to interoperability issues of the RF signal and therefore it can be classified as interference. When interference level is intended to disrupt, disorder, deny the service, etc. which can lead to serious failures into systems, then we are dealing with deliberate jamming. Jamming is perhaps the most serious threat of deliberate interference which can lead from serious system anomalies to the complete malfunction, disruption, or destruction of the entire system or systems. This is the reason why jamming must be fully understood, explained, accounted for, and eliminated to enable safe and normal operation of device or the system which can be an aircraft, destroyer, etc. *Sensitivity/harmfulness* seems like a broad term to cover both human safety concerns and interoperability. In order to avoid this ambiguity, ITU might decide to provide two subcategories for this term: one pertaining to human safety concerns and the other pertaining to interoperability. It seems that the interference issue, as it relates to receiver issues like *blocking and selectivity*, has more of a context within this book.

Interoperability: refers to a property of GRFS systems RF signal design's (or spectrum) ability to be nearly orthogonal in either time or frequency domain with other telecommunication devices signals taking into account RF signals spectrum as defined by the ITU Spectrum Monitoring handbook, all possible interfaces, and other factors that impact GRFS systems performance [11, 12]. This is a slightly different definition of interoperability from the one provided in Chap. 1 because here we have in view RF signals' definition and computation (or processing) and in Chap. 1, we have in view GRFS system's definition and functionality (design or simulation or test, etc.). Two RF signals then are said to be interoperable; if we were to take their product in either time or frequency domain, then this product should result in a term that is either zero or something very small and negligible whose tolerance level defines the allowable interoperability level of the two RF signals

[11, 12]. ITU Spectrum Monitoring handbook should have plenty of discussion on interoperability of RF signals and telecommunication devices and systems.

Integrity: it is a measure of trust that we place on the signal and its information. This is especially important when we rely on the signal's information for safety transportation such as aviation, automotive, rail, navy, public safety, search, and rescue operations, etc. More discussions on integrity are given in RF signals for satellite GRFS systems in Sect. 3.7.

Compatibility: refers to the property of the signal design of the GRFS systems that ensures coexistence of GRFS systems, principles, signal designs, components, technologies, and enables as little changes as possible to ensure that one GRFS system is capable of performing the same tasks as another GRFS system. This aspect of the GRFS signal design became important for the design of Galileo; modernization of GPS and GLONASS, and with design of GPS-III [145–315]. Compatibility ensures coexistence, survivability, slow-fading, and smooth transition of the best and widely used techniques, signals, principles, components, until a new generation of RF signals, components, principles, etc. has taken over. Due to the novel nature of the RF signal design requirements, it is not clear what should the cycle of compatibility for RF signals and systems under consideration be; however, it is clear that compatibility will ensure a smoother and more cooperative environment. Moreover, it is neither the purpose nor the focus of this book to determine all the global factors that influence compatibility because the minute detailed study of compatibility might result in a book such as RF signal compatibility design.

Signal/spectrum protection: for all RF signals that are approved by the FCC in the United States, and ITU from the United Nations in ITU Spectrum Monitoring handbook, signal/spectrum protection is that authority and obligation that is given by ITU to enforce protection or legitimate use of RF signals as approved in the ITU spectrum monitoring handbook or during the World Radio Conference (WRC); during which changes of the ITU are proposed and approved.

Signal/source density: The number of independent, identical RF signals or sources in a particular area or volume. We could also define signal density as a function of RF signals coming from a particular direction, in this case will be a linear density. We are going to see later in the chapter that signal density is an important parameter in the Wireless Mobile Networks and perhaps in the case of multipath, interference, and jamming.

This completes the first description of the RF signals' main parameters. We should also mention that these are not all the parameters, only the most important one or at least for the purpose of this book they are considered the most important ones.

3.2.2 How Can We Best Describe RF Signals?

Now that we have given a definition of all the RF signal parameters, we can start discussing how we can best describe RF signals based on the information provided in Fig. 3.1. There are at least nine ways to describe RF signals using: (1) *symbolic math* or *notation*; (2) a real or complex *number*, *vector*, or a *matrix*; (3) *vector state diagram*; (4) *2D/3D power/gain, amplitude, frequency* plots; (5) *2D/3D complex diagrams*; (6) *propagation models, signal density, and absorption models*; (7) *tensors* with the following information, (*x*, *y*, *z*, *t*) should geospatial multidimensional be provided; (8) *higher order tensors*; and (9) *the effects they cause on secondary systems*.

First, we can describe RF signals using *symbolic math* or *notation*. This is perhaps the most common and the easiest of description for an RF signal. We can use any of the letters or symbols to describe RF signals.

Second, an RF signal can be described using a *real* or *complex number*, *vector*, or a *matrix*. A number is usually a numeric value associated with an RF signal parameter generally or typically employed in every numerical assessment or analysis section of real, *complex numbers*, *vectors* or *matrices*.

Third, RF signals can be represented by means of a *vector state diagram*. This form of representation is also very common because it represents the signal as having a magnitude and phase and since we are interested in finding out the angle of arrival for our RF signal sources we shall see in Chap. 4 that this form of representation will be fairly useful to us.

Fourth, RF signals are generally represented by *2D/3D power/gain, amplitude, frequency* plots. This form of representation, illustration, and interpretation is perhaps the most beloved form from all engineers and all professionals in the field. Chapters 4 and 6 contain many illustrations of signals which are in fact 2D/3D power, gain, amplitude, and frequency plots.

Fifth, RF signals are also represented as *2D/3D complex diagrams*. This form of representation is useful to enable a visual interpretation of complex waveforms. Generally speaking, this form of representation requires a lot of information about the signal structure or designs which is especially useful for designing a workable receiver. Chapter 4 will examine scenarios (or case studies) on information about 2D/3D complex diagrams of various RF signals.

Sixth, RF signals can be represented using *propagation models*, *signal density parameters*, *and absorption models*. These models can be discussed either directly into the category of tensor or higher order tensors which are treated next or as more simple models which are considered in a separate chapter in *Indoor Geolocation Systems: Theory and Applications*.

Seventh, RF signals can be represented as *tensors* with the following information (*x*, *y*, *z*, *t*) should geospatial multidimensional be provided. Because scalars and vectors are special cases of tensors, this form of representation is inferred for geospatial maps and solutions by the National Geospatial and Intelligence Agency (NGA) [134]. This form of representation of RF signals is more accurate than the previous forms that we have seen and therefore at specific cases it might represent information that might be considered classified. Again to the extent that this form of representation is used in the unclassified literature for commercially available RF signals or military signals, we will discuss, analyze, represent, and interpret them in Chaps. 4 and 6. Tensors are used quite a bit in general theory of relativity and electrodynamics, [2] so we might use them in Chaps. 4 and 6.

Eight, RF signals can also be represented using *higher order tensors*. This is perhaps the most accurate form of representation of RF signals that we are going to discuss or infer anywhere in this book because it utilizes information from the general theory of relativity and electrodynamics. Again to the extent that this form of representation is used in the unclassified literature for commercially available signals or military signals, we will discuss, analyze, represent, and interpret them in Chaps. 4 and 6. Higher order tensors are used quite a bit in the general theory of relativity [2], astronomy, cosmology, electrodynamics, radars, etc. so we might use them in Chaps. 4 and 6.

Ninth and final, RF signals are represented by *the effects they cause on secondary systems* which are receivers of any kind. For example, we are too familiar with FM signals and AM signals. Well, these RF signals are all different from the RF TV signals. RF signals that enable our communications via the mobile phones are different from RF signals that are used to do the same via the cordless phones and garage door openers, etc. Also, RF signals that are used by the GPS receiver are different from RF signals that are used by the radars, etc. US Air Force, Army, and Navy employs RF signals' special feature to fire off missiles and weapons, to guide munitions, communicate voice, video, data, to gather intelligence by scanning remote areas with radars; to perform reconnaissance and surveillance, etc. NASA uses RF signals: to guide space shuttles; to position geosatellites in geosynchronous orbit; to transmit and receive huge amounts of data, voice, video, live picture; to communicate with astronauts in and around the space shuttle or space stations, etc.; to geolocate stars, galaxies, star constellations, etc. As for the purposes of this book, we are going to give some detail to secondary effects on RF signals but we will try to maintain our focus on giving very detailed and accurate description of the principle simulation examples (or illustrations) on GRFS systems.

Chapter 3 is organized as follows: We briefly discuss RF signals for indoor GRFS systems in Sect. 3.3. Next, we investigate RF signals for urban GRFS systems in Sect. 3.4. In Sect. 3.5, we provide a proper explanation on RF signals for suburban GRFS systems. RF signals for global GRFS systems are discussed in Sect. 3.6. RF signals for satellite GRFS systems are assessed in Sect. 3.7. This is the reason why GRFS systems have maximum performance achievable of any system under normal conditions and yield up to 0° angle of arrival estimation precision accuracy. The best state of the art GRFS techniques (or algorithms) are discussed in Chap. 4. Applications of GRFS are discussed in Chaps. 4 and 6. Chapter 3 is concluded in Sect. 3.8.

3.3 RF Signals for Indoor GRFS Systems

RF signals for indoor GRFS systems are grouped into: (1) RF signals for wireless networks (which include indoor geolocation systems) GRFS systems which are discussed in Sect. 3.3.1.

3.3.1 RF Signals for Wireless Networks GRFS Systems

The Wireless signal structure (or spectrum) in the unlicensed 2.4 and 5 GHz frequencies is based on several standards such as IEEE 802.11 "A, B 'Wi-Fi,' E, and G" [106] and IEEE 802.15.3 [108], IEEE 802.15.4 [109], and other proprietary signal designs [135, 136].

First, the IEEE 802.11a specifications uses OFDM modulation and is capable of achieving data rates up to 54 Mb/s and its effective range is up to 100 m and its current drain is greater than 350 mA [106, 110, 111, 113]. There are technical needs for WLAN and Wireless Personal Area Networks (WPANs) technologies based on customer use models: (1) WPAN technologies replace wireless and docking station connections; and (2) WLAN provides Internet/enterprise connectivity [106, 110, 111, 113]. Appendix A includes a description of a baseband model of an IEEE® 802.11a physical layer WLAN.

Second, the IEEE 802.11b "Wi-Fi," operates in the 2.4 GHz with data rate up to 22 Mb/s and effective range up to 100 m. Two leading PHY candidates for the 802.11g standard are single-carrier trellis-coded 8-phase shift keying (PSK) modulation and OFDM schemes both of which offer much more costly radio and baseband implementations than the 802.15.3 PHY layer which is discussed next [106, 110, 111, 113]. Appendix A provides a description of the baseband Simulink block diagram of IEEE® 802.11b WLAN Physical Layer.

Third, the IEEE 802.15.3 layer operates in the unlicensed frequency band between 2.4 and 2.4835 GHz, and is designed to achieve data rates up to 11–55 Mb/s and symbol rate of 11 Mbaud (same as 802.11b); these rates are commensurate with the high definition (HD) video and audio [106, 110, 111, 113]. This standard includes five distinct modulation formats such as uncoded quadrature PSK (QPSK) modulation at 22 Mb/s and trellis coded QPSK; 16/32/64-QAM at 11, 33, 44, and 55 Mb/s, respectively TCM [106, 110]. This standard has the advantage of offering somewhat lower current drain on the order of less than 80 mA [106, 110, 111, 113]. Appendix A shows a description of the baseband Simulink block diagram of IEEE® 802.15.3 UWB Multiband OFDM Physical Layer.

Fourth, the IEEE 802.15.4 is designed to offer data rates at 250 kb/s at industrial, scientific, and medical (ISM) 2.4 GHz band, range 10–20 m, 16 channels with channel center frequency at $2{,}405 + 5(k - 11)$ for $k = 11, 12, \ldots, 26$; symbol rate at 62.5 kbaud and 16-ary orthogonal with chipping rate at 2 Mchips/s and O-QPSK modulation [109–113]. This standard often employs a 16-ary quasi-orthogonal modulation technique based on the direct sequence spread spectrum (DSSS) technique [109–113]. Binary data are grouped into the 4-bit symbols, and each symbol specifies one of 16 nearly orthogonal 32-chip pseudonoise (PN) sequences for transmission [109–113].

The signal structure for this type of standard enables PN sequences for successive data symbols to be concatenated (or linked together) and the aggregate chip sequence is modulated unto the carrier using minimum phase shift keying (MSK) which is equivalent to offset QPSK or O-QPSK with half-sine pulse shaping [109–113].

Fifth, we have several UWB protocols such as: (a) continuous wave UWB (C-UWB); (b) direct sequence UWB (DS-UWB); and (c) multiband-OFDM UWB (MB-OFDM UWB) [133]. C-UWB uses bursts of pulses and variable spreading codes to trade data rate for communication range which is specified as an alternative physical layer in IEEE 802.15.4a and IEEE 802.15.3a and which is proposed in the IEEE 802.15.3c [133]. DS-UWB is based on DSSS technology, and MB-OFDM UWB uses a combination of frequency hopping and OFDM technologies and both DS-UWB and MB-OFDMA-UWB are physical specifications for WPANs. DS-UWB is vulnerable to intersymbol-interference (ISI) and MB-OFDM-UWB is robust to ISA but requires higher computational power for FFTs [133]. The FCC PSD emission limit for devices operating in the UWB band is −41.3 dBm/MHz, with emission levels significantly lower in other parts of the spectrum which allows the UWB systems to coexist with other narrowband systems. Because UWB transmission power level cannot be adjusted; spreading technologies in both the time-domain and frequency domain are used to vary the data rates [133]. As we are going to see more extensively in Giftet Inc. 2006 and 2010 OFDMA signal designs is that UWB signal design can provide ideal high-precision ranging for real-time location systems (or the practice of indoor geolocation systems) [133].

Figure 3.2 illustrates a description of the present and future wireless technologies' 3D signal designs in the frequency (GHz), indoor range (m), and data rate (kb/s ~ Gb/s) based on IEEE 802.11 "A, B 'Wi-Fi,' E, and G," IEEE 802.15.3.1; IEEE 802.15.4; IEEE 802.15.3a (UWB); and Giftet Inc. 2006 OFDMA signal design and Giftet Inc. 2010 OFDMA signal design.

Figure 3.2 does not include the description of the present and future mobile phone 3D signal design such as 2G, 3G, 4G, and WPAN. For a detailed description of these signals the reader should refer to Sect. 3.4.1.

Note also that because this is an original chart and it is proprietary information of Giftet Inc, it is released with permission to the ION and IEEE and Springer. Duplication and reprint of this chart can be done only with a written permission to Giftet Inc. headquarters' office from www.giftet.com. While all the present wireless technology standards, as depicted in Fig. 3.2, are of interests, we are going to focus mainly on the description of Giftet Inc. 2010 3D OFDMA signal design and also provide some background on Giftet Inc. 2006 OFDMA signal design.

Giftet Inc. 2006 was designed to offer ~100 m indoor range and it was designed for commercial, university campuses, banks, hospitals, sport arenas that required submeter position accuracy 99.9% of the time (see [135]). Giftet Inc. 2010 signal design in [136] is produced to offer ~10 m indoor range especially for in home geolocation systems; offers over 1 GHz data rate in order to meet all future and next generation wireless local area networks needs for wireless communications connectivity, geolocation capability (Fig. 3.3).

It is without a doubt that Giftet Inc. 2010 offers the best choice for the highest performance possible in any indoor (in home networking and geolocation)

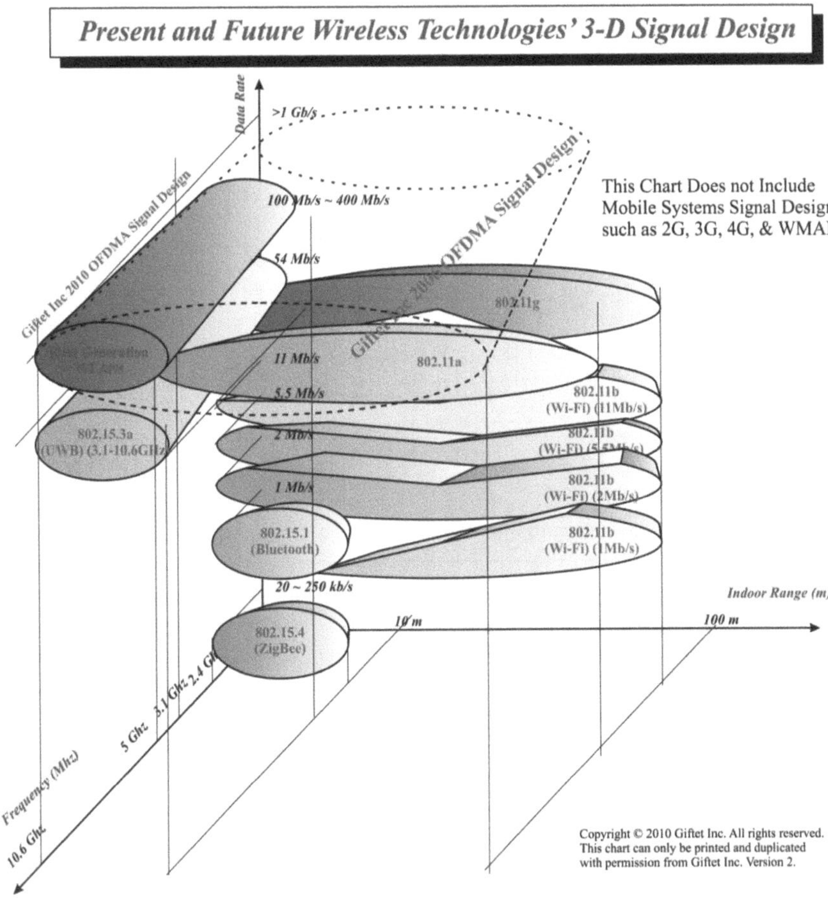

Fig. 3.2 A description of the present and future wireless technologies' 3D signal designs in the frequency (GHz), indoor range (m), and data rate (kb/s ~ Gb/s) based on IEEE 802.11 "A, B 'Wi-Fi,' E, and G"; IEEE 802.15.3.1; IEEE 802.15.4; IEEE 802.15.3a (UWB); and Giftet Inc. OFDMA signal design. Reprinted with permission, © 2010 Giftet Inc

environment due to super high bandwidth demand and also its ability to combat multipath and interference. However, that comes at a high price of available bandwidth and sophisticated hardware.

There are a number of standards and applications pertaining to in home and in office consumer wireless electronics such as a 60 GHz wireless network for enabling uncompressed video communications [127]. These systems are known as the mmWave WPANs and occupy frequencies in the range of 57–66 GHz also known as the millimeter-wave unlicensed spectrum which can support multigigabit transmissions such as uncompressed HD video up to 3 Gb/s [127]. The standards pertaining to these frequencies are WirelessHD, ECMA, TC48, and IEEE 802.15.3c.

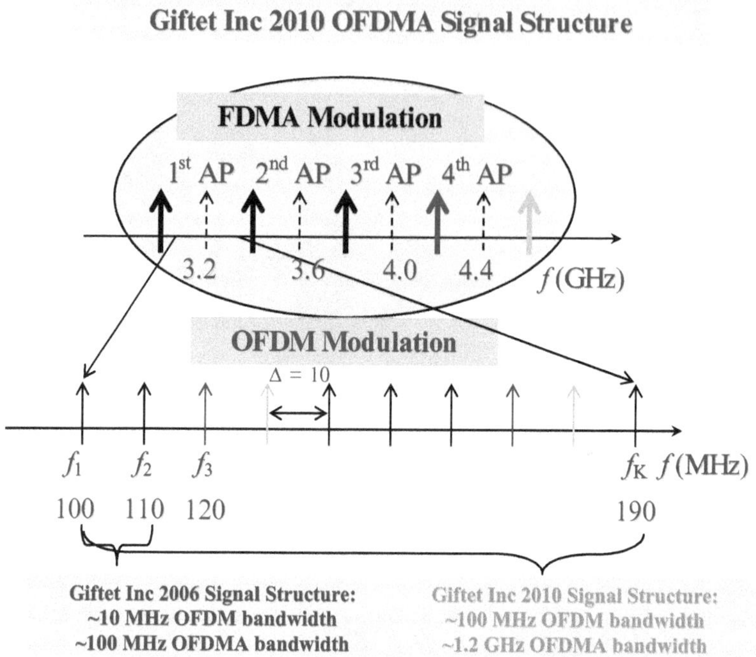

Fig. 3.3 A description of Giftet Inc. 2010 OFDMA signal design [136]. For a description of Giftet Inc. 2006 signal structure please refer to [136]. Reprinted with permission, © 2010 Giftet Inc

Let's underline some of the benefits of the 60 GHz mmWave band for supporting short range applications such as uncompressed video streaming:

1. *Bandwidth*: There is a huge spectrum that can be provided in the 57–64 GHz unlicensed band available in the United States which allows for higher data rates with lower spectral efficiencies because data rate is calculated as data rate (bits per second) = spectral efficiency (bits per second per Hertz) × bandwidth (Hertz) [127].
2. *Coverage*: Due to high attenuation of 60 GHz signals by obstacles, the effective range for indoor mmWave WPANs is on the order of 10 m [127].
3. Form factor: This technology actually uses multipatch beam forming technologies to create directivity, somewhat different, and much more complex than a directional antenna. Multipatch beam forming technologies to create directivity are far too easier to implement at 60 GHz band than at 2.4–5 GHz because of the smaller wavelength which reduces the antenna size and improves the directional transmission and reception which simplifies the transceiver design by significantly reducing the delay spread and intersymbol interference [127].

Simulation examples on this standard may be provided on future editions of this book due to complex signal waveforms and complex GRFS systems receiver design.

3.4 RF Signals for Urban GRFS Systems

RF signals for urban GRFS systems include: (1) RF signals for mobile systems and metropolitan area networks (MAN) in Sect. 3.4.1; (2) RF signals for FM and TV stations in Sect. 3.4.2.

3.4.1 *RF Signals for Mobile Systems and Metropolitan Area Networks (MAN)*

In this subsection of the RF signals for urban GRFS systems, we will discuss several mobile system designs such as 2G, 3G, 4G, etc., and the MAN; and complete the discussion of other RF signals of cellular networks in Sect. 3.5.2 since the effective range of this networks fall within the suburban GRFS systems.

Before we go and consider various wireless systems, signal designs we should first consider that there exists fundamental limitations on increasing data rate in wireless systems [126]. Reference [126] and Fig. 1 in reference [126] provide very helpful information in a cellular system; the range (m) vs. data rate (b/s) dependency as a function of $1/d^m$ where d is the transmitter–receiver distance and the exponent m is empirically determined to be in the range from 2 to 5 or 6 in different environments with typical values ranging from 3 to 4. Shannon-Hartley Capacity Theorem vs. SNR, some basic principles of propagation theory vs. frequency and many other channel features will be discussed more extensively in a separate chapter in the *Indoor Geolocation Systems: Theory and Applications* book. If it is possible, we may need to reproduce Fig. 1 in reference [126] adding also a third axis – the frequency. So hopefully, the purpose of this work in Chap. 3 on *Geolocation of RF Signals: Principles and Simulations* is to come up for the first time with a very unique chart that takes into consideration three main parameters: (1) range (m); (2) data rate (b/s); and (3) frequency (Hz).

One example is the long-term evolution (LTE) as defined by the 3rd Generation Partnership Project (3GPP) is highly flexible radio interface with the following characteristics: (1) peak rates of 300 Mb/s; (2) radio-network delay of less than 5 ms; (3) a significant increase in spectrum efficiency; (4) it supports both FDD and TDD and a wide range of system bandwidths in order to operate in a large number of different spectrum allocations; (5) it aims for a smoother evolution (which requires a ton of investments to replace everything in the ground) from earlier 3GPP systems such as time division-synchronous code division multiple access (TD-SCDMA) and wide-band CDMA/high speed packet access (HSPA), as well as 3GPP2 systems such as CDMA 2000; (6) it already includes many of the features originally considered for fourth-generation systems; (7) it uses OFDM with data transmitted on a large number of parallel, narrow band subcarriers as the core of the LTE downlink radio transmission; (8) it uses single-carrier frequency-division multiple access (SC-FDMA) which is used for the LTE uplink; (9) in the physical layer, the transmitted data is turbo coded and modulated using QPSK, 16-QAM or 64-QAM, followed by OFDM modulation; (10) 2 GHz is the center frequency; (11) it allows for an overall system bandwidth as small as 1.4 MHz up to 20 MHz with a range from 280/3 subcarriers to 4000/3 subcarriers and 500 m inner cell site distance, indoor terminals; (12) base station/terminal power 46/23 dBm, respectively [132]. Appendix A depicts a description of the baseband Simulink block diagram of CDMA 2000 Physical Layer.

Next, the IEEE 802.16 which corresponds to the wireless metropolitan area networks (WirelessMAN™) includes both licensed and licensed-exempt 2–11 GHz band specifications which are driven by the need for the NLOS operation due to significant multipath propagation in three different formats: (1) WirelessMAN-SC2: which uses single carrier modulation format; (2) WirelessMAN-OFDM: which uses OFDM with a 256-point transform and achieves multiple access via time division multiple access (TDMA) which is mandatory for license-exempt bands; (3) WirelessMAN-OFDMA: which uses OFDMA with a 2048-point transform [108, 110, 111, 113]. Appendix A indicates a description of the baseband Simulink block diagram of IEEE 802.16-2004 OFDM Physical Layer Including Space–Time Block Coding.

3.4.2 RF Signals for FM and TV Stations

The RF signals for the FM broadcast band, throughout the world, falls within the VHF part of the radio spectrum. Usually 87.5–108.0 MHz is used, or some portion thereof, with few exceptions [142].

Most of the FM modulation and signal design is part of most "communications engineering" textbooks; therefore, it offers little advantage for us to reiterate information that is quite available in many textbooks.

The range of an FM monotransmission is related to the transmitter RF power, the antenna gain, and antenna height. The FCC (USA) publishes curves that aid in calculation of this maximum distance as a function of signal strength at the receiving location [142].

For FM stereo, the maximum distance covered is significantly reduced. This is due to the presence of the 38 kHz subcarrier modulation. Vigorous audio processing improves the coverage area of an FM stereo station.

For our everyday experience we can hear an FM radio well within 50 km. Hence, it is safe to assume that for most FM broadcast stations 50 km is their effective range which means that FM radio can be treated as either urban or suburban GRFS system.

While many people have seen many AM and FM receivers and many mobile wireless handhelds transceivers because of our interaction with them daily, not many people have seen an AM transmitter antenna, an FM transmitter antenna, or a mobile antenna array that are shown in Figs. 3.4 and 3.5.

Fig. 3.4 A description of an AM transmitter antenna and AM main and backup transmitters from the Holden, Massachusetts site. Reprinted with Permission © 2010, Ilir Progri

Fig. 3.5 A description of the first FM transmitter antenna and mobile wireless antenna array in Paxton, Massachusetts site. Reprinted with Permission © 2010, Ilir Progri

There are several *broadcast television systems* in use in the world today. An analog television system includes several components: a set of technical parameters for the broadcast signal, a system for encoding color, and possibly a system for encoding multichannel audio. In digital television, all of these elements are combined in a single digital transmission system [143].

Terrestrial television is a mode of television broadcasting which does not involve satellite transmission or underground cables – typically using radio waves through transmitting and receiving antennas or aerials. The term is more common in Europe, while in the United States it is referred to as *broadcast television* or sometimes *over-the-air television* [144].

In addition to the threat from cable television, analog terrestrial television is now also subjected to competition from satellite television and distribution of video and film content over the Internet. The technology of digital terrestrial television has been developed as a response to these challenges. The rise of digital terrestrial television, especially HDTV, may mark an end to the decline of broadcast television reception via traditional receiving antennas, which can receive over-the-air HDTV signals [144]. Terrestrial television and broadcast television are subject to the same propagation and other GRFS systems; therefore, their signals can be used for many GRFS applications such as future handsets which will have TV capability.

In North America, terrestrial broadcast television operates on TV channels 2 through 6 (VHF-low band, known as band I in Europe), 7 through 13 (VHF-high band, known as band III elsewhere), and 14 through 69 (UHF television band, elsewhere bands IV and V). Channel numbers represent actual frequencies used to broadcast the television signal. Additionally, television translators and boosters can be used to rebroadcast a terrestrial TV signal using an otherwise unused channel to cover areas with marginal reception. A chart showing the North American television bandplan can be found here [144].

In December 2005, the European Union decided to cease all analog television transmissions by 2012 and switch all terrestrial television broadcasting to digital (all EU countries have agreed on using DVB-T). The Netherlands completed the transition in December 2006, and some EU member states have decided to complete this switchover as early as 2008 (Sweden), and (Denmark) in 2009. While the UK began the switch in late 2007, it will not be complete until mid 2012. Norway ceased all analog television transmissions on December 1, 2009. Two member states (not specified in the announcement) have expressed concerns that they might not be able to proceed to the switchover by 2012 due to technical limitations; the rest of the EU member states are expected to stop analog television transmissions by 2012 [144].

Australia has adopted the DVB-T standard and the government's industry regulator, the Australian Communications and Media Authority, has mandated that all analog transmissions will cease by 2012. Mandated digital conversion commenced early in 2009 with a graduated program. The first center to experience analog switch-off will be the remote Victorian regional town of Mildura, in 2010. The government will supply underprivileged houses across the nation with free digital set-top DTV converter boxes in order to minimize any conversion disruption. Australia's major free-to-air television networks have all been granted digital transmission licenses and are each required to broadcast at least one high-definition and one standard-definition channel into all of their markets [144].

In North America, a specification laid out by the ATSC has become the standard for digital terrestrial television. In the United States the FCC has set a final deadline for the switchoff of analog service for June 12, 2009. All television receivers must now include a digital tuner. In Canada, the Canadian Radio-television and Telecommunications Commission (CRTC), has set August 31, 2011 as the date that over-the-air analog transmission service will cease in most parts of the country except in Northern Canada [144].

3.5 RF Signals for Suburban GRFS Systems

RF signals for suburban GRFS systems include: (1) RF signals for two-way radio as discussed in Sect. 3.5.1; and (2) RF signals for cellular network as presented in Sect. 3.5.2.

3.5.1 RF Signals for Two-Way Radio

One perfect example of indoor/urban/suburban range applications is the two way radio. We are going to discuss the two-way radio RF signals more as a suburban GRFS type of application than urban or indoor due to its extended effective range.

A *two-way radio* is a radio that can both transmit and receive RF signals (a transceiver) that contain voice or data, unlike a broadcast RF receiver which only receives RF signals that contain voice or data. A two-way radio is also known as point-to-point communications because it enables communications generally between two parties while others may be listening (i.e., in the listening mode); it supports both FDD and TDD and a wide range of system bandwidths in order to operate in a large number of different spectrum allocations and in different environments [139].

First, FDD historically has been used in voice-only applications; FDD supports two-way radio communications by using two distinct radio channels (or two different frequencies); unlike TDD which uses a single frequency to transmit signals in both the downstream (receive mode) and upstream (transmit mode) directions [316].

Second, consider for example suburban GRFS systems that use FDD, one frequency channel is employed to transmit downstream from a radio X to radio Y. In this case, radio Y and other radios such as Z, W, etc. are in the listening or receive mode. A second frequency is utilized in the upstream direction and supports transmission (and reception) from radio Y to radio X; thus, enabling simultaneous transmission (and reception) in both directions. However, a minimum amount of frequency separation (also known as FDD minimum bandwidth on the order of ~40 kHz for voice) must be maintained between the frequency pair to mitigate self-interference (or cochannel interference) between upstream and downstream transmissions [316].

Third, TDD operates by toggling transmission directions over a time interval. This toggling takes place very rapidly and is imperceptible to the user. Unlike FDD which support only symmetrical applications, TDD can support voice, symmetrical communication services, and more importantly *asymmetric data* services; i.e., TDD can handle a dynamic mix of several traffic types. Moreover, in TDD the relative capacity of the downstream and upstream links can be altered in favor of one direction over the other, i.e., in the case of one base station over a remote base station (or commanding unit), which will be favored over other remote stations in its transmissions. This is accomplished by giving a greater time allocation through time-slots to downstream transmission intervals than upstream. This asymmetry is useful for communication processes characterized by unbalanced information flow. An obvious civil application for this technique is Internet access (especially in combat, or search and rescue, firefighter, or other operations) in which a user enters a short message upstream and receives large information payloads downstream [316].

Fourth, FDD can also be used for asymmetric traffic; however, in order to be spectrally efficient, the downstream and upstream channel bandwidths must be matched precisely to the asymmetry which is a very hard thing to do since Internet traffic is bursty by nature and the asymmetry is always changing; therefore, the channel bandwidth cannot be precisely matched in FDD. This makes TDD more efficient than FDD for Internet type traffic bursty asymmetric applications. Although channel bandwidths typically are set by the FCC or limited by the functionality of available equipment and although users of FDD systems do not have the option to vary channel bandwidths dynamically in the upstream and downstream directions as perhaps suggested by [316], FDD can still fulfill all the requirements of a highly dynamic Internet Traffic by having channel bandwidth that allow for maximum available bandwidth and offers more redundancy than TDD will ever offer.

Fifth, frequency spectrum is an increasingly scarce commodity. This scarcity drives the need to optimize the use of available bandwidth. FDD systems operate on the principle of paired frequencies. A channel plan is devised that is comprised of downstream and upstream channels, typically defined by the FCC, ITU, or other governing body [316]. FDD channel plans maintain a guardband between the downstream and upstream channels. The guardband is required to avoid self-interference and, since it is unused and essentially is a wasted spectrum [316, 317].

Sixth, in contrast to FDD, TDD systems require a guard time (instead of a guardband) between transmit and receive streams. The TX/RX Transition Gap (TTG) is a gap between downstream transmission and the upstream transmission. This gap allows time for the base station to switch from transmit mode to receive mode and subscribers to switch from receive mode to transmit mode. During this gap, the base station and subscriber are not transmitting modulated data but are simply allowing the base station transmitter carrier to ramp down, the TX/RX antenna switch to actuate, and the base station receiver section to activate [316].

Based on these six arguments from the above discussion that has highlighted the differences, *it appears* that TDD has some *significant advantages* over FDD. These advantages can be summarized as follows:

1. It appears that FDD is an older scheme that is best suited for applications, such as voice, that generate symmetric traffic, while TDD is best suited for bursty, asymmetric traffic, such as Internet, or other datacentric services [316]. My question to Ref. [316] is that what happens for applications that require both voice and highly bursty asymmetric Internet traffic? In this case, a combination of FDD and TDD would be more appropriate than using TDD alone?
2. In TDD, both the transmitter and receiver operate on the same frequency but at different times. Therefore, TDD systems reuse the filters, mixers, frequency sources and synthesizers, thereby eliminating the complexity and costs associated with isolating the transmit antenna and the receive antenna. An FDD system uses a duplexer and/or two antennas that require spatial separation and, therefore, cannot reuse the resources. The result is more costly hardware of FDD vs. TDD GRFS systems [316]. However, an FDD can offer the advantage of redundancy in case one channel is jammed or needs repair. In this case, one FDD channel can be used in TDD mode without the interruption of communication.
3. TDD utilizes the spectrum more efficiently than FDD. FDD is not recommended in environments where the service provider does not have enough bandwidth to provide the required guardband between transmit and receive channels [316]. However, my question to Ref. [316] is that can it be designed to handle the maximum observable bandwidth for both voice and bursty asymmetric Internet data traffic. Bursty asymmetric Internet traffic occurs only at certain times of the day. Perhaps some parts of the spectrum can be licensed or shared with others so that when there is a need for more bandwidth that can be accomplished during those times. However, for some of the UWB technologies in the unlicensed band of 3.1–10.6 GHz this is not an issue at all.
4. TDD is more flexible than FDD in meeting the need to dynamically reconfigure the allocated upstream and downstream bandwidth in response to customer needs [316]. However, if TDD channel is jammed then the customer needs are not met at all. There are no data from [316] suggesting that TDD throughput is better than FDD when interference is present or against interference level in dB. This need to be quantified better and needs to be characterized better.
5. TDD allows interference mitigation via proper frequency planning. TDD requires only one interference-free channel compared with FDD, which requires two interference-free channels [316]. I think Ref. [316] meant cochannel (or self induced) interference as opposed to in-band (or outside) interference coming from other RF signals or sources. While this argument may be OK normal operations in interference-free environments, there are no data that suggests that this is the case for heavily jammed environments. I would think that TDD will perform worse than FDD because for a fixed channel bandwidth based on Shannon's

theorem or limit, TDD will not have enough bandwidth to mitigate interference vs. FDD which will have twice and much bandwidth to mitigate interference.

6. In summary, it appears that from [316] that TDD is a more desirable duplexing technology that allows system operators to receive the most from their investment in spectrum and telecom equipment, while meeting the needs of each individual customer [316]. However, there are no quantifiable data that support those claims such as graphs, charts, worst case scenarios from prestigious journals or magazines; therefore, most of arguments in [316] are only based on marketing or sales such as the statement that "In a data-driven world, FDD broadband wireless is a waste of space" [317].
7. There are a lot more issues with TDD than those presented in [316]. I think perhaps a combination of TDD and FDD would be a better choice than TDD or FDD used separately to handle a variety of bursty, voice, Internet video, etc. applications with detailed discussion on capacity, throughput, multipath, interference levels, signal models, and receiver design; a discussion that will exceed the scope of this chapter and this material and perhaps could be a focus for a future journal paper with the *IEEE Transactions on Wireless Communications* or a separate chapter for our second edition of this book. It could turn out that for most commercial usage, TDD might be the most preferred duplex mode than FDD; and for most military applications a combination of FDD with TDD might turn out to be the preferred mode of operations, at least that appears to be the trend that is driven by LTE applications based on customers' needs.

Two-way radios are available in at least three configurations: (1) mobile, (2) stationary base, and (3) hand-held portable configurations. Outside the defense electronics industry, mobile and hand-held are pretty much equivalent. Hand-held radios are often called walkie-talkies or handie-talkies because they contain a push-to-talk or press-to-transmit button that is often present to activate the transmitter [139]. Walkie-talkies generate a signal which is often known as burst RF signal because it is ON for a certain period of time and then it is OFF for another period of time via either TDD or FDD.

A mobile phone or cellular telephone is an example of a two-way radio that both transmits and receives RF signals with information (usually voice, video, and data) at the same time (or full-duplex). It uses either FDD or TDD to carry the two directions of the conversation simultaneously [139]. In this chapter we are only interested in the RF signal structure of Walkie-talkies which for the most part is used to carry only voice so their modulation is no different than those of AM radios; i.e., a voice modulated on a carrier.

One example of analog radio is AM aircraft radio, which is used to communicate with control towers and air traffic controllers. Another is a Family Radio Service walkie-talkie. In the case of analog two-way radio equipment, it is less complex than digital [139].

Examples of digital communication are: APCO Project 25, a standard for digital public safety radios, Nextel's iDEN, Motorola's MOTOTRBO, and NXDN implemented by Icom as IDAS and by Kenwood as NEXEDGE [139].

RF analog systems may communicate a single condition, such as water level in a livestock tank. A transmitter at the tank site continually sends a signal with a constant tone. The tone would change in pitch to indicate the tank's water level. A meter at the remote end would vary, corresponding to the pitch, to indicate the amount of water present in the livestock tank [139]. Sensor networks are a good topic, especially with water, gas, electric moving toward the use of these systems in urban and/or suburban environments.

Similar methods can be used to telemeter any RF analog condition. This type of radio system serves a purpose equivalent to a four-to-twenty milliampere (mA) loop. In the US, midband 72–76 MHz or UHF 450–470 MHz interstitial (or *situated among cells*) channels are often used for these systems. Some systems multiplex telemetry of several analog conditions by limiting each to a separate range of tone pitches, see for example [139].

Engineered systems are designed to perform close to a specification or standard. They are designed as systems with all equipment matched to perform together. For example, a modern, local government two-way radio system in the US may be designed to provide 95% area coverage in an urban area [139].

Two-way radios can operate on many different frequencies, and these frequencies are assigned differently in different countries. Typically channelized operations are used, so that operators need not tune equipment to a particular frequency but instead can use one or more preselected frequencies, easily chosen by a pushbutton or other means. For example, in the United States, there is a block of 22 channels (preselected radio frequencies) assigned, collectively, to the General Mobile Radio Service and Family Radio Service [139].

A channel number is just a shorthand notation for a frequency. It is, for instance, easier to remember "Channel 1" than to remember "26.965 MHz" (CB Channel 1) or "462.5625 MHz" (FRS/GMRS channel 1), or "156.05 MHz" (Marine channel 1). It is necessary to identify which radio service is under discussion when specifying a frequency by its channel number. Organizations, such as electric power utilities or police departments, may have several assigned frequencies in use with arbitrarily assigned channel numbers. For example, one police department's "Channel 1" might be known to another department as "Channel 3" or may not even be available. Public service agencies have an interest in maintaining some common frequencies for interarea coordination in emergencies [139].

The wavelength of a UHF and VHF signal plays a big role in which radio technology to use. UHF has a shorter wavelength which makes it easier for the signal to find its way through rugged terrain or the inside of a building. The longer wavelength of VHF means it can transmit further under ideal conditions. For most applications, lower radio frequencies are better for longer range than higher radio frequencies, which are without the use of up-converters and downconverter and guided ways such as parabolic or dish antennas. For example, a broadcasting TV station illustrates that on the one hand, a typical VHF station operates at about 100,000 W (or 100 kW) and has a coverage radius range of about 60 miles, on the other hand a UHF station with a 60-mile coverage radius requires transmitting at 3,000,000 W (or 3 MW) or three times as much power as the VHF station for equivalent effective range [139]. However, for satellite communications the homogeneity and heterogeneity of the environment and more importantly the absorption of certain frequencies from different layers of the atmosphere should be taken into consideration and also the availability of spectrum.

If an application requires working mostly outdoors, a VHF radio is probably the best choice, especially if a base station radio indoors is used and an external antenna is added. The higher the antenna is placed, the further the radio can transmit and receive. One exception to using a VHF radio outdoors is if it is used in a heavily wooded or rugged area. Under these conditions, a UHF radio may be able to transmit better through the terrain (unless the VHF antenna is raised above the terrain) [139].

If the radios are used mainly inside buildings, i.e., for indoor and urban applications, then UHF is likely the best solution since its shorter wavelength travels through the building better; which is known as RF penetration. There are also repeaters that can be installed that relay a UHF signal to increase the communication distance to enable better suburban signal reception [139].

We remind the reader that more discussions on: (1) propagation properties by frequency; (2) multipath/MIMO and spatial diversity principles; (3) absorption properties of water and other materials as they relate to RF signals, will be found on a separate chapter on channel models on *Indoor Geolocation Systems: Theory and Applications*.

3.5.2 RF Signals for Cellular Network GRFS Systems

Cellular network is another perfect example of a suburban GRFS system [140].

A *cellular network* is a radio network made up of a number of cells, each served by at least one fixed-location transceiver known as a cell site or base station. When joined together, these cells provide radio coverage over a wide geographic area. This enables a large number of portable transceivers (mobile phones, pagers, etc.) to communicate with each other and with fixed transceivers and telephones anywhere in the network, via base stations, even if some of the transceivers are moving through more than one cell during transmission [140].

To distinguish signals from several different transmitters, frequency division multiple access (FDMA), CDMA, and whole new subcategories such as OFDMA, W-CDMA and others were developed [140].

In FDMA, the transmitting and receiving frequencies used in each cell are different from the frequencies used in each neighboring cell. In a simple taxi system, the taxi driver manually tuned to a frequency of a chosen cell to obtain a strong signal and to avoid interference of signals from other cells [140]. However, FDMA allows for frequency reuse from no-adjacent cells.

The principle of CDMA is more complex which is based on using different codes for every transmitter and receiver, but achieves the same result; the distributed transceivers can select one cell and listen to it based on a unique code also known as a pseudorandom sequence [140]. More discussion on pseudorandom sequences can be found on signal model on *Indoor Geolocation Systems: Theory and Applications*.

Other available methods of multiplexing such as polarization division multiple access (PDMA) and TDMA cannot be used to separate signals from one cell to the next since the effects of both vary with position and this would make signal separation practically impossible. TDMA, however, is used in combination with either FDMA or CDMA in a number of systems to give multiple channels within the coverage area of a single cell [140].

The effect of frequency on cell coverage means that different frequencies serve better for different uses. Low frequencies, such as 450 MHz NMT, serve very well for countryside coverage. GSM 900 (900 MHz) is a suitable solution for light urban coverage. GSM 1800 (1.8 GHz) starts to be limited by structural walls. UMTS, at 2.1 GHz is quite similar in coverage to GSM 1800 [140].

Higher frequencies are a disadvantage when it comes to coverage, but it is a decided advantage when it comes to capacity. Pico cells, covering e.g., one floor of a building, become possible, and the same frequency can be used for cells which are practically neighbors [140].

RF signals for cellular network GRFS systems are a good application that discusses signal density as it relates to frequency and cell size. Following table shows the dependency of frequency on coverage area of one cell of a CDMA 2000 network [140]:

Frequency (MHz)	Cell radius (km)	Cell area (km^2)	Relative cell count
450	48.9	7,521	1
950	26.9	2,269	3.3
1,800	14.0	618	12.2
2,100	12.0	449	16.2

3.6 RF Signals for Global GRFS Systems

Most of the RF signals and devices designed for Satellite range are also designed for global range; therefore, the RF signals for global GRFS systems will be considered as RF signals for Satellite GRFS systems in Sect. 3.7.

3.7 RF Signals for Satellite GRFS Systems

For the purposes of this book and for the purposes of this chapter, the satellite signals of interests are those used for: (1) positioning, navigation, and timing as discussed in Sect. 3.7.1 as part of RF signals for GNSS; (2) communications connectivity for voice, data, video, and picture as treated in Sect. 3.7.2 and part of the RF Signals for Satellite Television Technology (STT); (3) those that are specifically designed for DVB and Digital Video Broadcasting–Satellite–Second Generation (DVB-S2) as presented in Sect. 3.7.3.

3.7.1 RF Signals for Global Navigation Satellite Systems (GNSS)

RF Signals for the GNSS are: (1) the Global Positioning System carriers in the L band, centered at 1176.45 MHz (L5), 1227.60 MHz (L2), 1381.05 MHz (L3), and 1575.42 MHz (L1) frequencies; (2) The Galileo Navigation System uses the L-band similarly to GPS; (3) The GLONASS System uses the L-band similarly to GPS.

These signals are employed in great detail in Chap. 6. However, due to the enormous similarities with Indoor Geolocation Systems' signal design, it is more convenient to investigate the signal design of GNSS in *Indoor Geolocation Systems: Theory and Applications.*

However, a number of issues related to GNSS RF signals are more appropriate to discuss them here such as: (1) *interoperability*; (2) institutional models for cooperation; (3) *integrity provision*; (4) *spectrum protection*; (5) safety and security; (6) liability; (7) user support within developing nations than in *Indoor Geolocation Systems: Theory and Applications.*

The ones that are really of interests in this chapter are: (1) interoperability; (2) integrity provision; and (3) spectrum protection.

Issues related to *interoperability*, *compatibility*, etc. have a long history in the GPS community going from 1975 [143] until the present time 2010 [315]. They cover all aspects of system design and operation, segment design and architecture, user design, operation, test, production, etc. but it was not until January 1998 when these issues really become the focus of the GNSS community with the Galileo signal design in the GNSS community [173], which is reflected form the proliferation of research, development, analysis, assessment, workshops, meetings, session, etc. dedicated to these very important issues [145–173, 318–344]. In order to understand these issues we need to understand how the GNSS architecture has evolved. These issues exist because we have competing requirements for spectrum, signal design, space, markets, etc. between GPS, GLNASS, Galileo and currently from Compass, MT-SAT, etc. So the question posed in 2001 was what should be focus on one GNSS system or global interoperability architecture of separate GNSS systems, as opposed to competing of regional or national systems [173]. The 2001 architecture comprises the U.S. GPS and Russian GLONASS as core systems along with satellite- and ground-based augmentation systems that either have been developed or under development in the United States, Europe, Russia, Japan, or elsewhere aiming to improve the basic GPS and GLONASS services such as accuracy, integrity, and availability [173].

The arrival of proposed Galileo European system led to the proposal for truly "seamless" global interoperability among independent global interoperable satellite navigation systems from April 1999 in Bermuda [173]. Recommendations aimed an insuring interoperability including the establishment of common definitions for open systems architecture and basic civil and public safety GNSS services, spectrum protection, liability, security on GNSS operations, etc. [173].

The discontinue of the GPS selective availability on May 1, 2000 was discontinued which resulted in immediate and significant benefits to GPS users worldwide with offered position accuracy to within 10 m.

The 2001 findings are as follows: "The European Union, the United States, and Russia have embarked on the development of future autonomous GNSS architectures that should also be complementary and interoperable if they are to provide the maximum benefit to users worldwide. Achieving this result requires a level of understanding among the parties that is greater than currently exists" [173, 178, 179].

GNSS interoperability can occur on a number of different levels, ranging from simple signal noninterference of autonomous systems to a complementary and interoperable "system-of-systems" [173]. These increasing levels of interoperability can not only yield increasing benefits to users of GNSS services, but also require increasing levels of technical, managerial, and operational cooperation on the part of the nations developing the systems [173, 179].

The first level (or instance) of interoperability (or noninterference) occurred between GPS and GLONASS which were designed independently with little consideration given to interoperability other than selecting operating frequencies which will not result in mutual interference [173]. Also in 2001, the results of the first interoperability flight testing conducted between a military Joint Precision Approach and Landing System (JPALS) ground station and a civil Local Area Augmentation System (LAAS) equipped commercial aircraft [175, 177]. For a detailed discussion, Galileo Interface Interoperability (GII) 6-Level diagram is used by an EC-funded study into interoperability in 2001 provided in [179].

The next level of interoperability occurs at the system level and can yield improved user benefits from multiple autonomous GNSS systems without extensive technical coordination or cooperation among the nations developing the systems. What is required is *an appropriate selection of frequency plans* and *noninterfering signal structures*, and *provision of time and geodetic corrections* by *either the core systems* or by *coordinated augmentation systems* [173].

However, a system-of-systems approach to developing and operating fully interoperable and complementary next-generation systems is needed as the optimum (or the best) level of interoperability would allow GNSS receivers to generate *high quality, high integrity position,* or *velocity solutions* using a *composite mix of viewable satellites from different systems*, and also require a much greater level of understanding than what currently exists between the United States, the European Union, and Russia [173, 178, 179].

In 2001, the United States had perhaps the most detailed and thoughtful vision for the future cooperation of the major policy principles [173]. Some that really relate to this chapter are the following:

1. Open signal structure for all civil services to promote equal access for applications development and value-added services (Indoor Geolocation Systems: Theory and Applications).

2. Protection of the current radionavigation spectrum from disruption and interference (Chap. 6).
3. Use of GPS time, geodesy, and signal structure standards (new chapter for second edition of the book).
4. Seamless, global interoperability of future systems with GPS (new chapter for second edition of the book).

The 2001, final recommendation was: "The European Union, the United States, and Russia should reach a common view on system interoperability that is consistent with the design schedules of the Galileo, GPS-III, and GLONASS-K programs" [173, 178, 179].

Galileo must be an open, global system, fully compatible with GPS, but independent from it [157].

A primary objective for GALILEO is to provide its services autonomously, thus avoiding any form of technical dependence or common modes of failure with other satellite navigation systems [178].

GALILEO and GPS Reference Co-ordinate and Time Systems will be interoperable, thus leading to an efficient combined use at receiver level [178].

Second, we discuss *integrity*! Multiple levels of GNSS integrity services are currently provided by separate augmentation systems designed to meet the specific needs of the various transport sectors, and are tied to the liability regimes established by national agencies responsible for these sectors. *Future GNSS architectures may provide these integrity services as part of the core GNSS design* [173]!

The 2001, recommendation on integrity was: "International transportation standards organizations such as ICAO, IMO, and others should develop a common understanding of integrity requirements, and determine if common global integrity standards are feasible and/or desirable" [173].

The third one is *spectrum protection*! The 2001 finding was: "GNSS service providers need support from user nations to protect the spectrum used by GNSS signals from interference and reallocation for other uses" [173].

There was explicitly expressed that "the need for vigilance in protecting against harmful interference and reallocation still exists and is a responsibility that requires the support of all nations with a vote in the ITU process and users who require assured access to GNSS signals" [173].

The 2001 recommendation on spectrum protection was "The U.N. Office of Outer Space Affairs should emphasize the need for support in protecting GNSS spectrum in its GNSS educational workshops in developing nations" [173].

The second finding on spectrum protection in 2001 was "Recent investigations in the United States have shown that proposed ultrawideband (UWB) systems can cause harmful interference to GNSS signals" [173].

The 2001 recommendation on spectrum protection was "Measures should be taken to protect the frequency bands allocated to GNSS from interference caused by UWB operations" [173].

2002 and 2003 were marked by a series of publications and advancements in the analysis, methods, clarification of concepts, simulation, test, and verification of interoperability, compatibility, design, etc. of RF GNSS signals [184–218].

It was generally proposed that compatibility is more related to the user equipment as oppose to interoperability which is more related to the system level [202].

Compatibility as the capability of two or more systems to operate simultaneously without interference causing a significant degradation of the system requirements with respect to the situation in which each system is operated alone [202].

Interoperability was defined as the capability of two or more systems to operate simultaneously, providing the user, while operating in this mode, with a service exceeding in a significant way the requirements and-or specifications of each system when operated stand-alone [202].

The benefit of the discussion on interoperability and compatibility as far as signal design is concerned can be summarized in two main scenarios: (1) in which the frequencies are different and (2) in which the frequencies are the same.

First, while signal structure and frequencies, in principle, need not to be the same; from the frequency allocation point of view, operating on the same frequency with a CDMA scheme may be highly desirable in order to reduce the band allocation in an already crowded frequency spectrum, make easier for the regulating bodies to defend the allocated band to the navigation services, and to lower the cost for the manufacturers of the user equipment by simplifying the RF front-end [202]. It would be very hard for GPS to share L1 and L2 with GLONASS, Galileo, and others. However, L3, L4, and L5 could be shared with other systems. So far, provision is only made for L5 and perhaps L3 and L4 could be considered for GPS IV or GPS V.

Second, if the same frequencies are used in a CDMA scheme the ranging codes have to be orthogonal, noninterfering, and mutually agreed. The above choices, if implemented, will produce user equipment with the lowest possible cost, allowing at the same time maximum flexibility for special categories of users, such as those using GNSS systems in differential mode [202]. This could be the case of the GPS IV or GPS V. This might require much more integration and cooperation among the nations. While something that is technically advantages and superior maybe unattainable for a while due to limitations driven by policy and as long as policies change so will interoperability and compatibility. Hopefully the LTE of GNSS systems will drive the interoperability and compatibility of GNSS systems towards the perfect case scenario which is a unique system with the same technical requirements but with different service requirements. While technical requirements will ensure optimum interoperability and compatibility, different service requirements will ensure national security and individual protection.

References [184–218] consider many interoperability and compatibility studies of varied applications among GPS, Galileo (in particular), and other GNSS signals.

In 2004, interoperability and compatibility studies among current and future GPS, Galileo, GLONASS, and other GNSS signals continued with means, methods, analysis, numerical results on how to compare and contrast current and future

GNSS systems, components, and services to achieve greater interoperability and as much compatibility among these signals and systems as possible [219–236].

Although 2005 marked the year that the Galileo requirements were frozen, the more advanced and refined interoperability and compatibility studies on GNSS signals continued [237–251]. Especially important is the development with the US and Russia on GPS and Galileo that "both sides will work together to the maximum extent practicable to maintain compatibility and promote interoperability of GPS and GLONASS for civil user benefits worldwide," and "intend to establish working groups on matters of development and use of GLONASS and GPS and their respective augmentations" [244].

In 2006, there were also many studies of interoperability and compatibility of GNSS signals [252–263]; however, of particular interests are the development on "Handheld, low cost situational awareness using existing military fielded equipment" by Kelly et al. [259] on issues related to interoperability and compatibility of current military handheld with future GPS signals via primarily a software upgrade; "The benefits of multiconstellation GNSS augmentations" by [260] and "Description of the L1C signal" by [262]. "The benefits of multiconstellation GNSS augmentations" [260] related interoperability with higher traffic density (or higher signal density) and "Description of the L1C signal" by [262] discusses the development of L1C as a new stage in international GNSS; not only is the signal being designed for transmission from GPS, it is designed for interoperability with Galileo's Open Service signal and for virtually seamless interoperability with signals from Japan's Quazi-Zenith Satellite System (QZSS).

From 2007 onwards, discussions on GNSS signals' interoperability and compatibility [264–315] are discussed on Indoor Geolocation Systems: Theory and Applications and perhaps in a future edition of this material.

3.7.2 RF Signals for Satellite Television Technology (STT)

Satellite television is television delivered by the means of communications satellite and received by a satellite dish and set-top box. In many areas of the world it provides a wide range of channels and services, often to areas that are not serviced by terrestrial or cable providers [141].

Satellites used for television signals are generally in either naturally highly elliptical (with inclination of $\pm 63.4^\circ$ and orbital period of about 12 h, also known as Molniya orbit) or geostationary orbit 37,000 km (22,300 miles) above the earth's equator [141].

Satellite television, like other communications relayed by satellite, starts with a transmitting antenna located at an uplink facility. Uplink satellite dishes are very large, as much as 9–12 m (30–40 ft) in diameter. The increased diameter results in more accurate aiming and increased signal strength at the satellite. The uplink dish is pointed toward a specific satellite and the uplinked signals are transmitted within a specific frequency range, so as to be received by one of the transponders tuned to that frequency range aboard that satellite. The transponder "retransmits" the signals back to Earth but at a different frequency band (a process known as translation, used to avoid interference with the uplink signal), typically in the C-band (4–8 GHz) or K_u-band (12–18 GHz) or both. The leg of the signal path from the satellite to the receiving Earth station is called the downlink [141].

A typical satellite has up to 32 transponders for K_u-band (12–18 GHz) and up to 24 for a C-band (4–8 GHz) and more for hybrid satellites. Typical transponders each have a bandwidth between 27 and 50 MHz. Each geostationary C-band (4–8 GHz) satellite needs to be spaced 2° from the next satellite (to avoid interference). For K_u, the spacing can be 1°. This means that there is an upper limit of $360/2 = 180$ geostationary C-band satellites and $360/1 = 360$ geostationary K_u-band satellites. C-band transmission is susceptible to terrestrial interference while K_u-band transmission is affected by rain (as water is an excellent absorber of microwaves at this particular frequency) [141, 345]. Chapter 5 of [345] provides an excellent description of atmospheric emission sources including water, their absorption characteristics, and the semiempirical equation for water vapor absorption as given by Waters (1976).

The downlinked satellite signal, quite weak after traveling the great distance (see inverse-square law), is collected by a parabolic receiving dish, which reflects the weak signal to the dish's focal point. Mounted on brackets at the dish's focal point is a device called a feedhorn. This feedhorn is essentially the flared front-end of a section of waveguide that gathers the signals at or near the focal point and "conducts" them to a probe or pickup connected to a low-noise block downconverter or LNB. The LNB amplifies the relatively weak signals, filters the block of frequencies in which the satellite TV signals are transmitted, and converts the block of frequencies to a lower frequency range in the L-band (~1–2 GHz) range. The evolution of LNBs was one of necessity and invention [141].

The original C-Band satellite TV systems used a Low Noise Amplifier connected to the feedhorn at the focal point of the dish. The amplified signal was then fed via very expensive and sometimes 50 Ω impedance gas filled hardline coaxial cable to an indoor receiver or, in other designs, fed to a downconverter (a mixer and a voltage tuned oscillator with some filter circuitry) for downconversion to an intermediate frequency. The channel selection was controlled, typically by a voltage tuned oscillator with the tuning voltage being fed via a separate cable to the headend. But this design evolved [141].

Designs for microstrip based converters for Amateur Radio frequencies were adapted for the 4 GHz C-Band. Central to these designs was concept of block downconversion of a range of frequencies to a lower, and technologically more easily handled block of frequencies (intermediate frequency) [141].

The advantages of using an LNB are that cheaper cable could be used to connect the indoor receiver with the satellite TV dish and LNB, and that the technology for handling the signal at L-Band and UHF was far cheaper than that for handling the signal at C-Band frequencies. The shift to cheaper technology from the 50 Ω impedance cable and N-Connectors of the early C-Band systems to the cheaper

75 Ω technology and F-Connectors allowed the early satellite TV receivers to use, what in reality were, modified UHF TV tuners which selected the satellite television channel for down conversion to another lower intermediate frequency centered on 70 MHz where it was demodulated. This shift allowed the satellite television DTH industry to change from being a largely hobbyist one where receivers were built in low numbers and complete systems were expensive (costing thousands of Dollars) to a far more commercial one of mass production [141].

In the United States, service providers use the intermediate frequency ranges of 950–2,150 MHz to carry the signal to the receiver. This allows for transmission of UHF band signals along the same span of coaxial wire at the same time. In some applications, (e.g., DirecTV AU9-S and AT-9) ranges – the lower B-Band and upper 2,250–3,000 MHz, are used. Newer LNBFs in use by DirecTV referred to as SWM, use a more limited frequency range of 950–1,800 MHz [141].

The satellite receiver demodulates and converts the signals to the desired form (outputs for television, audio, data, etc.). Sometimes, the receiver includes the capability to unscramble or decrypt; the receiver is then called an Integrated receiver/decoder or IRD. The cable connecting the receiver to the LNBF or LNB should be of the low loss type RG-6, quad shield RG-6, or RG-11, etc. RG-59 is not recommended for this application as it is not technically designed to carry frequencies above 950 MHz, but will work in many circumstances, depending on the quality of the coaxial wire [141].

Analog television distributed via satellite is usually sent scrambled or unscrambled in National Television System Committee (NTSC), phase alternate line (PAL), or "Sequential Color with Memory" SECAM television broadcast standards. The analog signal is frequency modulated and is converted from an FM signal to what is referred to as baseband. This baseband comprises the video signal and the audio subcarrier(s). The audio subcarrier is further demodulated to provide a raw audio signal [141].

If the signal is a digitized television signal or multiplex of signals, it is typically QPSK [141].

In general, digital television, including that transmitted via satellites, are generally based on open standards such as "the generic coding of moving pictures and associated audio information" MPEG and Digital Video Broadcasting-Satellite (DVB-S) or Integrated Services Digital Broadcasting (ISDB-S) [141]. Appendix A depicts a description of a baseband Simulink block diagram of the RF Satellite Link.

3.7.3 *RF Signals for Digital Video Broadcasting (DVB) and Digital Video Broadcasting–Satellite–Second Generation (DVB-S2)*

The outstanding spectrum efficiency of DVB-S2 along with adaptability and configurability makes it a promising technology for next-generation satellite communications [128].

DVB-S2 defines the forward link transmission format for next generation satellite services and has already generated significant industry activity since it was adopted by the European Telecommunications Standard Institute (ETSI) as a European Standard in March 2005 with these main features:

1. As previously mentioned, the DVB-S2 specification originated from work on Turbo Code error correction schemes and the desire for improved efficiency [138].
2. Ironically, having evaluated all the error correction schemes submitted to the DVB-S2 group the best performing scheme of a low density parity check (LDPC) code concatenated with a Bose-Chaudhuri-Hocquenghem (BCH) code was chosen [138].
3. This new FEC scheme can be thought of as a replacement of the DVB-S convolutional coding with LDPC coding and Reed-Solomon encoding with a different BCH encoding. It is this new FEC scheme that forms the heart of the DVB-S2 standard [138].
4. Direct input of one or more MPEG-2 Transport Streams (TS). MPEG-TS is supported using a compatibility mode [137].
5. The native stream format for DVB-S2 is called Generic Stream (GS), and can be used to efficiently carry IP-based data, including MPEG-4 AVC/H.264 services [137].
6. Backward compatibility to DVB-S, intended for end users, and DVB-DSNG, used for backhauls and electronic news gathering [137].
7. Variable coding and modulation (VCM) to optimize bandwidth utilization based on the priority of the input data, e.g., SDTV could be delivered using a more robust setting than the corresponding HDTV service [137].
8. Adaptive coding and modulation (ACM) to allow flexibly adapting transmission parameters to the reception conditions of terminals, e.g., switching to a lower code rate during fading [137].
9. Four modulation modes:
   (a) QPSK and 8PSK are proposed for broadcast applications, and can be used in nonlinear transponders driven near to saturation.
   (b) 16APSK and 32APSK are used mainly for professional, semilinear applications, but can also be used for broadcasting though they require a higher level of available C/N and an adoption of advanced predistortion methods in the uplink station in order to minimize the effect of transponder linearity.
10. Improved roll-off: $\alpha = 0.20$ and 0.25 in addition to the roll-off of DVB-S $\alpha = 0.35$ [137].
11. Improved coding: a modern large LDPC code is concatenated with an outer BCH code to achieve quasierror-free (QEF) reception conditions on an AWGN channel. The outer code is introduced to avoid error floors at low bit-error rates. A single FEC frame may have either 64,800 bits (normal) or 16,200 bits (short). If VCM or ACM is used, the broadcast can be a combination of normal and short frames.

12. Several code rates for flexible configuration of transmission parameters: 1/4, 1/3, 2/5, 1/2, 3/5, 2/3, 3/4, 4/5, 5/6, 8/9, and 9/10. Code rates 1/4, 1/3, and 2/5 have been introduced for exceptionally poor reception conditions in combination with QPSK modulation. Encoding values 8/9 and 9/10 behave poorly under marginal link conditions (where the signal level is below the noise level). However, with targeted spot K_u (12–18 GHz) or Ka (26.5–40 GHz) band downlinks these code rates may be recommended to prevent out-of-region viewing for copyright or cultural reasons [137].
13. Optional input stream synchronization to provide a constant end-to-end delay [137].
14. Depending on code rate and modulation, the system can operate at a C/N between −2.4 dB (QPSK, 1/4) and 16 dB (32APSK, 9/10) with a QEF goal of a 10^{-7} TS packet error rate. Distance to the Shannon limit ranges from 0.7 to 1.2 dB [137].

With the exception of L-band Satellite signals which do not require a special antenna and hardware to receive, all the other satellite signals such as STT and DVB-S or DVB-S2 require a special satellite dish antenna LNB to be received and down-converted in the L-band.

3.8 Conclusions

In conclusion we have provided a detailed description (or qualitative study) of RF signals starting with (1) an introduction of RF signals in Sect. 3.2; (2) RF signals for indoor GRFS systems in Sect. 3.3; then with (3) RF signals for urban GRFS systems in Sect. 3.4; then with (4) RF signals for suburban GRFS systems in Sect. 3.5; and then continue with (5) RF signals for global GRFS systems in Sect. 3.6; and (6) conclude with RF signals for satellite GRFS systems in Sect. 3.7.

First, in the introduction of RF signals in Sect. 3.2, we discussed thirteen RF signals main parameters which are: (a) type; (b) center (or reference) frequency; (c) bandwidth; (d) gain/power; (e) modulation; (f) standard; (g) usability/usefulness; (h) sensitivity/harmfulness; (i) interoperability; (j) integrity; (k) compatibility; (l) signal/source density; and (m) signal/source protection. We also considered nine ways to describe RF signals which are (1) *symbolic math* or *notation*; (2) a real or complex *number*, *vector*, or a *matrix*; (3) *vector state diagram*; (4) *2D/3D power/gain, amplitude, frequency* plots; (5) *2D/3D complex diagrams*; (6) *propagation models, signal density, and absorption models*; (7) *tensors* with the following information (x, y, z, t) should geospatial multidimensional be provided; (8) *higher order tensors*; and (9) *the effects they cause on secondary systems*. If someone thinks that more RF signal parameters should be considered and more ways to describe them then by all means send your request to the address provided at the end of the book.

Second, in RF signals for indoor GRFS systems in Sect. 3.3, special consideration is given to a description of the present and future wireless technologies' 3D signal designs in the frequency (GHz), indoor range (m), and data rate (kb/s ~ Gb/s) based on IEEE 802.11 "A, B 'Wi-Fi,' E, and G"; IEEE 802.15.3.1; IEEE 802.15.4; IEEE 802.15.3a (UWB); and Giftet Inc. 2006 OFDMA signal design and Giftet Inc. 2010 OFDMA signal design.

Third, in RF signals for urban GRFS systems in Sect. 3.4, special attention is provided to several mobile system designs such as 2G, 3G, 4G, etc. and the MAN and also to FM and TV stations.

Fourth, in RF signals for suburban GRFS systems in Sect. 3.5, special consideration is given to RF signals for two-way radio with a super detailed description of FDD and TDD technologies and also with a detailed discussion on cellular networks with a detailed discussion on signal density and provided avenues for future research.

Sixth, RF signals for global GRFS systems in Sect. 3.6 are discussed as part of the RF signals for satellite GRFS systems in Sect. 3.7 which included GNSS RF Signals; RF Signals for STT; and DVB and DVB-S2. In RF GNSS signals, special consideration is given to interoperability and compatibility of GNSS signals from 1975 until 2007.

In Appendix A at the end of the book, we have also illustrated a good number of RF signal design with Simulink baseband demos. This provides an outstanding opportunity for the beginner RF engineer to test their signal designs and build more sophisticated signal designs and perform more sophisticated simulations.

There are many more RF signals which we have not considered because they are either not part of any standard or because there is no information about them in a public domain. Nevertheless, RF signals that we have considered here are sufficient to complete the initial study of Part II on *Geolocation of RF Signals: Principles and Simulations* which consists of Chaps. 4–6.

In Part II of the book, we will focus mainly on GRFS which require as little hardware as possible leaving the more sophisticated signals and the more sophisticated techniques for the second edition of the book.

References

1. Progri, I., An assessment of indoor geolocation systems, *Ph.D. Dissertation*, Worcester Polytechnic Institute, May 2003.
2. Arfken, G.B., and Weber, H.J., *Mathematical Methods for Physicists*, San Diego: Academic Press, 1995.
3. Davies, N., *Europe a History*, New York: Oxford University Press, 1996.
4. Butterfield, H., *The Origins of Modern Science, 1300–1800*. London: Routledge, 1947.
5. Nebeker, F., September and October in EE history, *IEEE Inst.*, vol. 25, p. 9, 2001.
6. Parkinson, B., Gromov, K., Stansell, T., and Beard, R., A history of satellite navigation, in *Proc. ION-AM*, pp. 17–65, Colorado Springs, June 1995.
7. Parkinson, B.W., Spilker, J.J., Axelrad, P., and Enge, P., *Global Positioning System: Theory and Applications*, vols. 1 and 2, Washington: AIAA, 1996.

8. Parkinson B., and Gilbert, S.W., NAVSTAR: global positioning system – ten years later, *IEEE*, vol. 71, pp. 1177–1186, 1983.
9. Parkinson, B.W., Spilker, J.J., Jr., Axelrad, P., and Enge, P., *The Global Positioning System-Theory and Applications*, Washington: American Institute of Aeronautics and Astronautics (vol. 1, chap. 2, *Overview of GPS operation and design*), 1996.
10. Misra, P., and Enge, P., *Global Positioning System – Signals Measurements and Performance*, Lincoln: Ganga-Jamuna Press, 2001.
11. International Telecommunications Union, *Spectrum Monitoring Handbook*, 2002 Edition. http://www.itu.int/publ/R-HDB-23/en
12. International Telecommunications Union, *Supplement to Handbook on Spectrum Monitoring*, 2008 Edition. http://www.itu.int/publ/R-HDB-53/en
13. Agnes, E.M., (Editor), *Webster's New World Dictionary of the American Language*. College Edition, Cleveland: Webster's New World, Aug., 1998.
14. Codara, L.C., *Smart Antennas*, Boca Raton: CRC Press, 2004.
15. Stapleton, S.P., and Costescu, F.C., An adaptive predistorter for a power amplifier based on adjacent channel emissions [mobile communications], *IEEE Trans. Veh. Tech.*, vol. 41, no. 1, pp. 49–56, 1992.
16. Belkerdid, M.A., Mears, T.J., and Weeter, H.T., UHF SATCOM adjacent channel emissions and modem implementation loss: predictions/measurements for the linear phase SBPSK modulation waveform family, in *Proc. MILCOM*, vol. 3, pp. 1970–1976, October 2005.
17. 8960 W-CDMA/HSPA Online User's Guide E1963A, E6703E. *Spectrum Emission Mask Measurement Description*, Agilent Technologies, January 2009. http://wireless.agilent.com/rfcomms/refdocs/wcdma/wcdma_meas_spec_em_mask_desc.php
18. Proakis, J.G., *Digital Communications*, 4th ed., Boston: McGraw-Hill, 2001.
19. Monmoiner, M., *Privacy and Technologies of Identity a Cross-Disciplinary Conversation* (book chapter 5 "Geolocation and location privacy: the 'inside' story on geospatial tracking"), Berlin: Springer, December 2006.
20. Boettcher, P.W., and Shaw, G.A., *Lecture Notes in Computer Science* (book chapter on "Energy-constrained collaborative processing for target detection, tracking, and geolocation"), Berlin: Springer, January 2003.
21. Nerguizian, C., Despins, C., and Affès, S., *Lecture Notes in Computer Science* (book chapter on "Indoor geolocation with received signal strength fingerprinting technique and neural networks"), Berlin: Springer, July 2004.
22. Sahi, P., *The Springer International Series in Engineering and Computer Science:* (book chapter 2 on "Geolocation on cellular networks"), Netherlands: Springer, April 2006.
23. Harris, M.P., Daunt, F., Newell, M. Phillips, R.A., and Wanless, S., Wintering areas of adult Atlantic puffins *Fratercula arctica* from a North Sea colony as revealed by geolocation technology, Berlin: Springer, December 2009.
24. Shaffer, S.A., Tremblay, Y., Awkerman, J.A., Henry, R.W., Teo, S.L.H., Anderson, D.J., Croll, D.A., Block, B.A., and Costa, D.P., Comparison of light- and SST-based geolocation with satellite telemetry in free-ranging albatrosses, Berlin: Springer, vol. 147, no. 4, August 2005.
25. Bassiouni, M., and Cui, W., *The Springer International Series in Engineering and Computer Science* (book chapter 7 on "Enhancing terminal coverage and fault recovery in configurable cellular networks using geolocation services"), Netherlands: Springer, vol. 4217/2006, pp. 231–254, April 2006.
26. Ki, Y.M., Kim, J.W., Kim, S.R., and Kim, D.K. (book chapter on "Modified RWGH and positive noise mitigation schemes for TOA geolocation in indoor multi-hop wireless networks"), *Lecture Notes in Computer Science*, Netherlands: Springer, vol. 4217/2006, pp. 231–254, April 2006.
27. Hunter, E., Aldridge, J.N., Metcalfe, J.D., and Arnold, G.P., Geolocation of free-ranging fish on the European continental shelf as determined from environmental variables, *J. Mar. Biol.*, vol. 142, no. 3, pp. 601–609, 2004.

28. Thygesen, U.H., and Nielsen, A., *Reviews: Methods and Technologies in Fish Biology and Fisheries Tagging and Tracking of Marine Animals with Electronic Devices* (book chapter on "Lessons from a prototype geolocation problem"), Netherlands: Springer, vol. 9, pp. 257–276, June 2009.
29. Liu, X., Ma, H., and Sun, W., *Lecture Notes in Computer Science: Advances in Machine Vision, Image Processing, and Pattern Analysis* (book chapter on "Study on the geolocation algorithm of space-borne SAR Image"), Berlin: Springer, pp. 270–280, August 2006.
30. Green, J.A., Wilson, R.P., Boyd, I.L., Woakes, A.J., Green, C.J., and Butler, P.J., Tracking macaroni penguins during long foraging trips using "behavioural geolocation," *J. Polar Biol.*, vol. 32, no. 4, pp. 645–653, 2009.
31. Mielke, C., and Chen, H., *Lecture Notes in Computer Science, Intelligence and Security Informatics: Intelligence and Security Informatics* (book chapter on "Mapping dark web geolocation"), Berlin: Springer, pp. 97–107, November 2008.
32. Kikiras, P., and Drakoulis, D., An integrated approach for the estimation of mobile subscriber geolocation, *J. Wireless Pers. Commun.*, vol. 30, no. 2–4, pp. 217–231, 2004.
33. Mielke, C., and Chen, H., *Ultra-Wideband, Short-Pulse Electromagnetics 5* (book chapter on "Experimental results from an ultra wideband precision geolocation system"), Berlin: Springer, pp. 97–107, 2002.
34. Luthcke, S.B., Carabajal, C.C., Rowlands, D.D., and Pavlis, D.E., Improvements in spaceborne laser altimeter data geolocation, *Surv. Geophys.*, vol. 22, no. 5–6, 549–559, 2001.
35. Hunter, E., Metcalfe, J.D., Holford, B.H., and Arnold, G.P., Geolocation of free-ranging fish on the European continental shelf as determined from environmental variables II. Reconstruction of plaice ground tracks, *J. Mar. Biol.*, vol. 144, no. 4, pp. 787–798, 2004.
36. Bargshady, N., Nayef, A., Alsindi, N.A., and Pahlavan, K., *Lecture Notes in Computer Science: Mobile Entity Localization and Tracking in GPS-less Environments* (book chapter on "Performance of TOA- and RSS-based indoor geolocation for cooperative robotic applications"), Berlin: Springer, pp. 255–266, September 2009.
37. Pahlavan, K., Li, X., Ylianttila, M., Chana, R., and Latva-aho, M., *Lecture Notes in Computer Science: Mobile and Wireless Communications Networks* (book chapter on "An overview of wireless indoor geolocation techniques and systems"), Berlin: Springer, pp. 1–13, January 2000.
38. Gueye, B., Uhlig, S., Ziviani, A., and Fdida, S., *Lecture Notes in Computer Science, Networking 2006. Networking Technologies, Services, and Protocols; Performance of Computer and Communication Networks; Mobile and Wireless Communications Systems* (book chapter on "Leveraging buffering delay estimation for geolocation of internet hosts"), Berlin: Springer, pp. 319–330, April 2006.
39. Gueye, B., Uhlig, S., Ziviani, A., and Fdida, S., *Reviews: Methods and Technologies in Fish Biology and Fisheries, Tagging and Tracking of Marine Animals with Electronic Devices* (book chapter on "Summary report of a workshop on geolocation methods for marine animals"), Netherlands: Springer, pp. 342–363, June 2009.
40. Palaniswami, M., Sundaram, B., Jayavardhana, R., and Shilton, A., *Informatics in Control, Automation and Robotics II* (book chapter on "Target localization using machine learning"), Netherlands: Springer, pp. 27–33, June 2007.
41. Rowlands, D.D., Lemoine, F. G., Chinn, D. S., and Luthcke, S.B., A simulation study of multi-beam altimetry for lunar reconnaissance orbiter and other planetary missions, *J. Geod.*, vol. 83, no. 8, pp. 709–721, 2008.
42. Choi, W.-J., and Tekinay, S., Location-based service provisioning for next generation wireless networks, *Int. J. Wireless Inform. Networks*, vol. 10, no. 3, pp. 127–139, 2003.
43. Thygesen, U.H., Pedersen, M.W., and Madsen, H., *Reviews: Methods and Technologies in Fish Biology and Fisheries, Tagging and Tracking of Marine Animals with Electronic Devices* (book chapter 114 on "Geolocating fish using hidden Markov models and data storage tags"), Netherlands: Springer, pp. 277–293, June 2009.

44. Lee, S.C., Lee, W.R., and You, K.H., *Communications in Computer and Information Science, Control and Automation* (book chapter on "TDoA based UAV localization using dual-EKF algorithm"), Berlin: Springer, pp. 47–54, 2009.
45. Tanner, S., Stein, C., and Graves, S.J., *Scientific Data Mining and Knowledge Discovery* (book chapter on "On-board data mining"), Berlin: Springer, pp. 345–376, 2009.
46. Jeske, D.R., *Statistics for Industry and Technology, Advances in Mathematical and Statistical Modeling* (book chapter on "Jackknife bias correction of a clock offset estimator"), Boston: Birkhäuser, pp. 245–254, 2008.
47. Arslan, H., and Celebi, H., *Cognitive Radio, Software Defined Radio, and Adaptive Wireless Systems* (book chapter 10 on "Location information management systems for cognitive wireless networks"), Netherlands: Springer, pp. 291–323, 2007.
48. Drent, R.H., Fox, A.D., and Stahl, J., Travelling to breed, *J. Ornithol.*, vol. 147, no. 2, pp. 185–188, 2006.
49. Riboni, D., Pareschi, L., and Bettini, C., *Lecture Notes in Computer Science, Privacy in Location-Based Applications* (book chapter 10 on "Privacy in georeferenced context-aware services: a survey"), Berlin: Springer, pp. 151–172, 2009.
50. Xiang-zheng, D., Jin-yan, Z., Ji-yuan, L., and Da-fang, Z., The global rainforest mapping project JERS-1: a paradigm of international collaboration for monitoring land cover change, *J. Geogr. Sci.*, vol. 12, no. 1, pp. 185–188, 2002.
51. Huang, L., Yamane, H., Matsuura, K., and Sezaki, K., *Advances in Information Security, Secure Localization and Time Synchronization for Wireless Sensor and Ad Hoc Networks* (book chapter on "Location privacy in wireless LAN"), Berlin: Springer, pp. 299–322, 2007.
52. Zhou, R., and Schneider, G., *Lecture Notes of the Institute for Computer Sciences, Social Informatics and Telecommunications Engineering, Mobile Wireless Middleware, Operating Systems, and Applications* (book chapter on "Location-based botany guide: a prototype of web-based tracking and guiding"), Berlin: Springer, pp. 72–86, 2007.
53. DongChen, E., Shen, Q., Xu, Y., and Chen, G., High-accuracy topographical information extraction based on fusion of ASTER stereo-data and ICESat/GLAS data in Antarctica, *Sci. China Ser. D Earth Sci.*, vol. 52, no. 5, pp. 714–722, 2009.
54. Taheri, J., and Zomaya, A.Y., A combined genetic-neural algorithm for mobility management, *JMMA*, vol. 6, no. 3, pp. 714–722, 2007.
55. Yang, J., Zhang, C., Li, X., Huang, Y., Fu, S., and Acevedo, M.F., Integration of wireless sensor networks in environmental monitoring cyber infrastructure, *Wireless Networks*, vol. 6, no. 3, pp. 714–722, 2007.
56. Villafuerte, F.L., and Schiller, J., *Lecture Notes in Computer Science, Ad-Hoc, Mobile and Wireless Networks* (book chapter on "An ad-hoc algorithm to estimate distances in wireless sensor networks"), Berlin: Springer, pp. 162–175, 2008.
57. Gardiner, K., Yin, J., and Carswell, J.D., *Lecture Notes in Computer Science, Web and Wireless Geographical Information Systems* (book chapter on "EgoViz – a mobile based spatial interaction system"), Berlin: Springer, pp. 135–152, 2009.
58. Yang, F., Zhang, Z., and Chen, Y., Maximum-likelihood estimator for coarse carrier frequency offset estimation in OFDM systems, *J. Wireless Pers. Commun.*, vol. 49, no. 1, pp. 55–66, 2009.
59. Musick, S., *Foundations and Applications of Sensor Management* (book chapter 11 on "Defense applications"), Berlin: Springer, pp. 257–268, 2008.
60. Frattasi, S., Fitzek, P.H.P., and Prasad, R., *IFIP International Federation for Information Processing, Intelligence in Communication Systems* (book chapter on "A look into the 4G crystal ball"), Berlin: Springer, pp. 281–290, 2005.
61. Dil, B., Dulman, S., and Havinga, P., *Lecture Notes in Computer Science, Wireless Sensor Networks* (book chapter on "Range-based localization in mobile sensor networks"), Berlin: Springer, pp. 164–178, 2006.
62. Shen, X., Mark, J.W., and Ye, J., Mobile location estimation in CDMA cellular networks by using fuzzy logic, *J. Wireless Pers. Commun.*, vol. 22, no. 1, pp. 57–70, 2002.

63. Mineno, H., Hida, K., Mizutani, M., Miyauchi, N., Kusunoki, K., Fukuda, A., and Mizuno, T., *Lecture Notes in Computer Science, Knowledge-Based Intelligent Information and Engineering Systems* (book chapter on "Position estimation for goods tracking system using mobile detectors"), Berlin: Springer, pp. 431–437, 2005.
64. Buschmann, C., Hellbrück, H., Fischer, S., Kröller, A., and Fekete, S., *Lecture Notes in Computer Science, Wireless Sensor Networks* (book chapter on "Radio propagation-aware distance estimation based on neighborhood comparison"), Berlin: Springer, pp. 325–340, 2007.
65. Li, J., Li, J., Guo, L., and Wang, P., *Lecture Notes in Computer Science, Advanced Web and Network Technologies, and Applications* (book chapter on "Power-efficient node localization algorithm in wireless sensor networks"), Berlin: Springer, pp. 420–430, 2006.
66. Muthukrishnan, K., and Hazas, M., *Lecture Notes in Computer Science, Location and Context Awareness* (book chapter on "Position estimation from UWB pseudorange and angle-of-arrival: a comparison of non-linear regression and Kalman filtering"), Berlin: Springer, pp. 222–239, 2009.
67. Kontkanen, P., Myllymäki, P., Roos, T., Tirri, H., Valtonen, K., and Wettig, H., *Emerging Location Aware Broadband Wireless Ad Hoc Networks* (book chapter on "Probabilistic methods for location estimation in wireless networks"), Berlin: Springer, pp. 173–188, 2005.
68. Doherty, P., and Rudol, P., *Lecture Notes in Computer Science, AI 2007: Advances in Artificial Intelligence* (book chapter on "A UAV search and rescue scenario with human body detection and geolocalization"), Berlin: Springer, pp. 1–13, 2007.
69. De la Cruz, A., Laneve, G., Cerra, D., Mielewczyk, M., Garcia, M., Santilli, G., Cadau, E., and Joyanes, G., *Lecture Notes in Geoinformation and Cartography, Geomatics Solutions for Disaster Management* (book chapter on "On the application of nighttime sensors for rapid detection of areas impacted by disasters"), Berlin: Springer, pp. 17–36, 2007.
70. Martínez-de-Dios, J.R., Merino, L., Ollero, A., Ribeiro, L.M., and Viegas, X., *Springer Tracts in Advanced Robotics, Multiple Heterogeneous Unmanned Aerial Vehicles* (book chapter on "Multi-UAV experiments: application to forest fires"), Berlin: Springer, pp. 207–228, 2007.
71. Sun, G., and Guo, W., A novel indoor geo-location method using MIMO array, *J. Electron. (China)*, vol. 23, no. 6, pp. 810–813, 2003.
72. Pahlavan, K., Li, X., Ylianttila, M., and Latva-aho, M., *The Springer International Series in Engineering and Computer Science, Wireless Communication Technologies: New Multimedia Systems* (book chapter on "Wireless data communications systems"), Berlin: Springer, pp. 201–214, 2000.
73. Merino, L., Caballero, F., Ferruz, J., Wiklund, J., Forssén, P.-E., and Ollero, A., *Springer Tracts in Advanced Robotics, Multiple Heterogeneous Unmanned Aerial Vehicles* (book chapter on "Multi-UAV cooperative perception techniques"), Berlin: Springer, pp. 67–110, 2007.
74. Sunay, M., *The Springer International Series in Engineering and Computer Science, Next Generation Wireless Networks* (book chapter on "Evaluation of location determination technologies towards satisfying the FCC E-911 Ruling"), Netherlands: Springer, pp. 157–192, 2002.
75. Yarlykov, M.S., and Yarlykova, S.M., Signal-detection and signal-processing algorithms for code-division multiple-access satellite mobile communications systems employed simultaneously with satellite radio navigation systems, *J. Comm. Tech. Electron.*, vol. 51, no. 8, pp. 874–894, 2006.
76. Amundson, I., Manish Kushwaha, M., and Koutsoukos, X.D., *Lecture Notes in Computer Science, Mobile Entity Localization and Tracking in GPS-Less Environments* (book chapter on "On the feasibility of determining angular separation in mobile wireless sensor networks"), Berlin: Springer, pp. 115–127, 2009.
77. Richton, B., Vannucci, G., and Wilkus, S., *The Springer International Series in Engineering and Computer Science, Next Generation Wireless Networks* (book chapter on "Assisted GPS

for wireless phone location – technology and standards"), Netherlands: Springer, pp. 129–155, 2002.
78. Howard, A., and Tunstel, F., *Frontiers of Geographic Information Technology* (book chapter on "Using geospatial information for autonomous systems control"), *Berlin: Springer*, pp. 63–84, 2006.
79. Hoeher, P., and Schmeink, K., *Lecture Notes in Electrical Engineering, Multi-Carrier Spread Spectrum 2007* (book chapter on "Joint navigation & communication based on interleave-division multiple access"), Netherlands: Springer, pp. 97–106, 2007.
80. Howard, A., and Tunstel, F., *Lecture Notes in Computer Science, Algorithmic Aspects of Wireless Sensor Networks* (book chapter on "Algorithms for location estimation based on RSSI Sampling"), Berlin: Springer, pp. 72–86, 2008.
81. Cianca, E., Sanctis, M.D., Araniti, G., Molinaro, A., Iera, A., Torrisi, M., and Ruggieri, M., *Signals and Communication Technology, Satellite Communications and Navigation Systems* (book chapter on "Integration of navigation and communication for location and context aware RRM"), Berlin: Springer, pp. 25–50, 2008.
82. Song, L., Adve, R., and Hatzinakos, D., *Lecture Notes in Computer Science, Wireless Sensor Networks* (book chapter on "Matrix pencil for positioning in wireless ad hoc sensor network"), Berlin: Springer, pp. 18–27, 2004.
83. Chen, Y., and Rapajic, P., Human respiration rate estimation using ultra-wideband distributed cognitive radar system, *Institute of Automation, Chinese Academy of Sciences, co-published with Springer-Verlag GmbH, International Journal of Automation and Computing*, vol. 4, no. 5, pp. 325–333, October 2008.
84. Hämäläinen, M., Saloranta, J., Mäkelä, J.-P., Oppermann, I., and Patana, T., Ultra-wideband signal impact on the performances of IEEE 802.11b and bluetooth networks, *International Int. J Wireless Inform. Networks*, vol. 10, no. 4, pp. 201–210, 2003.
85. Guvenc, I., Sahinoglu, Z., Orlik, P., and Arslan, H., Searchback algorithms for TOA estimation in non-coherent low-rate IR-UWB systems, *J. Wireless Pers. Commun.*, vol. 48, no. 4, pp. 585–603, 2009.
86. Zeger, L.M., and Kobayashi, H., A simplified EM algorithm for detection of CPM signals in a fading multipath channel, *J. Wireless Networks*, vol. 8, no. 6, pp. 649–658, 2002.
87. Barber, B.D., Redding, J.D., McLain, T.W., Beard, R.W., and Taylor, C.N., Vision-based target geo-location using a fixed-wing miniature air vehicle, *J. Intell. Robot. Syst.*, vol. 47, no. 4, pp. 361–382, 2006.
88. Gezici, S., A survey on wireless position estimation, *J. Wireless Pers. Commun.*, vol. 44, no. 3, pp. 263–282, 2008.
89. Ye, L., Geng, Z., Xue, L., and Liu, Z., *Lecture Notes in Computer Science,* Computational Science and its Applications – ICCSA 2007 (book chapter on "A novel real time method of signal strength based indoor localization"), Berlin: Springer, pp. 678–688, 2007.
90. Guoqiang, M., Barş, F., and Anderson, D., *Sensor Networks and Configuration* (book chapter on "Localisation"), Berlin: Springer, pp. 281–315, 2007.
91. Khokhar, S., and Nilsson, A.A., *Lecture Notes in Computer Science,* Wireless Algorithms, Systems, and Applications (book chapter on "Introduction to mobile trajectory based services: a new direction in mobile location based services"), Berlin: Springer, pp. 398–407, 2009.
92. Zàruba, G.V., Huber, M., Kamangar, F.A., and Chlamtac, I., Indoor location tracking using RSSI readings from a single Wi-Fi access point, *J. Wireless Networks*, vol. 13, no. 2, pp. 221–235, 2007.
93. Gu, Z., and Gunawan, E., Radiolocation in CDMA Cellular system based on joint angle and delay estimation, *J. Wireless Pers. Commun.*, vol. 23, no. 3, pp. 297–309, 2002.
94. Arslan, H., and Celebi, H., *Cognitive Radio, Software Defined Radio, and Adaptive Wireless Systems* (book chapter on "Software defined radio architectures for cognitive radios"), Netherlands: Springer, pp. 109–144, 2007.
95. Fontana, R.J., *Ultra-Wideband, Short-Pulse Electromagnetics 5* (book chapter on "Recent applications of ultra wideband radar and communications systems"), Berlin: Springer, pp. 225–234, 2002.

96. Boryssenko, A.O., and Schaubert, D.H., Electromagnetics-related aspects of signaling and signal processing for UWB short range radios, *J. VLSI Signal Process.*, vol. 43, no. 1, pp. 89–104, 2006.
97. Mazuelas, S., Lago, F.A., Fernandez, P., Bahillo, A., Blas, J., Lorenzo, R.M., and Abril, E.J., Ranking of TOA Measurements Based on the Estimate of the NLOS Propagation Contribution in a Wireless Location System, *J. Wireless Pers. Commun.*, vol. 53, nr. 1, pp. 35–52, 2009.
98. Manodham, T., Loyola, L., and Miki, T., A novel wireless positioning system for seamless internet connectivity based on the WLAN infrastructure, *J. Wireless Pers. Commun.*, vol. 44, no. 3, pp. 295–309, 2008.
99. Frattasi, S., and Monti, M., *Cognitive Wireless Networks* (book chapter on "Cooperative mobile positioning in 4G wireless networks"), Netherlands: Springer, pp. 213–233, 2007.
100. Artieda, J., Sebastian, J.M., Campoy, P., Correa, J.F., Mondragón, I.F., Martínez, C., and Olivares, M., Visual 3-D SLAM from UAVs, *J. Intell. Robot. Syst.*, vol. 55, no. 4–5, pp. 299–321, 2009.
101. *Principles of Spread-Spectrum Communication Systems* (book chapter on "Code-Division Multiple Access"), Berlin: Springer, pp. 293–386, 2005.
102. Webb, W., Broadband fixed wireless access as a key component of the future integrated communications environment, *IEEE Com. Mag.*, vol. 39, no. 9, pp. 115–121, 2001.
103. Danesh, M., Zuniga, J.-C., Concilio, F., Fixed low-frequency broadband wireless access radio systems, *IEEE Com. Mag.*, vol. 39, no. 9, pp. 134–138, 2001.
104. Izadpanah, H., A military-wave broadband wireless access technology demonstrator for the next-generation internet network reach extension, *IEEE Com. Mag.*, vol. 39, no. 9, pp. 140–145, Sep. 2001.
105. Heegard, C., Coffey, J.T., Gummadi, S., Murphy, P.A., Provencio, R., Rossin, E.J., Schrum, S., and Shoemake, M.B., High-performance wireless Ethernet, *IEEE Com. Mag.*, vol. 39, no. 11, pp. 64–73, 2001.
106. Karaguz, J., High-rate wireless personal area network, *IEEE Com. Mag.*, vol. 39, no. 12, pp. 96–102, 2001.
107. Smulders, P., Exploiting the 60 GHz band for local wireless multimedia access: prospects and future directions, *IEEE Com. Mag.*, vol. 40, no. 1, pp. 140–147, 2002.
108. Eklund, C., Marks, R.B., Stanwood, K.L., and Wang, S., IEEE standard 802.16: a technical overview of the WirelessMAN™ air interface for broadband wireless access, *IEEE Com. Mag.*, vol. 40, no. 6, pp. 90–96, 2002.
109. Callaway, E., Gorday, P., Hester, L., Gutierrez, J.A., Naeve, M., Heile, B., and Bahl, V., Home networking with IEEE 802.15.4: a developing standard for low-rate wireless personal area networks, *IEEE Com. Mag.*, vol. 40, no. 8, pp. 70–77, 2002.
110. Porcino, D., and Hirt, W., Ultra-wideband radio technology: potential and challenges ahead, *IEEE Com. Mag.*, vol. 41, no. 7, pp. 66–74, 2003.
111. Mahonen, P., Riihijarvi, J., Petrova, M., and Shelby, Z., Hop-by-hop toward future mobile broadband IP, *IEEE Com. Mag.*, vol. 42, no. 3, pp. 138–146, 2004.
112. Zheng, J., and Lee, M.J., Will IEEE 802.15.4 make ubiquitous networking a reality?: a discussion on a low power, low bit rate standard, *IEEE Com. Mag.*, vol. 42, no. 6, pp. 140–146, 2004.
113. Williams, C., Beach, M., Neirynick, D., Nix, A., Chen, K., Morris, K., Kitchener, D., Presser, M., Li, Y, and McLauchglin, S., Personal area technologies for internetworked services, *IEEE Com. Mag.*, vol. 42, no. 12, pp. S15–S26, 2004.
114. Ghosh, A., Wolter, D.R., Andrews, J.G., and Chen, R., Broadband wireless access with WiMax/802/16: current performance benchmarks and future potential, *IEEE Com. Mag.*, vol. 43, no. 2, pp. 129–136, 2005.
115. Tardy, I., and Grondalen, O., On the role of future high-frequency BFWA systems in broadband communication networks, *IEEE Com. Mag.*, vol. 43, no. 2, pp. 138–144, 2005.
116. Chan, S., Shared spectrum access for the DoD, *IEEE Com. Mag.*, vol. 45, no. 6, pp. 58–66, 2007.

117. Weerackody, V., and Gonzales, L., Mobile small aperture satellite terminals for military communications, *IEEE Com. Mag.*, vol. 45, no. 10, pp. 70–75, 2007.
118. Li, B., Yang, Q., Low, C.P., and Gwee, C.L., A survey on mobile WiMAX, *IEEE Com. Mag.*, vol. 45, no. 12, pp. 70–75, 2007.
119. Guiliano, R., Luglio, M., and Mazzenga, F., Interoperability between WiMAX and broadband mobile space networks, *IEEE Com. Mag.*, vol. 46, no. 3, pp. 50–57, 2008.
120. Cheng, L., Henty, B.E., Cooper, R., Stancil, D.C., and Bai, F., A measurement study of time-scaled 802.11a waveforms over the mobile-to-mobile vehicular channel at 5.9 GHz, *IEEE Com. Mag.*, vol. 46, no. 5, pp. 84–91, 2008.
121. Stotts, L., Seidel, S., Krout, T., and Kolodzy, P., MANET gateways: radio interoperability via the internet, not the radio, *IEEE Com. Mag.*, vol. 46, no. 6, pp. 51–59, 2008.
122. Greenspan, A., Klerer, M., Tomcik, J., Canchi, R., and Wilson, J., IEEE 802.20: mobile broadband wireless access for the twenty-first century, *IEEE Com. Mag.*, vol. 46, no. 7, pp. 56–63, 2008.
123. Etemad, K., Overview of mobile WiMAX technology and evolution, *IEEE Com. Mag.*, vol. 46, no. 10, pp. 31–40, 2008.
124. Wang, F., Chosh, A., Sankaran, C., Fleming, P.J., Hsieh, F., and Benes, S.J., Mobile WiMAX systems: performance and evolution, *IEEE Com. Mag.*, vol. 46, no. 10, pp. 41–49, 2008.
125. Yen, S.-P., Talwar, S., Lee, S.-C., and Kim, H., WiMAX femtocells: a perspective on network architecture, capacity, and coverage, *IEEE Com. Mag.*, vol. 46, no. 10, pp. 58–65, 2008.
126. Cox, D.C., Fundamental limitations on increasing data rate in wireless systems, *IEEE Com. Mag.*, vol. 46, no. 12, pp. 16–17, 2008.
127. Singh, H., Oh, J., Kweon, C.Y., Qin, X., Shao, H.-R., and Ngo, C., A 60 GHz wireless network for enabling uncompressed video communication, *IEEE Com. Mag.*, vol. 46, no. 12, pp. 71–78, 2008.
128. Gardikis, G., and Kourtis, A., Using DVB-S2 adaptive coding and modulation for provision of Satellite triple play services, *IEEE Com. Mag.*, vol. 46, no. 12, pp. 128–135, 2008.
129. Peters, S.W., and Hearth, R.W., Jr., The future of WiMAX: multihop relaying with IEEE 802.16j, *IEEE Com. Mag.*, vol. 47, no. 1, pp. 104–111, 2009.
130. Stevenson, C.R., Chouinard, G., Lei, Z., Hu, W., Shellhammer, S.J., and Caldwell, W., IEEE 802.22: the first cognitive radio wireless regional area network standard, *IEEE Com. Mag.*, vol. 47, no. 1, pp. 130–138, 2009.
131. Bazelon, C., Licensed or unlicensed: the economic considerations in incremental spectrum allocations, *IEEE Com. Mag.*, vol. 47, no. 3, pp. 110–116, 2009.
132. Astely, D., Dahlman, E., Furuskar, A., Jading, Y., Lindstrom, M., and Parkvall, S., LTE: the evolution of mobile broadband, *IEEE Com. Mag.*, vol. 47, no. 4, pp. 44–51, 2009.
133. Cai, L.X., Shen, X., and Mark, J.W., Efficient MAC protocol for ultra-wideband networks, *IEEE Com. Mag.*, vol. 47, no. 6, pp. 179–185, 2009.
134. National Geospatial and Intelligence Agency. http://www1.nga.mil/Pages/Default.aspx
135. Progri, I.F., Ortiz, W., Michalson, W.R., and Wang, J., The Performance and simulation of an OFDMA pseudolite indoor geolocation system, in *Proc. ION-GNSS 2006*, Fort Worth, pp. 3149–3162, September 2006.
136. Progri, I.F., Wireless-enabled GPS indoor geolocation system, in *Proc. IEEE/ION-PLANS 2010*, Palm Springs, pp. 526–538, May 2010.
137. Digital Video Broadcasting – Satellite – Second Generation (DVB-S2). http://en.wikipedia.org/wiki/DVB-S2
138. A COMPANION GUIDE TO DVB-S2. http://www.arctekhd.com/news/stories/DVBS-2_guide.pdf, © 2004 TANDBERG Television Ltd. All rights reserved.
139. Two-way Radio. http://en.wikipedia.org/wiki/Two-way_radio
140. Cellular Network. http://en.wikipedia.org/wiki/Cellular_network
141. Satellite Television. http://en.wikipedia.org/wiki/Satellite_television
142. FM Broadcasting. http://en.wikipedia.org/wiki/FM_broadcasting

143. Broadcast Television Systems. http://en.wikipedia.org/wiki/Broadcast_television_systems
144. Terrestrial Television. http://en.wikipedia.org/wiki/Terrestrial_television
145. Betz, J.W., Effect of narrowband interference on GPS code tracking accuracy, in *Proc. 2000 Nat. Tech. Mtg. ION*, Anaheim, pp. 16–27, January 2000.
146. Wallace, K., and Schwoerer, B., Concept for automatic reporting of Global Positioning System radio frequency interference, in *Proc. 2000 Nat. Tech. Mtg. ION*, Anaheim, pp. 9–15, January 2000.
147. Betz, J.W., Analysis of M code signal interference with C/A code receivers, in *Proc. 2000 Nat. Tech. Mtg. ION*, Anaheim, pp. 1–8, January 2000.
148. Khan, M., and Lazar, S., GPS Radio Frequency (RF) spectrum defense, in *Proc. IAIN World Cong. 56th Ann. Mtg. ION*, The Catamaran Resort Hotel, San Diego, pp. 200–208, June 2000.
149. Lazar, S., Raghavan, S., and Turner, D., Issues in sharing and encroachment: a hierarchical approach to like-system spectrum sharing, in *Proc. IAIN World Cong. 56th Ann. Mtg. ION*, The Catamaran Resort Hotel, San Diego, pp. 209–216, June 2000.
150. Winkel, J.O., Konig, D., Oehler, V., Eissfeller, B., Hein, G., Positioning and navigation using mobile communication: alternative or supplement to a future GNSS-2?, in *Proc. IAIN World Cong. 56th Ann. Mtg. ION*, San Diego, pp. 451–460, June 2000.
151. Hegarty, C., Van Dierendonck, A.J., Bobyn, D., Tran, M., and Grabowski, J., Suppression of pulsed interference through blanking, in *Proc. IAIN World Cong. 56th Ann. Mtg. ION*, San Diego, pp. 399–408, June 2000.
152. Peiro, A.B.M., Beech, T.W., Garcia, A.M., and Merino, M.M.R., Galileo in-orbit control strategy, in *Proc. IAIN World Cong. 56th Ann. Mtg. ION*, San Diego, pp. 469–480, June 2000.
153. Hein, G., Eissfeller, B., Oehler, V., and Winkel, J.O., Synergies between satellite navigation and location services of terrestrial mobile communication, in *Proc. 13th Inter. Tech. Mtg. Sat. Div. ION (ION GPS 2000)*, Salt Lake City, pp. 535–544, September 2000.
154. Kim, H.S., Hyun, M.P., Park, S.J., Jee, G.-I., Lee, Y.J., You, H.R., and Woo, S.S., Performance analysis of position location methods based on IS-801 Standard, in *Proc. 13th Inter. Tech. Mtg. Sat. Div. ION (ION GPS 2000)*, Salt Lake City, pp. 545–553, September 2000.
155. Lazar, S., Raghavan, S., and Turner, D., Issues in sharing and encroachment: a hierarchical approach to like-system spectrum sharing, in *Proc. 13th Inter. Tech. Mtg. Sat. Div. ION (ION GPS 2000)*, Salt Palace Convention Center, Salt Lake City, pp. 973–980, September 2000.
156. Lortie, J.P., Jr., Future GNSS architecture: interoperability or compatibility? between systems what is the prudent course to pursue, in *Proc. 13th Inter. Tech. Mtg. Sat. Div. ION (ION GPS 2000)*, Salt Palace Convention Center, Salt Lake City, pp. 1332–1339, September 2000.
157. Eissfeller, B., Hein, G.W., Winkel, J., and Hartl, P.H., Requirements on the Galileo signal structure, in *Proc. 13th Inter. Tech. Mtg. Sat. Div. ION (ION GPS 2000)*, Salt Lake City, pp. 1772–1781, September 2000.
158. Godet, J., GPS/GALILEO radio frequency compatibility analysis, in *Proc. 13th Inter. Tech. Mtg. Sat. Div. ION (ION GPS 2000)*, Salt Lake City, pp. 1782–1790, September 2000.
159. Arif, M.Z., Brami, S., Haro, P., and Pasquali, R., The GEMINUS proposed Galileo services: evolution of GNSS, in *Proc. 13th Inter. Tech. Mtg. Sat. Div. ION (ION GPS 2000)*, Salt Lake City, pp. 1798–1806, September 2000.
160. Dafesh, P.A., Cooper, L., and Partridge, M., Compatibility of the interplex modulation method with C/A and P(Y) code signals, in *Proc. 13th Inter. Tech. Mtg. Sat. Div. ION (ION GPS 2000)*, Salt Lake City, pp. 2122–2129, September 2000.
161. Bull, B., A real time differential GPS tracking system for NASA sounding rockets, in *Proc. 13th Inter. Tech. Mtg. Sat. Div. ION (ION GPS 2000)*, Salt Lake City, pp. 2028–2037, September 2000.
162. Silva, J.M., The nationwide differential global positioning system and electric power line interference, in *Proc. 13th Inter. Tech. Mtg. Sat. Div. ION (ION GPS 2000)*, Salt Lake City, pp. 489–498, September 2000.
163. Klukas, R., Lachapelle, G., and Fattouche, M., Estimation noise of a cellular telephone positioning system, *Navigation*, vol. 47, no. 3, pp. 167–174, fall 2000.

164. Ross, J.T., Leva, J.L., and Yoder, S., Effect of partial-band interference on receiver estimation of C/N0: measurements, in *Proc. 2001 Nat. Tech. Mtg. ION*, Long Beach, pp. 829–838, January 2001.
165. Guggenbuehl, P., and Boggs, M., Global positioning system advanced targeting system, in *Proc. 2001 Nat. Tech. Mtg. ION*, Long Beach, pp. 426–431, January 2001.
166. Martin, S., Usage of terrestrial DGPS Services for high precision aircraft applications, in *Proc. 2001 Nat. Tech. Mtg. ION*, Long Beach, pp. 310–318, January 2001.
167. Lee, R., Slattery, R., Kovach, K., Thompson, R., and Kuhlmann, K.R., Improving GPS for aviation navigation: GPS operational requirements document (ORD) aviation annex, in *Proc. 2001 Nat. Tech. Mtg. ION*, Long Beach, pp. 124–132, January 2001.
168. Dafesh, P.A., Wong, R.L., Partridge, M.D., and Fan, T., Measurements of GPS receiver compatibility with interplex-modulated M-code signals, in *Proc. 57th Ann. Mtg. ION*, Albuquerque, pp. 100–108, June 2001.
169. Tetewsky, A., Soltz, A., Bogner, T., and O'Neil, E., The key to unlocking the emerging market for civil (and DoD) applications of navigation technology: cost scalable interface standardization, in *Proc. 57th Ann. Mtg. ION*, Albuquerque, pp. 722–732, June 2001.
170. Holmes, J.K., and Dafesh, P.A., Matched filter mean acquisition time performance for P(Y), BOC(10,5), and MAN(10) codes with FFT aiding and noncoherent combining, in *Proc. 57th Ann. Mtg. ION*, Albuquerque, pp. 121–139, June 2001.
171. Luo, M., Akos, D., Koenig, M., Opshaug, G., Pullen, S., Enge, P., Erlandson, B., and Frodge, S., Testing and research on interference to GPS from UWB transmitters, in *Proc. 14th Inter. Tech. Mtg. Sat. Div. ION (ION GPS 2001)*, Salt Lake City, pp. 1–13, September 2001.
172. Erlandson, R.J., and Van Dierendonck, A.J., GPS receiver susceptibility to ultra-wideband RFI: test results and RFI link analyses, in *Proc. 14th Inter. Tech. Mtg. Sat. Div. ION (ION GPS 2001)*, Salt Lake City, pp. 14–25, September 2001.
173. Turner, D., Report of the AIAA international space cooperation working group on global navigation satellite systems, in *Proc. 14th Inter. Tech. Mtg. Sat. Div. ION (ION GPS 2001)*, Salt Lake City, pp. 81–89, September 2001.
174. Lavrakas, J.W., A systems approach to a national position, navigation and timing service, in *Proc. 14th Inter. Tech. Mtg. Sat. Div. ION (ION GPS 2001)*, Salt Lake City, pp. 703–736, September 2001.
175. Andreone, V., Mediavilla, R., and Broughton, C., North slope 2000: flight testing Military/Civil GPS receivers against north warning system radars, in *Proc. 14th Inter. Tech. Mtg. Sat. Div. ION (ION GPS 2001)*, Salt Lake City, pp. 912–923, September 2001.
176. Bett, C.J., Simon, S.A., Farnworth, B.J., Boyd, R.W., and Brewer, J.J., Flight test of the JPALS LDGPS demonstration system, in *Proc. 14th Inter. Tech. Mtg. Sat. Div. ION (ION GPS 2001)*, Salt Lake City, pp. 1159–1169, September 2001.
177. Katanik, T., Simon, S., Bett, C., Driscoll, B., Tsamis, D., Flemming, J., Norwood, R., and Barry, J., Interoperability between civil LAAS and military JPALS precision approach and landing systems, in *Proc. 14th Inter. Tech. Mtg. Sat. Div. ION (ION GPS 2001)*, Salt Lake City, pp. 1179–1189, September 2001.
178. Tytgat, L., and Campagne, P., GALILEO: taking the next step, in *Proc. 14th Inter. Tech. Mtg. Sat. Div. ION (ION GPS 2001)*, Salt Lake City, pp. 1255–1263, September 2001.
179. Dutton, L., Ahmed, R., and Bou, J.-F., The commercialisation of GNSS services, in *Proc. 14th Inter. Tech. Mtg. Sat. Div. ION (ION GPS 2001)*, Salt Lake City, pp. 1283–1290, September 2001.
180. Hein, G.W., Godet, J., Issler, J.-L., Martin, J.-C., Lucas-Rodriguez, R., and Pratt, T., The GALILEO frequency structure and signal design, in *Proc. 14th Inter. Tech. Mtg. Sat. Div. ION (ION GPS 2001)*, Salt Lake City, pp. 1273–1282, September 2001.
181. Pasquali, R., Navigation related communication services, the case of a GALILEO service option, in *Proc. 14th Inter. Tech. Mtg. Sat. Div. ION (ION GPS 2001)*, Salt Lake City, pp. 2067–2085, September 2001.
182. Barker, B.C., III, and Straton, J.R., GPS military signal modernization: updates to design and characteristics, in *Proc. 14th Inter. Tech. Mtg. Sat. Div. ION (ION GPS 2001)*, Salt Lake City, pp. 2716–2721, September 2001.

183. McKendree, T.L., Ideas for GPS IV, in *Proc. 14th Inter. Tech. Mtg. Sat. Div. ION (ION GPS 2001)*, Salt Lake City, pp. 2745–2750, September 2001.
184. Dafesh, P.A., Bow, R.T., Partridge, M.D., and Fan, T. Simulation and hardware demonstration of new L2 civilian GPS ranging codes, in *Proc. 2002 Nat. Tech. Mtg. ION*, The Catamaran Resort Hotel, San Diego, pp. 925–930, January 2002.
185. Hegarty, C., and Tran, M., Compatibility of the new military GPS signals with civil aviation receivers, in *Proc. 58th Ann. Mtg. ION CIGTF 21st Guid. Test Sym.*, Hyatt Regency Hotel, Albuquerque, pp. 555–568, June 2002.
186. Lee, Y.C., Compatibility of the new military GPS signals with non-aviation receivers, in *Proc. 58th Ann. Mtg. ION CIGTF 21st Guid. Test Symp.*, Hyatt Regency Hotel, Albuquerque, pp. 581–597, June 2002.
187. Hein, G.W., Godet, J., Issler, J.-L., Martin, J.-C., Erhard, P., Lucas-Rodriguez, R., and Pratt, T., Status of Galileo frequency and signal design, in *Proc. 15th Inter. Tech. Mtg. Sat. Div. ION (ION GPS 2002)*, Portland, pp. 266–277, September 2002.
188. Fyfe, P., Davis, K., Jeng, I., Kelley, C., and Mosley, C., GPS and Galileo – interoperability for civil aviation applications, in *Proc. 15th Inter. Tech. Mtg. Sat. Div. ION (ION GPS 2002)*, Portland, pp. 289–302, September 2002.
189. Bou, J.F., Dussurgey, C., Leblond, V., and Panefieu, B., Galileo PRS service, in *Proc. 15th Inter. Tech. Mtg. Sat. Div. ION (ION GPS 2002)*, Portland, pp. 303–319, September 2002.
190. Leonard, A., Blomenhofer, H., and Izquierdo, I., GPS and GALILEO interoperability and synergies, in *Proc. 15th Inter. Tech. Mtg. Sat. Div. ION (ION GPS 2002)*, Portland, pp. 330–341, September 2002.
191. Hegarty, C., Tran, M., and Lee, Y., Simplified techniques for analyzing the effects of non-white interference on GPS receivers, in *Proc. 15th Inter. Tech. Mtg. Sat. Div. ION (ION GPS 2002)*, Portland, pp. 620–629, September 2002.
192. Cabler, H., and DeCleene, B., LPV: new, improved WAAS instrument approach, in *Proc. 15th Inter. Tech. Mtg. Sat. Div. ION (ION GPS 2002)*, Portland, pp. 1013–1021, September 2002.
193. Van Dierendonck, A.J., and Hegarty, C., Methodologies for assessing intrasystem and intersystem interference to satellite navigation systems, in *Proc. 15th Inter. Tech. Mtg. Sat. Div. ION (ION GPS 2002)*, Portland, pp. 1241–1250, September 2002.
194. Titus, B.M., Dafesh, P.A., Wong, R., Maine, K.P., Jr., and Stansell, T.A., Assessing ultra wide band (UWB) interference to GPS receivers, in *Proc. 15th Inter. Tech. Mtg. Sat. Div. ION (ION GPS 2002)*, Portland, pp. 1251–1259, September 2002.
195. Godet, J., de Mateo, J.C., Erhard, P., and Nouvel, O., Assessing the radio frequency compatibility between GPS and Galileo, in *Proc. 15th Inter. Tech. Mtg. Sat. Div. ION (ION GPS 2002)*, Oregon Convention Center, Portland, pp. 1260–1269, September 2002.
196. Soellner, M., Kohl, R., Luetke, W., and Erhard, Ph., The impact of linear and non-linear signal distortions on Galileo code tracking accuracy, in *Proc. 15th Inter. Tech. Mtg. Sat. Div. ION (ION GPS 2002)*, Portland, pp. 1270–1285, September 2002.
197. Lionel R., Macabiau, C., Nouvel, O., Jeandel, Q., Vigneau, W., Calmettes, V., and Issler, JL., A software receiver for GPS-IIF L5 signal, in *Proc. 15th Inter. Tech. Mtg. Sat. Div. ION (ION GPS 2002)*, Portland, pp. 1540–1553, September 2002.
198. Grabowski, J., and Hegarty, C., Characterization of L5 receiver performance using digital pulse blanking, in *Proc. 15th Inter. Tech. Mtg. Sat. Div. ION (ION GPS 2002)*, Portland, pp. 1630–1635, September 2002.
199. Gao, Y., and Shen, X., A new method for carrier-phase-based precise point positioning, *Navigation*, vol. 49, no. 2, pp. 109–116, sum. 2002.
200. Sousa, P., and Wellons, L., Results of shipboard relative GPS (SRGPS) testing for the pegasus X-47A unmanned combat air vehicle (UCAV), in *Proc. 59th Ann. Mtg. ION CIGTF 22nd Guid. Test Sym.*, Albuquerque, pp. 291–300, June 2003.
201. Kovach, K., Possibilities for a 27 optimized constellation, in *Proc. 59th Ann. Mtg. ION CIGTF 22nd Guid. Test Symp.*, Albuquerque, pp. 22–56, June 2003.
202. Dellago, R., Detoma, E., and Luongo, F., Galileo-GPS interoperability and compatibility: a synergetic viewpoint, in *Proc. 16th Inter. Tech. Mtg. Sat. Div. ION (ION GPS/GNSS 2003)*, Portland, pp. 542–548, September 2003.

203. Bastide, F., Macabiau, C., Akos, D.M., and Roturier, B., Assessment of L5 receiver performance in presence of interference using a realistic receiver simulator, in *Proc. 16th Inter. Tech. Mtg. Sat. Div. ION (ION GPS/GNSS 2003)*, Portland, pp. 142–152, September 2003.
204. Jones, R., and Price, D., GALILEO augmentation for a precise positioning application, in *Proc. 16th Inter. Tech. Mtg. Sat. Div. ION (ION GPS/GNSS 2003)*, Portland, pp. 579–592, September 2003.
205. Ganguly, S., and Jovancevic, A., Interoperability study between GPS and Galileo signals, in *Proc. 16th Inter. Tech. Mtg. Sat. Div. ION (ION GPS/GNSS 2003)*, Portland, pp. 1023–1034, September 2003.
206. Van Dierendonck, A.J., Coker, R., Razumovsky, O., Bobyn, D., and Kroon, H., Inmarsat-4 navigation transponder test equipment and control software for in-orbit tests, in *Proc. 16th Inter. Tech. Mtg. Sat. Div. ION (ION GPS/GNSS 2003)*, Portland, pp. 1307–1319, September 2003.
207. Kim, T., and Grabowski, J., Validation of GPS L5 coexistence with DME/TACAN and link-16 systems, *Proc. 16th Inter. Tech. Mtg. Sat. Div. ION (ION GPS/GNSS 2003)*, Portland, pp. 1455–1469, September 2003.
208. Goodrich, B., Gonzales, B., Hughes, L.H., Weaver, R., and Deary, B., Signal generation, verification, and interaction analysis of legacy and modernized GNSS systems, in *Proc. 16th Inter. Tech. Mtg. Sat. Div. ION (ION GPS/GNSS 2003)*, Portland, pp. 1494–1501, September 2003.
209. Jurek, M.H., and Franco, G.A., GPS modernized block IIR system test program and anomaly reporting process, in *Proc. 16th Inter. Tech. Mtg. Sat. Div. ION (ION GPS/GNSS 2003)*, Portland, pp. 1515–1520, September 2003.
210. Hammock, A., Maj Anderson, J., Maj Hebert, J., Jokerst, S., Emery, A., Stroing, C., and Torres, J., Anechoic chamber testing of M-Code interference with legacy military receivers, in *Proc. 16th Inter. Tech. Mtg. Sat. Div. ION (ION GPS/GNSS 2003)*, Oregon Convention Center, Portland, pp. 1521–1525, September 2003.
211. Luba, O., Boyd, L., and Gower, A., GPS III system operations concepts, in *Proc. 16th Inter. Tech. Mtg. Sat. Div. ION (ION GPS/GNSS 2003)*, Portland, pp. 1561–1570, September 2003.
212. Swann, J., Chatre, E., and Ludwig, D., Galileo: benefits for location-based services, in *Proc. 16th Inter. Tech. Mtg. Sat. Div. ION (ION GPS/GNSS 2003)*, Portland, pp. 1603–1612, September 2003.
213. Normark, P.-L., Ståhlberg, C., Seco-Granados, G., Interference study – processing semi-live Galileo L1 and live GPS L1 data in a high-performance GNSS software receiver, in *Proc. 16th Inter. Tech. Mtg. Sat. Div. ION (ION GPS/GNSS 2003)*, Portland, pp. 1965–1972, September 2003.
214. Titus, B.M., Betz, J., Hegarty, C., and Owen, R., Intersystem and intrasystem interference analysis methodology, in *Proc. 16th Inter. Tech. Mtg. Sat. Div. ION (ION GPS/GNSS 2003)*, Portland, pp. 2061–2069, September 2003.
215. Ernou, Y., Renard, A., and Kirby, E., ATC radar interference impact on air receiver, in *Proc. 16th Inter. Tech. Mtg. Sat. Div. ION (ION GPS/GNSS 2003)*, Portland, pp. 2093–2101, September 2003.
216. Bastide, F., Akos, D., Macabiau, C., and Roturier, B., Automatic gain control (AGC) as an interference assessment tool, in *Proc. 16th Inter. Tech. Mtg. Sat. Div. ION (ION GPS/GNSS 2003)*, Portland, pp. 2042–2053, September 2003.
217. Stevens, J.R.A., Lage, M.E., Nam, Y.S., and Peterson, B.R., JPALS availability modeling and assessment in nominal and jamming environments, in *Proc. 16th Inter. Tech. Mtg. Sat. Div. ION (ION GPS/GNSS 2003)*, Portland, pp. 2337–2348, September 2003.
218. Miller, J.J., GPS & Galileo: evolution towards GNSS, in *Proc. 2004 Nat. Tech. Mtg. ION*, The Catamaran Resort Hotel, San Diego, pp. 73–91, January 2004.
219. Erlandson, R.J., Kim, T., Hegarty, C., and Van Dierendonck, A.J., Pulsed RFI effects on aviation operations using GPS L5, in *Proc. 2004 Nat. Tech. Mtg. ION*, San Diego, pp. 1063–1076, January 2004.

220. Bastide, F., Chatre, E., Macabiau, C., and Roturier, B., GPS L5 and GALILEO E5a/E5b signal-to-noise density ratio degradation due to DME/TACAN signals: simulations and theoretical derivation, in *Proc. 2004 Nat. Tech. Mtg. ION*, San Diego, pp. 1049–1062, January 2004.
221. Carroll, K., Celano, T., Biggs, C., and Weeks, G.K., Jr., Enhanced Loran for timing and frequency, in *Proc. 2004 Nat. Tech. Mtg. ION*, San Diego, pp. 735–739, January 2004.
222. Hegarty, C., Tran, M., and Betz, J.W., Multipath performance of the new GNSS signals, in *Proc. 2004 Nat. Tech. Mtg. ION*, San Diego, pp. 333–342, January 2004.
223. DiEsposti, R., DiLellio, J., Kelley, C., Dorsey, A., Fliegel, H., Berg, J., Edgar, C., McKendree, T., and Shome, P., The proposed state vector representation of broadcast navigation message for user equipment implementation of GPS satellite ephemeris propagation, in *Proc. 2004 Nat. Tech. Mtg. ION*, San Diego, pp. 294–312, January 2004.
224. Lee, Y.C., Performance of receiver autonomous integrity monitoring (RAIM) in the presence of simultaneous multiple satellite faults, in *Proc. 60th Ann. Mtg. ION*, Dayton, pp. 687–697, June 2004.
225. Peterson, B.R., Johnson, G., and Stevens, J., Feasible architectures for joint precision approach and landing system (JPALS) for land and sea, in *Proc. 17th Inter. Tech. Mtg. Sat. Div. ION (ION GNSS 2004)*, Long Beach, pp. 544–554, September 2004.
226. Simonsen, K., Suycott, M., Crumplar, R., and Sloat, S., LOCO GPSI: detection and location of GPS interference/jamming, in *Proc. 17th Inter. Tech. Mtg. Sat. Div. ION (ION GNSS 2004)*, Long Beach, pp. 555–560, September 2004.
227. Rodríguez, J.Á.Á., Irsigler, M., Hein, G.W., and Pany, T., Combined Galileo/GPS frequency and signal performance analysis, in *Proc. 17th Inter. Tech. Mtg. Sat. Div. ION (ION GNSS 2004)*, Long Beach, pp. 632–649, September 2004.
228. Ganguly, S., Jovancevic, A., and Noronha, J., Interoperability between GPS and Galileo, in *Proc. 17th Inter. Tech. Mtg. Sat. Div. ION (ION GNSS 2004)*, Long Beach, pp. 670–680, September 2004.
229. Bedrich, S., and Müncheberg, S., GNSS-based sensor fusion for telematic applications in railway traffic, in *Proc. 17th Inter. Tech. Mtg. Sat. Div. ION (ION GNSS 2004)*, Long Beach, pp. 681–691, September 2004.
230. Lee, Y.C., Investigation of extending receiver autonomous integrity monitoring (RAIM) to combined use of Galileo and modernized GPS, in *Proc. 17th Inter. Tech. Mtg. Sat. Div. ION (ION GNSS 2004)*, Long Beach, pp. 1691–1698, September 2004.
231. Blomenhofer, H., Ehret, W., Leonard, A., and Blomenhofer, E., GNSS/Galileo global and regional integrity performance analysis, in *Proc. 17th Inter. Tech. Mtg. Sat. Div. ION (ION GNSS 2004)*, Long Beach, pp. 2158–2168, September 2004.
232. Christie, J.R.I., Bentley, P.B., Ciboci, J.W., Gellrich, C.A., Gondek, J. M., Knoth, B.K., Ressler, M.B., Oetzel, G.N., GPS signal quality monitoring system, in *Proc. 17th Inter. Tech. Mtg. Sat. Div. ION (ION GNSS 2004)*, Long Beach, pp. 2239–2245, September 2004.
233. Bastide, F., Galileo E5a/E5b and GPS L5 acquisition time statistical characterization and application to civil aviation, in *Proc. 17th Inter. Tech. Mtg. Sat. Div. ION (ION GNSS 2004)*, Long Beach, pp. 2623–2635, September 2004.
234. Luo, M., Pullen, S., Ene, A., Qiu, D., Walter, T., and Enge, P., Ionosphere threat to LAAS: updated model, user impact, and mitigations, in *Proc. 17th Inter. Tech. Mtg. Sat. Div. ION (ION GNSS 2004)*, Long Beach, pp. 2771–2785, September 2004.
235. Gupta, S.K., and Louie, B.M., LADO (Launch, anomaly, and disposal operations), a new challenge for GPS, in *Proc. 17th Inter. Tech. Mtg. Sat. Div. ION (ION GNSS 2004)*, Long Beach, pp. 2219–2227, September 2004.
236. Monnerat, M., Couty, R., Vincent, N., Huez, O., and Chatre, E., The assisted GNSS, technology and applications, in *Proc. 17th Inter. Tech. Mtg. Sat. Div. ION (ION GNSS 2004)*, Long Beach, pp. 2479–2488, September 2004.
237. Fan, T., Lin, V.S., Wang, G.H., Maine, K.P., and Dafesh, P.A., The RF compatibility of flexible navigation signal combining methods, in *Proc. 2005 Nat. Tech. Mtg. ION*, San Diego, pp. 810–818, January 2005.

238. Partridge, M.D., and Dybdal, R.B., Design/calibration of GPS on-orbit monitoring capability, in *Proc. 2005 Nat. Tech. Mtg. ION*, San Diego, pp. 358–364, January 2005.
239. Pratt, A.R., and Owen, J.I.R., Galileo signal optimisation In L1, in *Proc. 2005 Nat. Tech. Mtg. ION*, San Diego, pp. 332–345, January 2005.
240. Van Dierendonck, A.J., GNSS user assessment of the plans and the benefits of GNSS modernized signals and services, in *Proc. 61st Ann. Mtg. ION*, Royal Sonesta Hotel, Cambridge, pp. 201–209, June 2005.
241. Nunes, F.D., Sousa, F.M.G., and Leitao, J.M.N., Improving multipath mitigation in GPS/Galileo BOC signals with gating functions, in *Proc. 61st Ann. Mtg. ION*, Cambridge, pp. 498–507, June 2005.
242. Shmihluk, K., Carroll, K., Celano, T., Biggs, C., and Powers, E., Enhanced LORAN implementation and evaluation for timing and frequency, in *Proc. 61st Ann. Mtg. ION*, Cambridge, pp. 379–385, June 2005.
243. Peterson, B.R., Pullen, S., Pervan, B., McGraw, G., Skidmore, T., and Anderson, S., Investigation of common architectures for land- and sea-based JPALS, in *Proc. 18th Inter. Tech. Mtg. Sat. Div. ION (ION GNSS 2005)*, Long Beach, pp. 26–37, September 2005.
244. Ripple, M.U., and Vidal, A., New U.S. GPS policy from a European perspective, in *Proc. 18th Inter. Tech. Mtg. Sat. Div. ION (ION GNSS 2005)*, Long Beach, pp. 407–413, September 2005.
245. Wallner, S., Hein, G.W., Pany, T., Avila-Rodriguez, J.-A., and Posfay, A., Interference computations between GPS and GALILEO, in *Proc. 18th Inter. Tech. Mtg. Sat. Div. ION (ION GNSS 2005)*, Long Beach, pp. 861–876, September 2005.
246. Hein, G.W., Avila-Rodriguez, J.-A., Ries, L., Lestarquit, L., Issler, J.-L., Godet, J., and Pratt, T., A candidate for the Galileo L1 OS optimized signal, in *Proc. 18th Inter. Tech. Mtg. Sat. Div. ION (ION GNSS 2005)*, Long Beach, pp. 833–845, September 2005.
247. Normark, P.-L., and Ståhlberg, C., Hybrid GPS/Galileo real time software receiver, in *Proc. 18th Inter. Tech. Mtg. Sat. Div. ION (ION GNSS 2005)*, Long Beach, pp. 1906–1913, September 2005.
248. Blunt, P., Ebinuma, T., Hodgart, S., and Unwin, M., A demonstration of Galileo transmitter / receiver architecture for space applications, in *Proc. 18th Inter. Tech. Mtg. Sat. Div. ION (ION GNSS 2005)*, Long Beach, pp. 1914–1921, September 2005.
249. Paynter, G., and Banas, V., Compliance of NovAtel's GPS-702L antenna to the European Union's new WEEE (Waste Electrical and Electronic Equipment) and RoHS (Restriction of the use of Certain Hazardous Substances) Directives, in *Proc. 18th Inter. Tech. Mtg. Sat. Div. ION (ION GNSS 2005)*, Long Beach, pp. 2253–2262, September 2005.
250. Li, B., Dempster, A., Rizos, C., Lee, H.K., and Li, D., Application of interpolation to mobile phone fingerprinting for positioning, in *Proc 18th Inter. Tech. Mtg. Sat. Div. ION (ION GNSS 2005)*, Long Beach, pp. 2268–2276, September 2005.
251. Pratt, A.R., and Owen, J.I.R., Signal multiplex techniques in satellite channel availability possible applications to Galileo, in *Proc. 18th Inter. Tech. Mtg. Sat. Div. ION (ION GNSS 2005)*, Long Beach, pp. 2448–2460, September 2005.
252. Rebeyrol, E., Macabiau, C., Ries, L., Issler, J.-L., Bousquet, M., Boucheret, M.-L., Interplex modulation for navigation systems at the L1 Band, in *Proc. 2006 Nat. Tech. Mtg. ION*, Monterey, pp. 100–111, January 2006.
253. Peng, C., Intelligence vehicle-base station navigation system VSB/BSV, in *Proc. 2006 Nat. Tech. Mtg. ION*, Monterey, pp. 813–824, January 2006.
254. Shen, X., and Gao, Y., Analyzing the impacts of Galileo and modernized GPS on precise point positioning, in *Proc. 2006 Nat. Tech. Mtg. ION*, Hyatt Regency Hotel, Monterey, pp. 837–846, January 2006.
255. Siu, P., Apker, C., McMillen, J., and Sorber, S., G-STAR GPS anti-jam technology, in *Proc. 2006 Nat. Tech. Mtg. ION*, Hyatt Regency Hotel, Monterey, pp. 1057–1063, January 2006.
256. Sousa, F.M., Nunes, F.D., and Leitao, J.M., Strobe pulse design for multipath mitigation in BOC GNSS receivers, in *Proc. IEEE/ION PLANS 2006*, San Diego, pp. 348–355, April 2006.

257. Ho, C.-S., Weng, C.-T., and Guan, M., A crosswind hazard study with using FDR Data, in *Proc. IEEE/ION PLANS 2006*, San Diego, pp. 818–822, April 2006.
258. Hein, G.W., Avila-Rodriguez, J.-A., Wallner, S., Pratt, A.R., Owen, J., Issler, J.-L., Betz, J.W., Hegarty, C.J., Lenahan, S., Rushanan, J.J., Kraay, A.L., Stansell, T.A., MBOC: the new optimized spreading modulation recommended for GALILEO L1 OS and GPS L1C, in *Proc. IEEE/ION PLANS 2006*, San Diego, pp. 883–892, April 2006.
259. Kelly, J.T., Chapman, M.D., Barton, B., Handheld, low cost situational awareness using existing military fielded equipment, in *Proc. 19th Inter. Tech. Mtg. Sat. Div. ION (ION GNSS 2006)*, Fort Worth, pp. 182–188, September 2006.
260. Alcantarilla, I., Porras, D., Tajdine, A., Zarraoa, N., and Lévy, J.C., The benefits of multi-constellation GNSS augmentations, in *Proc. 19th Inter. Tech. Mtg. Sat. Div. ION (ION GNSS 2006)*, Fort Worth, pp. 930–938, September 2006.
261. Lee, Y.-W., Suh, Y., and Shibasaki, R., Simulation-based estimation of multipath mitigation using 3D-GIS and spatial statistics, in *Proc. 19th Inter. Tech. Mtg. Sat. Div. ION (ION GNSS 2006)*, Fort Worth, pp. 1778–1783, September 2006.
262. Betz, J., Blanco, M.A., Cahn, C.R., Dafesh, P.A., Hegarty, C.J., Hudnut, K.W., Kasemsri, V., Keegan, R., Kovach, K., Lenahan, L.S., Ma, H.H., Rushanan, J., Rushanan, J.J., Sklar, D., Stansell, T.A., Wang, C.C., and Yi, S.K., Description of the L1C signal, in *Proc. 19th Inter. Tech. Mtg. Sat. Div. ION (ION GNSS 2006)*, Fort Worth, pp. 2080–2091, September 2006.
263. Mongrédien, C., Lachapelle, G., and Cannon, M.E., Testing GPS L5 acquisition and tracking algorithms using a hardware simulator, in *Proc. 19th Intern. Tech. Mtg. Sat. Div. ION (ION GNSS 2006)*, Fort Worth, pp. 2901–2913, September 2006.
264. Leite, N.P.O., and Walter, F., Static and dynamic performance of a novel GPS attitude determination algorithm (GADA), in *Proc. 2007 Nat. Tech. Mtg. ION*, San Diego, pp. 161–171, January 2007.
265. Sakai, T., Fukushima, S., Takeichi, N., and Ito, K., Augmentation performance of QZSS L1-SAIF signal, in *Proc. 2007 Nat. Tech. Mtg. ION*, San Diego, pp. 411–421, January 2007.
266. DiEsposti, R., and Hsu, K., Proposed operations concepts and flexible UE architectures for modernized user equipment SIS utilization for transition from test mode to IOC through FOC, in *Proc. 2007 Nat. Tech. Mtg. ION*, San Diego, pp. 548–560, January 2007.
267. Vo, A., Falchetti, C.R., Morrison, A.W., ADAP: enhancing GPS protection for Navwar, in *Proc. 2007 Nat. Tech. Mtg. ION*, San Diego, pp. 990–997, January 2007.
268. Rushanan, J.J., The spreading and overlay codes for the L1C signal, *Navigation*, vol. 54, no. 1, pp. 43–51, spring 2007.
269. Hein, G.W., Towards a GNSS system of systems, in *Proc. 20th Inter. Tech. Mtg. Sat. Div. ION (ION GNSS 2007)*, Fort Worth Convention Center, Fort Worth, pp. 22–30, September 2007 (Plenary Presentation).
270. Avila-Rodriguez, J.-A., Wallner, S., Hein, G.W., Eissfeller, B., Irsigler, M., and Issler, J.-L., A vision on new frequencies, signals and concepts for future GNSS systems, *Proc. 20th Inter. Tech. Mtg. Sat. Div. ION (ION GNSS 2007)*, Fort Worth, pp. 517–534, September 2007.
271. Soellner, M., Briechle, C., Hechenblaikner, G., Kaindl, M., Kohl, R., Lindemer, W., Middendorf, M., and Zecha, C., The BayNavTech(TM) signal experimentation facility (BaySEF(TM)) is ready for assessing GNSS signal performance, in *Proc. 20th Inter. Tech. Mtg. Sat. Div. ION (ION GNSS 2007)*, Fort Worth, pp. 1065–1072, September 2007.
272. Soualle, F., and Burger, T., Radio frequency compatibility criterion for code tracking performance, in *Proc. 20th Inter. Tech. Mtg. Sat. Div. ION* (*ION GNSS 2007*), Fort Worth Convention Center, Fort Worth, pp. 1201–1210, September 2007.
273. Brown, A., and Ganguly, S., Advanced family of software simulators, in *Proc. 20th Inter. Tech. Mtg. Sat. Div. ION (ION GNSS 2007)*, Fort Worth, pp. 1265–1276, September 2007.
274. Avila-Rodriguez, J.-A., Hein, G.W., Wallner, S., Issler, J.-L., Ries, L., Lestarquit, L., de Latour, A., Godet, J., Bastide, F., Pratt, T., and Owen, J., The MBOC modulation: the final touch to the Galileo frequency and signal plan, in *Proc. 20th Inter. Tech. Mtg. Sat.*

Div. ION (ION GNSS 2007), Fort Worth Convention Center, Fort Worth, pp. 1515–1529, September 2007 (Published in NAVIGATION).

275. Amin, B., Jitter analysis of QPSK and BOC(n,n) GNSS signals, in *Proc. 20th Inter. Tech. Mtg. Sat. Div. ION (ION GNSS 2007)*, Fort Worth, pp. 1543–1548, September 2007.
276. Wallner, S., Avila-Rodriguez, J.-A., Hein, G.W., and Rushanan, J.J., Galileo E1 OS and GPS L1C pseudo random noise codes – requirements, generation, optimization and comparison, in *Proc. 20th Inter. Tech. Mtg. Sat. Div. ION (ION GNSS 2007)*, Fort Worth, pp. 1549–1563, September 2007.
277. Kaindl, M., Soellner, M., Zecha, C., and Kohl, R., Performance analysis of GIOVE-A signals in comparison with GPS based on wideband measurements with the BayNavTechTM signal evaluation facility (BaySEFTM), in *Proc. 20th Inter. Tech. Mtg. Sat. Div. ION (ION GNSS 2007)*, Fort Worth, pp. 1588–1595, September 2007.
278. Kogure, S., Kishimoto, M., Sawabe, M., and Terada, K., Introduction of IS-QZSS (interface specification for Quasi Zenith Satellite System), in *Proc. 20th Inter. Tech. Mtg. Sat. Div. ION (ION GNSS 2007)*, Fort Worth, pp. 3008–3014, September 2007.
279. Grelier, T., Ghion, A., Dantepal, J., Ries, L., DeLatour, A., Issler, J.-L., Avila-Rodriguez, J.A., Wallner, S., and Hein, G.W., Compass signal structure and first measurements, in *Proc. 20th Inter. Tech. Mtg. Sat. Div. ION (ION GNSS 2007)*, Fort Worth, pp. 3015–3024, September 2007.
280. Chiu, D.S., MacGougan, G., and O'Keefe, K., UWB assisted GPS RTK in hostile environments, in *Proc. 2008 Nat. Tech. Mtg. ION*, San Diego, p. 532, January 2008.
281. Sakai, T., Fukushima, S., Takeichi, N., and Ito, K., Implementation of the QZSS L1-SAIF message generator, in *Proc. 2008 Nat. Tech. Mtg. ION*, San Diego, p. 464, January 2008.
282. Renaudin, V., Merminod, B., and Kasser, M., Optimal data fusion for pedestrian navigation based on UWB and MEMS, in *Proc. IEEE/ION PLANS 2008*, Monterey, pp. 341–349, May 2008.
283. Kim, J.I., Lee, J.G., and Gee, G.I., Location determination in WiBro (Wireless Broadband) system, in *Proc. IEEE/ION PLANS 2008*, Monterey, pp. 832–837, May 2008.
284. Nosenko, Y., GLONASS in a multi-GNSS world, in *Proc. 21st Inter. Tech. Mtg. Sat. Div. ION (ION GNSS 2008)*, Savannah, pp. 7–19, September 2008.
285. Verhoef, P., European GNSS programmes and the future of GNSS, in *Proc. 21st Inter. Tech. Mtg. Sat. Div. ION (ION GNSS 2008)*, Savannah, pp. 20–29, September 2008.
286. Kim, D., and Langley, R.B., Toward the ultimate RTK: the last challenges in long-range real-time kinematic applications, in *Proc. 21st Inter. Tech. Mtg. Sat. Div. ION (ION GNSS 2008)*, Savannah, pp. 385–396, September 2008.
287. Terada, K., Quasi-Zenith Satellite System (QZSS) program update, in *Proc. 21st Inter. Tech. Mtg. Sat. Div. ION (ION GNSS 2008)*, Savannah, pp. 493–527, September 2008.
288. Wallner, S., Avila-Rodriguez, J.-A., Won, J.-H., Hein, G., and Issler, J.-L., Revised PRN code structures for Galileo E1 OS, in *Proc. 21st Inter. Tech. Mtg. Sat. Div. ION (ION GNSS 2008)*, Savannah, pp. 887–899, September 2008.
289. Lawrence, D., Wide area augmentation system (WAAS) program status, in *Proc. 21st Inter. Tech. Mtg. Sat. Div. ION (ION GNSS 2008)*, Savannah, pp. 1009–1031, September 2008.
290. Gadimova, S., International committee on global navigation satellite systems (ICG): 2008 activities, in *Proc. 21st Inter. Tech. Mtg. Sat. Div. ION (ION GNSS 2008)*, Savannah, pp. 1994–2003, September 2008.
291. Manandhar, D., Okano, K., Ishii, M., Torimoto, H., Kogure, S., and Maeda, H., Development of ultimate seamless positioning system based on QZSS IMES, in *Proc. 21st Inter. Tech. Mtg. Sat. Div. ION (ION GNSS 2008)*, Savannah, pp. 1698–1705, September 2008.
292. Schmitz-Peiffer, A., Stopfkuchen, L., Soualle, F., Floch, J-J., King, R., Fernandez, A., Jorgensen, R., Eissfeller, B., Rodriguez, J.A., Wallner, S., Won, J-H., Pany, T., Anghileri, M., Lankl, B., Schuler, T., and Colzi, E., Assessment on the use of C-Band for GNSS within the European GNSS evolution programme, in *Proc. 21st Inter. Tech. Mtg. Sat. Div. ION (ION GNSS 2008)*, Savannah, pp. 2189–2198, September 2008.

293. Avila-Rodriguez, J.-A., Wallner, S., Won, J.-H., Eissfeller, B., Schmitz-Peiffer, A., Floch, J.-J., Colzi, E., and Gerner, J.-L., Study on a Galileo signal and service plan for C-band, in *Proc. 21st Inter. Tech. Mtg. Sat. Div. ION* (*ION GNSS 2008*), Savannah, pp. 2515–2529, September 2008.
294. Catalán, C., Hernández, C., Mozo, A., Fernández, L., and Amarill, F., Improved integrity concept for Future GNSS evolutions, in *Proc. 21st Inter. Tech. Mtg. Sat. Div. ION (ION GNSS 2008)*, Savannah, pp. 2547–2557, September 2008.
295. Huang, Y.-S., and Tsai, M.-L., The impact of Compass/Beidou-2 on future GNSS: A perspective from Asia, in *Proc. 21st Inter. Tech. Mtg. Sat. Div. ION (ION GNSS 2008)*, Savannah, pp. 2227–2238, September 2008.
296. Avila-Rodriguez, J.-A., Hein, G.W., Wallner, S., Issler, J.-L., Ries, L., Lestarquit, L., de Latour, A., Godet, J., Bastide, F., Pratt, T., and Owen, J., The MBOC modulation: the final touch to the Galileo frequency and signal plan, *Navigation*, vol. 55, no. 1, pp. 14–28, Spring 2008.
297. Park, B., and Kee, C., Temporal and spatial decorrelation error reduction by a compact network RTK, in *Proc. 2009 Inter. Tech. Mtg. ION*, Anaheim, pp. 341–352, January 2009.
298. Sakai, T., Yoshihara, T., Fukushima, S., and Ito, K., The ionospheric correction processor for SBAS and QZSS L1-SAIF, in *Proc. 2009 Inter. Tech. Mtg. ION*, Anaheim, pp. 312–323, January 2009.
299. Cerruti, A.P., Rushanan, J.J., and Winters, D.W., Modeling C/A on C/A interference, in *Proc. 2009 Inter. Tech. Mtg. ION*, Anaheim, pp. 142–156, January 2009.
300. Simsky, A., De Wilde, W., Willems, T., Mertens, D., Koitsalu, E., and Sleewaegen, J-M., First field experience with L5 signals: DME Interference Reality Check, in *Proc. 22nd Inter. Tech. Mtg. Sat. Div. ION (ION GNSS 2009)*, Savannah, pp. 29–37, September 2009.
301. Savasta, S., Lo Presti, L., Dovis, F., and Margaria, D., Trustworthiness GNSS signal validation by a time-frequency approach, in *Proc. 22nd Inter. Tech. Mtg. Sat. Div. ION (ION GNSS 2009)*, Savannah, pp. 66–75, September 2009.
302. Raghavan, S., and Powell, T., A simple approach to obtain the CA code spectral separation coefficient, in *Proc. 22nd Int. Tech. Mtg. Sat. Div. ION (ION GNSS 2009)*, Savannah, pp. 86–94, September 2009.
303. Iwata, T., Toda, K., Kondoh, Y., Yamamoto, T., Kakinuma, M., and Kumagai, S., Dual-frequency spaceborne GPS receiver for the advanced land observing satellite (ALOS): design and flight results, in *Proc. 22nd Inter. Tech. Mtg. Sat. Div. ION (ION GNSS 2009)*, Savannah, pp. 1404–1412, September 2009.
304. Liu, W., Zhai, C., Zhang, Y., and Zhan, X., Simulation analysis of GPS/Galileo/Compass radio frequency compatibility, in *Proc. 22nd Inter. Tech. Mtg. Sat. Div. ION* (*ION GNSS 2009*), Savannah, pp. 1552–1560, September 2009.
305. Mateu, I., Boulanger, C., Issler, J.-L., Ries, L., Avila-Rodriguez, J.-A., Wallner, S., Kraus, T., Eissfeller, B., Mulassano, P., Caporale, M., Germaine, S., Guyomard, J.-Y., Bastide, F., Godet, J., Hayes, D., Serant, D., Thevenon, P., Julien, O., and Hein, G.W., Exploration of possible GNSS signals in S-band, in *Proc. 22nd Inter. Tech. Mtg. Sat. Div. ION* (*ION GNSS 2009*), Savannah, pp. 1573–1587, September 2009.
306. Yin, X., Kou, Y., and Zhang, Z., Design and implementation of a flexible software-based GNSS IF signal simulator, in *Proc. 22nd Inter. Tech. Mtg. Sat. Div. ION (ION GNSS 2009)*, Savannah, pp. 2229–2240, September 2009.
307. O'Brien, A.J., Hayhurst, K., and Gupta, I.J., Effects of rotor blade modulation on GNSS receiver measurements, in *Proc. 22nd Inter. Tech. Mtg. Sat. Div. ION (ION GNSS 2009)*, Savannah, pp. 2352–2361, September 2009.
308. Li, X., Zhang, X., and Guo, F., Study on precise point positioning based on combined GPS and GLONASS, in *Proc. 22nd Inter. Tech. Mtg. Sat. Div. ION (ION GNSS 2009)*, Savannah, pp. 2449–2459, September 2009.
309. Robertson, G.J., Kieffer, R., Malik, M., Gatti, G., Alpe, V., and Johansson, M., GIOVE-B satellite design and performance validation, in *Proc. 22nd Inter. Tech. Mtg. Sat. Div. ION (ION GNSS 2009)*, Savannah, pp. 3008–3016, September 2009.

310. Ohshima, Y., Kawaguchi, Y., Takahashi, T., Soga, H., Moriguchi, T., Noda, H., Kogure, S., and Kishimoto, M., Navigation payload test of proto-flight model (PFM) for the high-accuracy positioning experimental system (HIAPEX) for the Quasi-Zenith Satellite System (QZSS), in *Proc. 22nd Inter. Tech. Mtg. Sat. Div. ION (ION GNSS 2009)*, Savannah, pp. 3285–3289, September 2009.
311. Hama, S., Takahashi, Y., Amagai, J., Fujieda, M., Nakamura, M., Takahashi, T., and Horiuchi, S., Development of the time management system, in *Proc. 22nd Inter. Tech. Mtg. Sat. Div. ION (ION GNSS 2009)*, Savannah, pp. 3338–3343, September 2009.
312. Inaba, N., Quasi-Zenith satellite system program update, in *Proc. 22nd Inter. Tech. Mtg. Sat. Div. ION* (*ION GNSS 2009*), Savannah, pp. 3411–3436, September 2009.
313. Sakai, T., Fukushima, S., and Ito, K., Recent development of QZSS L1-SAIF master station, in *Proc. 2010 Inter. Tech. Mtg. ION*, San Diego, pp. 354–364, January 2010.
314. Shibata, T., and Maeda, H., Extended theory of spectral separation coefficient for GNSS signal interference, in *Proc. 2010 Inter. Tech. Mtg. ION*, San Diego, pp. 930–940, January 2010.
315. Paonni, M., Anghileri, M., Wallner, S., Avila-Rodriguez, J.-A., and Eissfeller, B., Performance assessment of GNSS signals in terms of time to first fix for cold, warm and hot start, in *Proc. 2010 Inter. Tech. Mtg. ION*, San Diego, pp. 1051–1066, January 2010.
316. Time Division Duplex (TDD) vs Frequency Division Duplex (FDD) in Wireless Backhauls, White Paper. http://www.netkrom.com/support/whitepapers/TDD_vs_FDD_in_wireless_backhaul_white_paper.pdf
317. MIMO: The Paper War. http://www.dailywireless.org/2010/05/13/mimo-the-paper-war/
318. Skalski, H., Mayne, M., and Fauver, B., Future worldwide position, velocity, and timing (PVT) requirements and acquisition strategies, in *Proc. 54th Ann. Mtg. ION*, Denver, pp. 53–63, June 1998.
319. Silvy, O., and Lavroff, J.L., MAGNET: a European multi-modal approach of space-based navigation, in *Proc. 54th Ann. Mtg. ION*, Denver, pp. 113–119, June 1998.
320. Kalafus, R.M., Interference to GPS receivers from mobile satellite emissions, in *Proc. 11th Inter. Tech. Mtg. Sat. Div. ION (ION GPS 1998)*, Nashville, pp. 827–832, September 1998.
321. Lucia, D.J., and Anderson, J., Analysis and recommendation for the reuse of the L1 and L2 GPS spectrum, in *Proc.11th Inter. Tech. Mtg. Sat. Div. ION* (ION GPS 1998), Nashville, pp. 1877–1886, September 1998.
322. Martens, D., and Latterman, D., Stewardship – a growing need to manage a national asset, in *Proc. 11th Inter. Tech. Mtg. Sat. Div. ION (ION GPS 1998)*, Nashville, pp. 1433–1448, September 1998.
323. Dai, D., Walter, T., Enge, P., and Powell, J.D., Interoperation of distributed SBASS : theory, experience and future perspectives, in *Proc. 11th Inter. Tech. Mtg. Sat. Div. ION (ION GPS 1998)*, Nashville, pp. 1355–1364, September 1998.
324. Dafesh, P.A., Parker, A., Holmes, J.K., and Raghavan, S.H., Simulation of the Code tracking performance of square-wave modulated ranging codes, in *Proc. 11th Inter. Tech. Mtg. Sat. Div. ION (ION GPS 1998)*, Nashville, pp. 633–641, September 1998.
325. Fuller, R., Dai, D., Walter, T., Comp, C., Enge, P., and Powell, J.D., Interoperation and integration of satellite based augmentation systems, in *Proc. 11th Inter. Tech. Mtg. Sat. Div. ION (ION GPS 1998)*, Nashville, pp. 121–130, September 1998.
326. Lucia, D.J., and Anderson, J.M., Analysis and recommendation for reuse of the L1 and L2 GPS spectrum, *Navigation*, vol. 45, no. 4, pp. 251–264, win. 1998–1999.
327. Kozlov, D., and Tkachenko, M., Centimeter-level, real time kinematic positioning with GPS + GLONASS C/A receivers, *Navigation*, vol. 45, no. 2, pp. 137–160, sum. 1998.
328. Dafesh, P.A., Nguyen, T.M., Lazar, S., Coherent adaptive subcarrier modulation (CASM) for GPS modernization, in *Proc. 1999 Nat. Tech. Mtg. ION*, San Diego, pp. 649–660, January 1999.
329. Schweikert, R., Woerz, T., and De Gaudenzi, R., On new signal structures for GNSS-2, in *Proc. 1999 Nat. Tech. Mtg. ION*, San Diego, pp. 109–118, January 1999.

330. Van Dierendonck, A.J., Spilker, J.J., Jr., Proposed civil GPS signal at 1176.45 MHz: In-phase/quadrature codes at 10.23 MHz chip rate, in *Proc. 55th Ann. Mtg. ION*, Cambridge, pp. 761–770, June 1999.
331. McDonald, K.D., Performance improvements to GPS in the decade 2000–2010, in *Proc. 55th Ann. Mtg. ION*, Cambridge, pp. 1–15, June 1999.
332. Hegarty, C., Kim, T., Ericson, S., Reddan, P., Morrissey, T., and Van Dierendonck, A.J., Methodology for determining compatibility of GPS L5 with existing systems and preliminary results, in *Proc. 55th Ann. Mtg. ION*, Royal Sonesta Hotel, Cambridge, pp. 635–644, June 1999.
333. Holmes, J.K., Raghavan, S., Dafesh, P., and Lazar, S., Effective signal to noise ratio performance comparison of some GPS modernization signals, in *Proc. 12th Inter. Tech. Mtg. Sat. Div. ION (ION GPS 1999)*, Nashville, pp. 1755–1762, September 1999.
334. Lucas, R., and Ludwig, D., GALILEO: system requirements and architectures, in *Proc. 12th Inter. Tech. Mtg. Sat. Div. ION (ION GPS 1999)*, Nashville, pp. 2097–2102, September 1999.
335. Lazar, S., Clark, J., and Turner, D., Signal design guidelines for navigation satellite system design, in *Proc. 12th Intern. Tech. Mtg. Sat. Div. ION (ION GPS 1999)*, Nashville, pp. 2079–2086, September 1999.
336. Wolfrum, J., Healy, M., Provenzano, J.P., and Sassorossi, T., Galileo – Europe's contribution to the next generation of GNSS, in *Proc. 12th Inter. Tech. Mtg. Sat. Div. ION (ION GPS 1999)*, Nashville, pp. 1381–1390, September 1999.
337. Kelley, C.W., Martoccia, D., and Pendley, R., A modernization deployment strategy to meet military and civil needs, in *Proc. 12th Inter. Tech. Mtg. Sat. Div. ION (ION GPS 1999)*, Nashville, pp. 1343–1352, September 1999.
338. Skidmore, T.A., Nyhus, O.K., and Wilson, A.A., An overview of the LAAS VHF data broadcast, in *Proc. 12th Inter. Tech. Mtg. Sat. Div. ION (ION GPS 1999)*, Nashville, pp. 671–680, September 1999.
339. Garin, L.J., Chansarkar, M., Miocinovic, S., Norman, C., and Hilgenberg, D., Wireless assisted GPS-SiRF architecture and field test results, in *Proc. 12th Inter. Tech. Mtg. Sat. Div. ION (ION GPS 1999)*, Nashville, pp. 489–498, September 1999.
340. Vincent, J.P., Brunel, F., Silvy, O., Bruce, A.S., Gu, X., Wippich, H.-G., Caignault, A., and Lavroff, J.-L., Testing MAGNET receiver under operational conditions, in *Proc. 12th Inter. Tech. Mtg. Sat. Div. ION (ION GPS 1999)*, Nashville, pp. 259–268, September 1999.
341. Hegarty, C., and Van Dierendonck, A.J., Civil GPS/WAAS signal design and interference environment at 1176.45 MHz: results of RTCA SC159 WG1 activities, in *Proc. 12th Inter. Tech. Mtg. Sat. Div. ION* (*ION GPS 1999*), Nashville, pp. 1727–1736, September 1999.
342. Guérin, A., Chouffot, M., and Lebée, J.P., Towards a dual civil/military and secure effective approach of Global Radio-Navigation by Satellite Systems, in *Proc. 12th Inter. Tech. Mtg. Sat. Div. ION (ION GPS 1999)*, Nashville, pp. 1369–1376, September 1999.
343. Simpson, P., Mac Millan, A., and Chasko, A., Inverted range for GPS modernization testing, in *Proc. 2000 Nat. Tech. Mtg. ION*, Pacific Hotel Disneyland, Anaheim, pp. 636–645, January 2000.
344. Correia, J.T., Blanchard, J.J., and Fine, P.B., A hardware testbed for evaluation of the GPS modernization modulation candidates, in *Proc. 2000 Nat. Tech. Mtg. ION*, Anaheim, pp. 594–604, January 2000.
345. Gary, B.L., Tutorial on airborne microwave temperature profilers, chapter 5, Atmospheric emission sources. http://brucegary.net/MTP_tutorial/MTP_ch5.html

Chapter 4
Adaptive Array Algorithms for Geolocation of RF Signals

4.1 Introduction

Adaptive array algorithms for geolocation of RF signals present an overview of the best practices and innovative techniques in the art and science of adaptive array algorithms for geolocation of RF signals over the last 20 years in the literature of adaptive array algorithms for beamforming, interference cancelation, source location, etc. [1–58]. It covers all insights and aspects including theoretical analysis, RF signals, signal techniques, key block diagrams, and practical principle signal interpretations in the frequency band from 100 MHz to 66 GHz. Dr. Progri reveals the research and development process by demonstrating how to understand and explain a good number of adaptive algorithms such as those used in wireless networks, mobile phones (or cellular networks), indoor geolocation systems, AM and FM radio, two-way radio, satellite radio, TV broadcasting, satellite TV broadcasting, digital video broadcasting, Global Navigation Satellite Systems (GNSS), etc., from basic diagrams to be utilized for the principle simulation examples and make recommendations for the future final products of geolocation of RF signals [1–58]. Starting with Geolocation of RF Signals Main Adaptive Algorithm Description in Sect. 4.2, the chapter progressively examines various geolocation of RF signals best adaptive algorithm practices in Sect. 4.3, and then continues with the best blind adaptive algorithm per band and per application to achieve required performance objectives of up to 0° precision. Next follows a step-by-step approach of algorithm description and principle simulation test cases (or scenarios) from indoor to satellite environment, and concludes with recommendations on state-of-the-art geolocation implementation as well as advanced features found in signal generator instruments to be discussed in Chap. 6. This chapter includes the best mathematical techniques employed for geolocation of RF signals from 100 MHz to 66 GHz. The principle simulation examples, discussed in great detail in this chapter during the second part of the book, utilize a great deal of signal design knowledge accumulated in Chap. 3. This chapter offers invaluable insights on adaptive algorithm description, development, analysis, and simulation that are not found in any other adaptive antenna array textbook or manual, all in one source for the beginner, the experienced, expert analysts and professionals. This chapter is also the beginning of part II of the book,

I. Progri, *Geolocation of RF Signals*,
DOI 10.1007/978-1-4419-7952-0_4, © Springer Science+Business Media, LLC 2011

which is on best mathematical techniques and methods on Geolocation of RF signals: Principles and Simulations.

A great deal of discussion on this chapter is dedicated to adaptive antenna array employing a blind adaptive algorithm which can be exploited to extract signals with unknown characteristics coming from unknown locations based only on very limited knowledge of the received signal properties. These signals may be RF sources of interference to a desired GPS signal, Mobile phone, wireless network, two-way radio, satellite TV, FM station etc. and whose locations might be determined once these signals are extracted.

In this chapter we present the advantages and disadvantages of exploiting a blind adaptive algorithm both in the time and frequency domains. Simulation results are grouped into three main sections: (1) 2001 Simulation Test Setup in Sect. 4.6; (2) 2002 Simulation Test Setup Sect. 4.7, and 2010 Simulation Test Setup in Sect. 4.8 illustrate the performance of the blind algorithm by comparing the extracted signals with the original signals for very simple signal designs in 2001 and 2002 and more contemporary signal designs in 2010 and the estimated signal locations with the corresponding actual signal locations up to 0° precision.

4.2 Geolocation of RF Signals Main Principles

This section is the core foundation of the entire book as illustrated in Fig. 4.1 and yet I had not seen until I developed this section's material that clearly explains the main principles of operations as presented in Sect. 4.2.1 and the Cramér-Rao lower

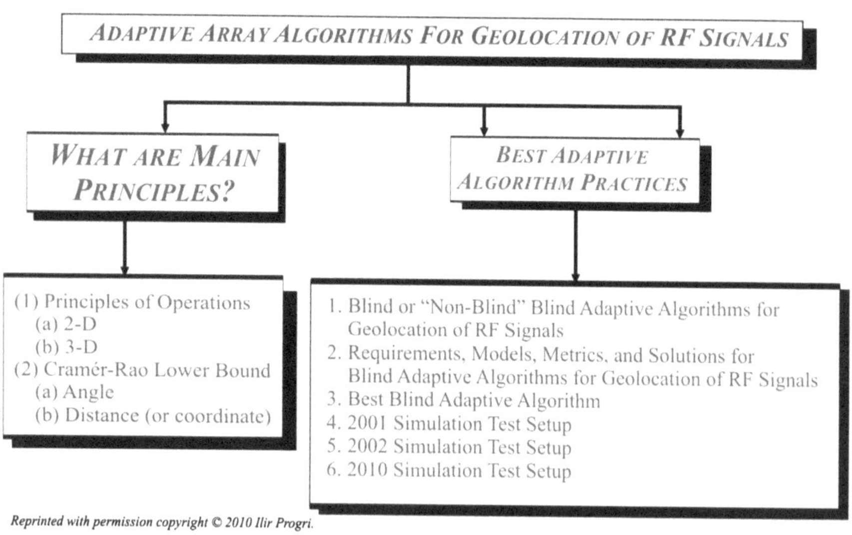

Fig. 4.1 Adaptive array algorithm for geolocation of RF signals. Reprinted with permission © 2010 Ilir Progri

bound, the Cramér-Rao lowest possible bound on angle accuracy as illustrated in Sect. 4.2.2 and the best adaptive algorithm practices Sect. 4.5 through Sect. 4.8.

We begin our discussion with the main principles of operations in Sect. 4.2.1.

4.2.1 Principles of Operations

In this section we are going to explain in detail the principles of operations. How we obtain an accurate elevation and azimuth angle of arrival estimation is the discussion that is derived later in Sect. 4.5.1. Therefore the details of the first and the second principle of operations that we initially discussed in Chap. 1 are found in Sect. 4.5.1. Here we are going to derive the third principle of operations which is that assuming that the azimuth and/or elevation angles are estimated very accurately then with the help of at least two antenna arrays located in a distance d from each-other in two configurations: (a) two dimensional (2D) principle of operations distance estimation in Sect. 4.2.1.1; (b) three-dimensional (3D) principle of operations distance estimation in Sect. 4.2.1.2.

4.2.1.1 (2D) Principle of Operations' Distance Estimation

The (2D) principle of operations' distance estimation is illustrated in Fig. 4.2. As shown in Fig. 4.2 the (2D) principle of operations' distance estimation we have two antenna arrays located in $A(A_x, A_y)$ and $B(B_x, B_y)$. Let us assume that we have two RF sources $C(C_x, C_y)$ and $D(D_x, D_y)$ which are not on the same line as A and B line as depicted in Fig. 4.2. Without loss of generality let us assume that the center of the 2D coordinate system is equally spaced between the A and B and that the x axis is the AB line as illustrated in Fig. 4.2. If the distance between the two antenna arrays is d then let us denote that half the distance by $x = d/2$. Therefore, the location of the two antenna arrays can be rewritten as $A(-x, 0)$ and $B(x, 0)$.

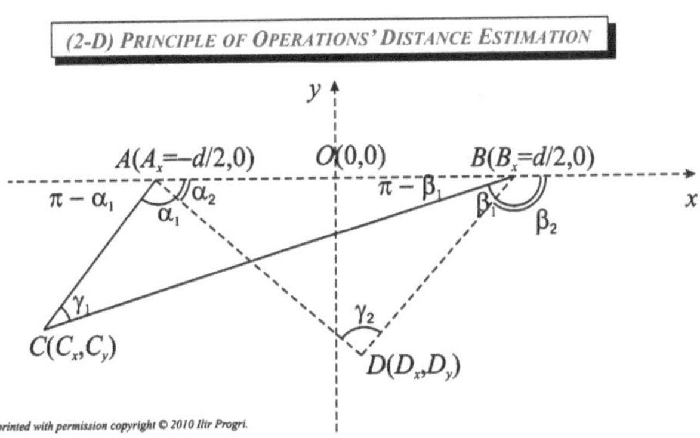

Fig. 4.2 (2D) principle of operations' distance estimation. Reprinted with permission © 2010 Ilir Progri

Because we know the direction of arrival from the two RF sources C and D that means that we know the angles (α_1, α_2) and the angles (β_1, β_2) which could be either azimuth or elevation angles depending on the application and on the configuration. Given this information we need to estimate the location of $C(C_x, C_y)$ and $D(D_x, D_y)$ which is simply a trigonometry solution. Let us develop then these solutions. Writing the sine law solution for the ACB triangle we have the following

$$\frac{d}{\sin\gamma_1} = \frac{AC}{\sin(\pi - \beta_1)} = \frac{BC}{\sin\alpha_1} \quad \text{which is the same as} \quad \frac{d}{\sin\gamma_1} = \frac{AC}{\sin\beta_1} = \frac{BC}{\sin\alpha_1}. \tag{4.1}$$

Therefore, the AC and BC distances are estimated as

$$AC = d\frac{\sin\beta_1}{\sin\gamma_1} = d\frac{\sin\beta_1}{\sin(\beta_1 - \alpha_1)} \quad \text{and} \quad BC = d\frac{\sin\alpha_1}{\sin\gamma_1} = d\frac{\sin\alpha_1}{\sin(\beta_1 - \alpha_1)}. \tag{4.2}$$

Therefore the coordinates for the RF source $C(C_x, C_y)$ are as follows:

$$\begin{aligned} C_x &= -AC\cos(\pi - \alpha_1) - \frac{d}{2} = AC\cos\alpha_1 - \frac{d}{2} = d\left[\frac{\sin\beta_1\cos\alpha_1}{\sin(\beta_1 - \alpha_1)} - \frac{1}{2}\right] \quad \text{and} \\ C_y &= -AC\sin(\pi - \alpha_1) = -AC\sin\alpha_1 = -d\frac{\sin\beta_1\sin\alpha_1}{\sin(\beta_1 - \alpha_1)}, \end{aligned} \tag{4.3}$$

$$\begin{aligned} C_x &= -BC\cos(\pi - \beta_1) + \frac{d}{2} = BC\cos\beta_1 + \frac{d}{2} = d\left[\frac{\sin\beta_1\cos\beta_1}{\sin(\beta_1 - \alpha_1)} + \frac{1}{2}\right] \quad \text{and} \\ C_y &= -BC\sin(\pi - \beta_1) = -BC\sin\beta_1 = -d\frac{\sin\beta_1\sin\beta_1}{\sin(\beta_1 - \alpha_1)} = -d\frac{\sin^2\beta_1}{\sin(\beta_1 - \alpha_1)} \end{aligned} \tag{4.4}$$

and $D(D_x, D_y)$

$$\begin{aligned} D_x &= AD\cos\alpha_2 - \frac{d}{2} = AD\cos\alpha_2 - \frac{d}{2} = d\left[\frac{\sin\beta_2\cos\alpha_2}{\sin(\beta_2 - \alpha_2)} - \frac{1}{2}\right] \quad \text{and} \\ D_y &= -AD\sin(\pi - \alpha_2) = -AD\sin\alpha_2 = -d\frac{\sin\beta_2\sin\alpha_2}{\sin(\beta_2 - \alpha_2)}, \end{aligned} \tag{4.5}$$

$$\begin{aligned} D_x &= -BD\cos(\pi - \beta_2) + \frac{d}{2} = BD\cos\beta_2 + \frac{d}{2} = d\left[\frac{\sin\beta_2\cos\beta_2}{\sin(\beta_2 - \alpha_2)} + \frac{1}{2}\right] \quad \text{and} \\ D_y &= -BD\sin(\pi - \beta_2) = -BD\sin\beta_2 = -d\frac{\sin\beta_2\sin\beta_2}{\sin(\beta_2 - \alpha_2)} = -d\frac{\sin^2\beta_2}{\sin(\beta_2 - \alpha_2)}. \end{aligned} \tag{4.6}$$

This concludes the (2D) principle of operations distance estimation problem. Next we exploit the (3D) principle of operations' distance estimation problem.

4.2.1.2 (3D) Principle of Operations' Distance Estimation

The (3D) principle of operations' distance estimation is illustrated in Fig. 4.3. As shown in Fig. 4.3 the (3D) principle of operations' distance estimation we have two antenna arrays located in $A(A_x, A_y, A_z)$ and $B(B_x, B_y, B_z)$. Let us assume that we have one RF source $C(C_x, C_y, C_z)$ which is not on the same line as A and B line as depicted in Fig. 4.3. Without loss of generality let us assume that the center of the 3D coordinate system is equally spaced between the A and B and that the x axis is the AB line as illustrated in Fig. 4.3. If the distance between the two antenna arrays is d then let us denote that half the distance by $x = d/2$. Therefore, the location of the two antenna arrays can be rewritten as $A(-x, 0, 0)$ and $B(x, 0, 0)$.

Because we know the direction of arrival from the RF source C that means that we know the angles (α_1, α_2) and the angles (β_1, β_2) which could be both azimuth and elevation angles depending on the application and on the configuration. Given this information we need to estimate the location of the projection $C_1(C_x, C_y, 0)$ and $C_2(C_x, 0, C_z)$ which is simply a trigonometry solution. We have already developed these solutions in Sect. 4.2.1.1 or the (2D) principle of operations' distance estimation. We are going to employ those equations to solve first for $C_1(C_x, C_y, 0)$ based on the

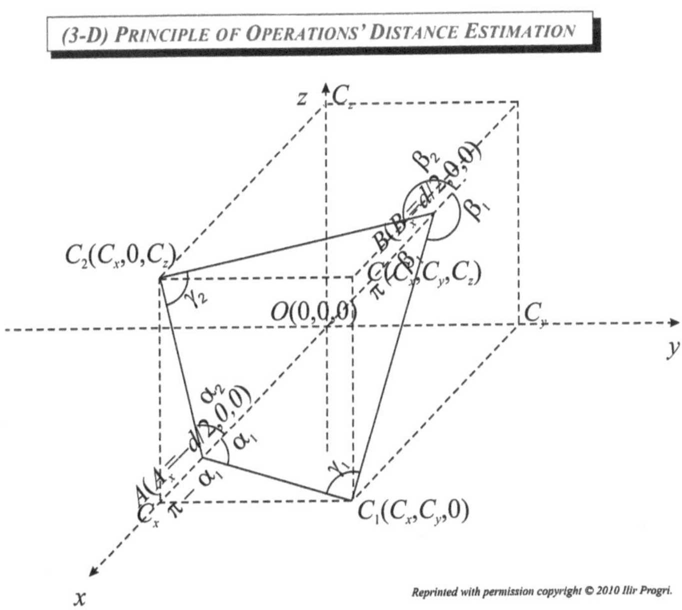

Fig. 4.3 (3D) principle of operations' distance estimation. Reprinted with permission © 2010 Ilir Progri

azimuth angles (α_1, α_2) in the azimuth plane $x0y$ and then solve for $C_2(C_x, 0, C_z)$ based on the elevation angles (β_1, β_2) in the vertical plane $x0z$.

$$\begin{aligned} C_x &= -AC_1\cos(\pi-\alpha_1) - \frac{d}{2} = AC_1\cos\alpha_1 - \frac{d}{2} = d\left[\frac{\sin\beta_1\cos\alpha_1}{\sin(\beta_1-\alpha_1)} - \frac{1}{2}\right] \quad \text{and} \\ C_y &= -AC_1\sin(\pi-\alpha_1) = -AC_1\sin\alpha_1 = -d\frac{\sin\beta_1\sin\alpha_1}{\sin(\beta_1-\alpha_1)}, \end{aligned} \tag{4.7}$$

$$\begin{aligned} C_x &= -BC_1\cos(\pi-\beta_1) + \frac{d}{2} = BC_1\cos\beta_1 + \frac{d}{2} = d\left[\frac{\sin\beta_1\cos\beta_1}{\sin(\beta_1-\alpha_1)} + \frac{1}{2}\right] \quad \text{and} \\ C_y &= -BC_1\sin(\pi-\beta_1) = -BC_1\sin\beta_1 = -d\frac{\sin\beta_1\sin\beta_1}{\sin(\beta_1-\alpha_1)} = -d\frac{\sin^2\beta_1}{\sin(\beta_1-\alpha_1)} \end{aligned} \tag{4.8}$$

and $C_2(C_x, 0, C_z)$

$$\begin{aligned} C_x &= AC_2\cos\alpha_2 - \frac{d}{2} = AC_2\cos\alpha_2 - \frac{d}{2} = d\left[\frac{\sin\beta_2\cos\alpha_2}{\sin(\beta_2-\alpha_2)} - \frac{1}{2}\right] \quad \text{and} \\ C_z &= -AC_2\sin(\pi-\alpha_2) = -AC_2\sin\alpha_2 = -d\frac{\sin\beta_2\sin\alpha_2}{\sin(\beta_2-\alpha_2)}, \end{aligned} \tag{4.9}$$

$$\begin{aligned} C_x &= -BC_2\cos(\pi-\beta_2) + \frac{d}{2} = BC_2\cos\beta_2 + \frac{d}{2} = d\left[\frac{\sin\beta_2\cos\beta_2}{\sin(\beta_2-\alpha_2)} + \frac{1}{2}\right] \quad \text{and} \\ C_z &= -BC_2\sin(\pi-\beta_2) = -BC_2\sin\beta_2 = -d\frac{\sin\beta_2\sin\beta_2}{\sin(\beta_2-\alpha_2)} = -d\frac{\sin^2\beta_2}{\sin(\beta_2-\alpha_2)}. \end{aligned} \tag{4.10}$$

The good thing is that we have a number of redundant equations for estimating the coordinates of the RF source $C(C_x, C_y, C_z)$ which leads to (4.7)–(4.10). Some of these equations can be also employed for integrity checks; however, we are not going to discuss which ones. In Sect. 4.7.3 we are going to employ these equations to obtain the accurate coordinates of various RF sources in principle simulation scenarios or (case studies).

This concludes the (3D) principle of operations distance estimation problem. Next we exploit the Cramér-Rao lower bound on angle accuracy for the maximum likelihood estimation method.

4.2.2 Cramér-Rao Lower Bound and Cramér-Rao Lowest Possible Bound on Angle Accuracy

The previous section provided the solution for the (2D) and (3D) principle of operations' distance estimation method. Accurate distance estimation clearly depends on the angle estimation (or direction of arrival estimation). Therefore, it is very important to provide the Cramér-Rao bound (CRB) which places a lower bound (or the minimum variance) on angle accuracy (or performance) of an unbiased estimator. The maximum likelihood estimator achieves the CRB [2]; while other estimators provide larger variance [3].

Although this derivation was initially derived in [2], we are re-deriving here because there are some essential differences that the readers should be aware of.

The equation that defines the beamformer for a universal linear array from the spatial matched filter that has its phase center at the center of the array

$$\mathbf{v}_{\mathrm{A}}(\phi) = e^{-j2\pi(M-1/2)(d/\lambda)}\sin\phi\mathbf{v}(\phi) \quad \text{and} \quad \mathbf{v}_{\mathrm{E}}(\theta) = e^{-j2\pi(M-1/2)(d/\lambda)}\sin\theta\mathbf{v}(\theta), \tag{4.11}$$

where $\mathbf{v}_{\mathrm{A}}(\phi)$ is referred to as the azimuth and $\mathbf{v}_{\mathrm{E}}(\theta)$ as the elevation beamformer. Due to geometry we have placed the two beam-formers orthogonal with each-other; therefore,

$$\mathbf{v}_{\mathrm{A}}^{\mathrm{H}}(\phi)\mathbf{v}_{\mathrm{E}}(\theta) = 0 \tag{4.12}$$

and they can make two independent measurements of the same signal which allow for the discrimination of both azimuth and elevation angles.

Employing the two steering vectors $\mathbf{v}_{\mathrm{A}}(\phi)$ and $\mathbf{v}_{\mathrm{E}}(\theta)$ we can form an adaptive, normalized azimuth and elevation beam-formers as follows

$$\mathbf{c}_{\mathrm{A}}(\phi) = \mathbf{R}^{-1}\mathbf{v}_{\mathrm{A}}(\phi) \quad \text{and} \quad \mathbf{c}_{\mathrm{E}}(\theta) = \mathbf{R}^{-1}\mathbf{v}_{\mathrm{E}}(\theta). \tag{4.13}$$

The output powers of the two beam-formers are computed as follows

$$P_{\mathrm{A}} = \mathbf{c}_{\mathrm{A}}^{\mathrm{H}}\mathbf{R}\mathbf{c}_{\mathrm{A}} \quad \text{and} \quad P_{\mathrm{E}} = \mathbf{c}_{\mathrm{E}}^{\mathrm{H}}\mathbf{R}\mathbf{c}_{\mathrm{E}}. \tag{4.14}$$

We can measure the normalized cross-correlation function

$$\rho_{\mathrm{AE}}^2 = \frac{|\mathbf{c}_{\mathrm{A}}^{\mathrm{H}}\mathbf{R}\mathbf{c}_{\mathrm{E}}|^2}{P_{\mathrm{A}}P_{\mathrm{E}}}. \tag{4.15}$$

The CRB on both azimuth and elevation angle accuracy can be found from

$$\sigma_{\phi}^2 \geq \frac{1}{2\pi\mathrm{SNR}_0 P_{\mathrm{A}}(1-\rho_{\mathrm{AE}}^2)\cos^2\phi} \quad \text{and} \quad \sigma_{\theta}^2 \geq \frac{1}{2\pi\mathrm{SNR}_0 P_{\mathrm{E}}(1-\rho_{\mathrm{AE}}^2)\cos^2\theta}, \tag{4.16}$$

where $\mathrm{SNR}_0 = M\sigma_{\mathrm{s}}^2/\sigma_{\mathrm{w}}^2$ is the signal to noise ratio for a spatial matched filter in the absence of interference; i.e., signal plus noise only (no interference).

No surprise that the formulas we have provided have very similar form with the formula (11.7.18) in [2] but with very essential differences. We have azimuth and elevation beamformers as opposed to sum and difference beamformers. In [2] the sum and difference beamformers are hypothetical as opposed to our azimuth and elevation beamformers which are much more realistic.

Let us go a bit further into our analyses which are not found anywhere or in any book or article which will help tremendously in the understanding of the material later in the chapter.

We want to find out what is the smallest possible Cramér-Rao Lower Bound by finding the minimum of (4.16).

Leaving the obvious details to the reader we find that

$$\begin{aligned}\sigma_\phi^2 &\geq \frac{1}{2\pi \mathrm{SNR}_0 P_\mathrm{A}(1-\rho_\mathrm{AE}^2)\cos^2\phi} \geq \frac{1}{2\pi \mathrm{SNR}_0 P_\mathrm{A}} \quad \text{and} \\ \sigma_\theta^2 &\geq \frac{1}{2\pi \mathrm{SNR}_0 P_\mathrm{E}(1-\rho_\mathrm{AE}^2)\cos^2\theta} \geq \frac{1}{2\pi \mathrm{SNR}_0 P_\mathrm{E}} \quad \text{(The angle is measured in radians).}\end{aligned} \tag{4.17}$$

If we wanted to measure the angle in degrees then we have the following

$$\begin{aligned}\sigma_\phi^2 &\geq \frac{1}{360 \mathrm{SNR}_0 P_\mathrm{A}} = \frac{1}{360}\frac{1}{M}\frac{\sigma_\mathrm{w}^2}{\sigma_\mathrm{s}^2}\frac{1}{P_\mathrm{A}}(\mathrm{deg}) \quad \text{and} \\ \sigma_\theta^2 &\geq \frac{1}{360 \mathrm{SNR}_0 P_\mathrm{E}} = \frac{1}{360}\frac{1}{M}\frac{\sigma_\mathrm{w}^2}{\sigma_\mathrm{s}^2}\frac{1}{P_\mathrm{E}}(\mathrm{deg}).\end{aligned} \tag{4.18}$$

Equation (4.18) is far easier to visualize than (4.16) because the smallest possible Cramér-Rao Lower Bound can be two to three orders of magnitude smaller than a degree. Therefore, theoretically it is possible to achieve angle estimation accuracy to within a 0° precision which means that we have for most practical purposes angle estimation accuracy to within two to three orders of magnitude smaller than a degree.

The rest of the discussion on the angle estimation accuracy is affected from lack of calibration and misalignment which we should consider in the second edition of the book or future R & D discussions or publications on the same subjects.

4.3 Geolocation of RF Signals Best Adaptive Algorithm Practices

Multiple antenna systems are currently the most important wireless systems in the forefront of the wireless research in three most important applications related in multiple access communications systems: (1) geolocation of RF signals via angle of arrival and distance estimation which is the grand theme of this chapter; (2) multiple access interference at the receiver via multi-user beamforming which are the grand themes of Chap. 6; and (3) and space-time modulation and coding for MIMO systems which will be discussed in *Indoor Geolocation Systems: Theory and Applications* as a novel technology in addition to other indoor geolocation systems [1–58].

Regardless of the multiple access communications system application, the intelligence or the heart of the multiple access communications system in this case GRFS is either "*blind*" or "*non-blind*" [1–58]. The "*blind*" adaptive algorithm is an iterative, very complex and sophisticated *intelligent* process which attempts to analyze, differentiate, discriminate, recognize, identify etc. all the unique RF signals (i.e., RF sources) in the environment only from exploiting properties of the received RF signal, wideband RF signal processing techniques, and standard RF waveforms or RF signal designs similar to those discussed in Chap. 3 [1–58]. Two promising blind adaptive-beamforming techniques widely available in the literature are: (1) Projection Approximation Subspace Tracking (PAST) and (2) Maximum Power (MP) which are based on the estimation of the dominant eigenvector of the received signal spatial-correlation matrix [24]. As we are going to see in Chap. 5, the iterative implementation of both algorithms guarantees linear computational complexity, thus avoiding the explicit calculation and storage of the spatial correlation matrix [24]. We should mention that the computational complexity of PAST and MP is linear in the number of sensors [24]. Because these algorithms are so important we are going to discuss the requirements for blind adaptive algorithms in the next Sect. 4.4.

"*Non-blind*" adaptive algorithms are employed in an environment in which we are concerned with (or interested in) a subset of signals in the environment which we know these signals characteristics and treat the remaining signals as interference or RF channel noise. These algorithms constitute more the traditional path of adaptive array processing or adaptive multiple access communications and geolocation systems and will be discussed in greater detail in Chap. 6.

Figure 4.4 illustrates the best adaptive array algorithm practices for geolocation of RF signals. In contrast to the block diagram of Fig. 4.1 which is the illustration of the best adaptive algorithms discussed in this chapter; this block diagram includes current and future work. The left hand-side of this block diagram includes the system configuration and architecture (or description). For the most part we have discussed system configuration in Chap. 1 with the general system description and requirements. System description (or architecture) and architecture patterns we have presented in Chap. 2. The third element of system configuration are system modeling and knowledge engineering which include RF signals discussed extensively in Chap. 3 and also anticipated to expand in the future and main principles of operations which are discussed in Chap. 4.

Therefore, the majority of the work in the system configuration and architecture is already covered. What remains to be added are more detailed descriptions of individual RF signals and more experimental system architecture pattern descriptions.

The left hand side of Fig. 4.4 includes the best adaptive algorithm practices which are discussed as follows. First we have the manual (or analyst in the loop) RF source identification, reliability and adaptive algorithm optimization methods. The following sections of this chapter and Chap. 6 will cover in great detail geolocation of RF signals based on properties of received signal which is still a manual (or requires the analyst in the loop) and is optimized to yield the best possible RF source geolocation based on the principles of operations discussed earlier. Second,

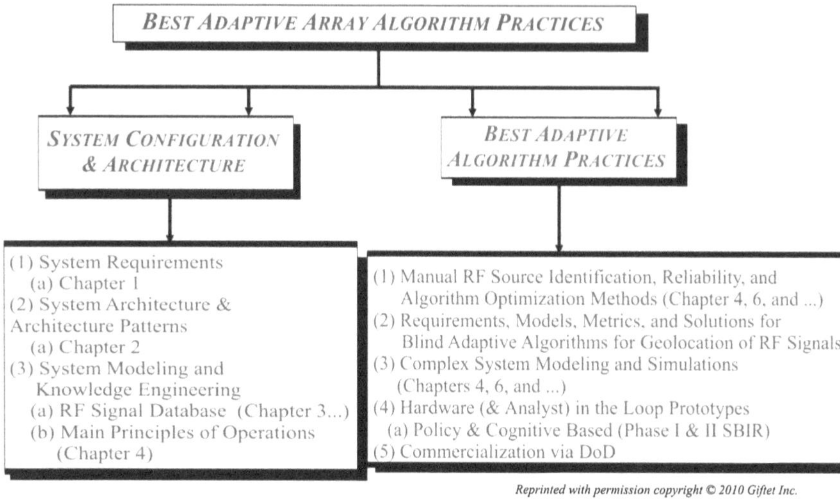

Fig. 4.4 Best adaptive array algorithm practices for geolocation of RF signals (this block diagram includes current and future work). Reprinted with permission © 2010 Giftet Inc

requirements, models, metrics, and solutions for best blind adaptive algorithms for geolocation of RF signals are discussed next. There is certainly more research needed in this area for more realistic system architecture and configurations. Third, complex system modeling and simulations are objects of Chaps. 4 and 6 and also future research and development programs. All the above is hoped to lead to hardware (and analyst) in the loop prototypes which are both policy and cognitive based in phases I and II of small business innovative research (SBIR) projects and programs. The final element of the best adaptive algorithms is implementation and commercialization within the Department of Defense (DoD) on various passive arrays or passive adaptive array systems.

The rest of Chap. 4 is an illustration of the best adaptive algorithm practices.

4.4 Requirements, Models, Metrics, and Solutions for Best Blind Adaptive Algorithms for Geolocation of RF Signals

While in Chap. 1 we discussed in general the requirements for geolocation of RF signals, here we focus on requirements, models, metrics, and solutions for blind adaptive algorithms for geolocation of RF signals. Because this chapter is the bed-rock of the entire book, the corner stone of everything we have discussed and we are going to analyze and discuss it is important to focus on requirements, models, metrics, and solutions for blind adaptive algorithms for geolocation of RF signals. The bottom line of this chapter and of this book can be understood by a series of questions such as: (1) what are we giving to the readers? (2) What are the requirements for blind adaptive algorithms for geolocation of RF signals? (3) How credible are our claims (or will the general audience perceive our claims as credible)? (4) How realistic are our models? (5) How sound is our solution and our approach? (6) Have we included enough scenarios (or principle simulation examples) in our study? (7) Can these algorithms be implemented in a real hardware? (8) How will the hardware look like?

First, at this point it is very important to clarify this question as to what are we giving to the readers? We are giving to the readers a methodical collection, refinement, interpretation of the best practices in the art and science of blind adaptive algorithms for geolocation of RF signals in the last 20 year or so of the published R & D research work; i.e., this book should serve as a manual for a student, developer, faculty member, etc. such as an R & D firm, university center, government agency etc. that if funding is provided to develop a test bed of such systems there is sufficient enough depth and knowledge in this book to guide through the analysis, simulations, design, and implementation phase. This book is not a replacement for the R & D work that needs to be done to analyze, simulate, design, and implement such systems; however, it is only meant to be an indispensable guide.

Second, what are the requirements for blind adaptive algorithms for geolocation of RF signals? This is a sufficiently large and important subject that will involve a subsection of its own; therefore, we are going to discuss this at the end.

Third, how credible are our claims (or will the general audience perceive our claims as credible)? What we would like to reader to understand is that we have made every attempt to properly characterize the work from the beginning to the end. We have made very attempt to include all the published materials from prestigious journal articles, conference proceedings, books, that seemed appropriate for the subject. We claim that we have made every attempt to crossed check every single possible idea, formula, or model, standard, reference, etc. in the context of theoretical work (or analysis) within the scope of the work. However, the credibility of this work does not mean that no more R & D work is required in the area and that there is other credible research work out there that we have not included. We are not saying that we have included every credible R & D work out there in the geolocation of RF signals: principles and simulations but we are saying that whatever material we have included, whatever assumptions we have made, and whatever organization we have provided, etc. are very credible work [37].

Fourth, now we attempt to address this question: how realistic are our models? This is probably the most difficult question to answer from both the philosophical and practical point of view. From the philosophical point of view ("While all the models are unrealistic, some of them are useful" – George Box and "Everything should be made as simple as possible but not simpler" – A. Einstein) we understand that there are philosophical differences that is neither the scope nor the objective of this book to clarify. This can be a very debatable issue because the technologies are new and innovative and as the first attempt to present this material all in one source is extremely hard and difficult work. From the practical point of view, we are not saying that the level of realism of this material is that of a finished product or of a product of products that have been tested over the years. However, we are saying that the level of realism is that of a guide (a tour guide or a road map) which again if funding is provided and proper design methods are used and as the result of this a prototype or a test-bed is built then it will certainly result in more realistic models other perhaps publishable material [37].

Fifth, how sound is our solution and our approach? Our approach is sound within the boundaries of assumptions that we have made. We can say for sure that all of our assumptions are not complete and we are not saying that this is "I am the best" approach because the literature review in this context is incomplete; but, we are saying that this is one of the very best books in the area and one of the very best guides and we are trying to demonstrate the best possible approach to illustrate out theory, assumptions we have made and principle simulation examples we have provided to fulfill the scope of a sound guided material [37].

Sixth, have we included enough scenarios (or principle simulation examples) in our study? We have included enough scenarios (or principle simulation examples) to illustrate the core principles of this book. In general test cases and realistic test scenarios are very examples; i.e., they cost a lot of money, they may not all be released to the public due to proprietary or other privileges or restrictions [37].

Seventh, can these algorithms be implemented in a real hardware? This is the ultimate goal of our research and of our work to secure funding to build prototypes to implement these algorithms and deploy them in real scenarios first for the interests of the DoD and perhaps several years later on some commercial products.

Eighth, how will the hardware look like? I think the real question at this point is will these algorithms be implemented in already built hardware; i.e., commercial-off-the shelf or will they require new hardware? Due to the originality of the work it may require custom made hardware especially new Field Programmable Gate Array (FPGA) and perhaps new antenna elements. Part of this question will also depend on the requirements for blind adaptive algorithms for GRFS systems which are discussed next.

4.4.1 Requirements for Blind Adaptive Algorithms for Geolocation of RF Signals Systems

Requirements for blind adaptive algorithms for geolocation of RF signals systems include an array of requirements that these systems should have in order to successfully perform geolocation of RF signals in the frequency band of 100 MHz to 66 GHz.

1. The first requirement is the availability of antenna elements that have sufficient enough gain the frequency band of 100 MHz to 66 GHz. It is well known that antenna elements have a non-flat gain as a function of frequency. If this is not achievable with the current antenna elements it might be achieved in the future or using segments of frequencies.
2. The second most important requirement is the availability of bandwidth in the RF frond end section of the receiver. GRFS systems require the highest possible bandwidth.

3. The third most important requirement is dynamic range available on the RF front end and baseband sections of the receiver. This is going to enable the receiver to detect the weakest signals in the presence of strong signals. This implies that when the received RF signal is converted in the baseband or IF frequency band, there is enough available dynamic range that enables to detect or discriminate all the strongest signals as well as the weakest signals.
4. The fourth most important requirement is intelligent blind adaptive signal processing for detection, discrimination, evaluation, and geolocation.
5. The fifth most important requirement is designing an intelligent blind adaptive algorithm that identifies all the signals in the environment vs. a blind adaptive algorithm that identifies the most important signals in the environment. Obviously designing a blind adaptive algorithm that identifies all the signals in the environment is neither feasible, nor practical, nor achievable without requiring tremendous resources, computational power, and other resources. Below we are going to see how to detect the most important signals in the environment.
6. The sixth most important requirement is the procedure that is used to identify all the most important signals in the environment. We would prefer to detect the signal (or the RF source) with the most dominant eigen-value first which could either be the signal that most dominant signal in power, closest in location, or the widest in bandwidth. Depending on the number of antenna elements we can only detect the number of independent RF sources that are less than the number of antenna elements.
7. Is it possible to detect and geolocate all the signals when the number of RF signals within the dynamic range in the environment is greater than the number of antenna elements? The answer to this question is yes. When signals with the largest eigen-value have been evaluated then perhaps those signals should be removed from the subspace and other signals should be attempted to be evaluated. It is an approach that needs to be evaluated and although it may suffer from limitations because we are reducing the eigen-value ratio.
8. What kind of processing should be followed to enable identification, detection, discrimination, and geolocation of all the signals in the environment? We should expect time domain processing, frequency domain processing, eigen-value decomposition, wavelet decomposing, and perhaps tensor analysis. We are going to exploit some of these options below leaving other options for future research and investigation.

4.5 Best Blind Adaptive Algorithm

In this section we present the advantages and disadvantages of exploiting a blind adaptive algorithm both in the time and frequency domains. Simulation results illustrate the performance of the blind algorithm by comparing the extracted signals with the original signals and the estimated signal locations with the corresponding actual signal locations.

Recently, there has been increased interest in blind adaptive algorithms to widen their applications [1–58]. Re-iterating some of the earlier works Xu [1] has proposed a blind adaptive algorithm for minimum variance receivers. Such receivers have very low complexity, which makes them very desirable. On the other hand, Martone [5] has proposed a blind adaptive detection scheme for DS/CDMA signals on time varying multipath channels with antenna arrays using high order statistics. This is especially important when the noise or interference is non-Gaussian. Blinder adaptive array algorithms are found in [1–58]. Two promising blind adaptive-beamforming techniques widely available in the literature are: (1) PAST and (2) MP which are based on the estimation of the dominant eigenvector of the received signal spatial-correlation matrix [24]. As we are going to see in Chap. 5, the iterative implementation of both algorithms guarantees linear computational complexity, thus avoiding the explicit calculation and storage of the spatial correlation matrix [24]. We should mention that the computational complexity of PAST and MP is linear in the number of sensors [24]. Because these algorithms are so important we are going to discuss the requirements for blind adaptive algorithms in Sect. 4.4.

Along these lines we have started to develop, and will present here, an approach for the blind detection of signals in the time and frequency domains, with very low receiver complexity exploiting only first and second order statistics.

First, we define *blind detection* as the extraction of a signal of interest in the presence of noise and interference, using only specific properties of the signal, without reliance on preambles or training sequences. Second, we present an approach for determining the direction of arrival of the corresponding signals based on Bromberg's approaches as presented in [6, 7]. Third, these approaches appear to be simpler than the approaches described by Gromov in [8, 9]. And fourth, this chapter begins to study the blind adaptive algorithm in the time and frequency domain based on the initial work of Bromberg [6, 7] and Progri [10].

This section is organized as follows: First, we present the blind adaptive algorithm. In this section we use the phrase "*blind*" to refer to the fact that the specific nature of the signal is not known *a priori*. Second, we perform simulation and describe the blind adaptive process both in the time and frequency domain for extracting four different signals and estimating the corresponding angles of arrival. Although the algorithm can adaptively identify and detect signals, these simulations are intended to show feasibility, rather than to fully study this adaptive property. Last, some conclusions are drawn which indicate that the blind adaptive algorithm in the frequency domain can successfully extract FM, CW, QAM, and BURST signals, as opposed to the blind adaptive algorithm in the time domain, which can only extract BURST signals. Moreover, the angles of arrival for these signals can be estimated at up to 0° precision. The reasons we have picked these signals are as follows: (1) FM is very typical RF signal for a sub-urban GRFS system; (2) CW is a very typical signal of a single carrier or multicarrier RF signal indoor or urban; (3) QAM is very widely used in a number of wireless networks IEEE 802.11a, IEEE

802.15.3, mobile network such as 2GPP, in RF Satellite downlink transmitter just to name a few; and (4) finally BURST signals represent two-way radio RF signals which are also discussed in Chap. 3.

4.5.1 A Blind Adaptive Array GRFS System Concept with an Analyst in the Loop

The generic block diagram of a blind adaptive array GRFS system concept with an analyst in the loop is shown in Fig. 4.5. It is assumed that there are J sources and that one source transmits one signal at some frequency or range of frequencies. It is further assumed that the number of antenna elements is equal to M. The user, which is identified with the receiver, is interested in detecting each signal and in estimating the angle of arrival for each individual source. In all of the cases studied in this work, we assume that the user and the sources are at fixed locations (or moving relatively slow for the window of time that we have extracted the data). This assumption is purely a matter of convenience, as we believe that this algorithm can be modified to handle source and receiver dynamics. Note that we may or may not have any information about the number of signals/sources, J. Nevertheless, this

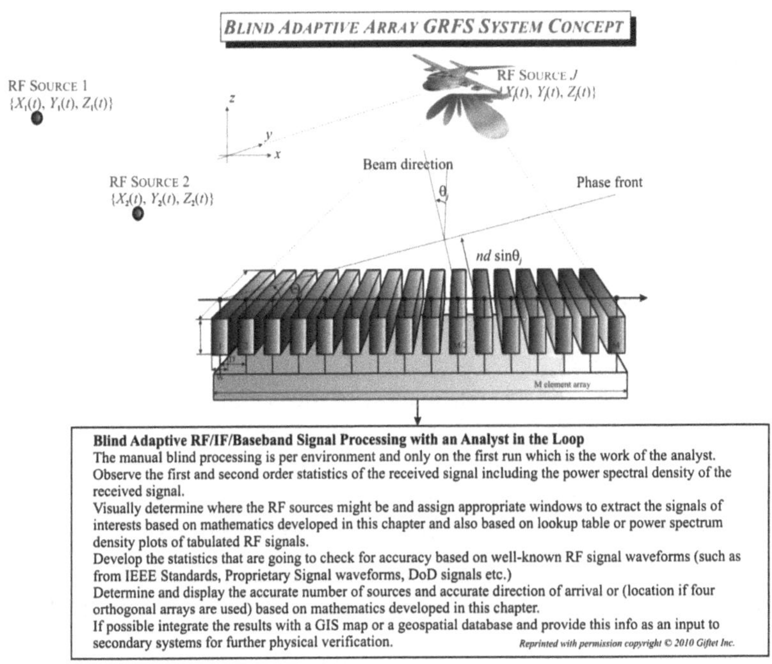

Fig. 4.5 Block diagram of a blind adaptive array GRFS system concept. Reprinted with permission © 2010 Giftet Inc

algorithm will be able to detect and extract $M - 1$ signals. Hence, for the purposes of our investigation the number of signals, J, is less than or equal to the number of antenna elements minus one; i.e., $J \leq M - 1$.

Perhaps the most important detail we have included in this revision/addition of the material is the role and the work of the analyst in the loop. An analyst is a trained individual with high graduate major degrees in electrical and computer engineering with major emphasis in signals (adaptive array signal, image, video processing, communications, RF engineering or related fields), and minor degrees in math, computer science, physics, or related fields and with extensive experience in system analysis, design, test, and evaluation. An analyst is someone who is knowledgeable in the signals waveform design, evaluation, interpretation, representation, standardization etc. We should mention that an analyst is only looking for *man-made signals or RF Sources* and his capabilities or abilities might be limited by the ability or capability of the state of the art laboratory equipment computer processing ability, training costs etc. The most important detail we are trying to emphasize here is that the work and the role of the analyst is indispensable for new signals or RF sources, for new environments, and in general for new configurations or policy changes or in general for new system upgrades etc. So the word adaptive or certain subsystems can be used adaptively to blind detect and geolocate certain signals of interest and for this kind of systems we can develop prototypes. Therefore, while the first role or work of the analyst is implied for future R & D work the secondary role is what is kept in mind throughout the treatment of this chapter as follows:

The manual blind processing is per environment and only on the first run which is the work of the analyst.

1. Observe the first and second order statistics of the received signal including the power spectral density of the received signal.
2. Visually determine where the RF sources might be and assign appropriate windows to extract the signals of interests based on mathematics developed in this chapter and also based on lookup table or power spectrum density plots of tabulated RF signals.
3. Develop the statistics that are going to check for accuracy based on well-known RF signal waveforms (such as from IEEE Standards, Proprietary Signal waveforms, DoD signals etc.).
4. Determine and display the accurate number of sources and accurate direction of arrival or (location if four orthogonal arrays are used) based on mathematics developed in this chapter.
5. If possible integrate the results with a GIS map or a geospatial database and provide this info as an input to secondary systems for further physical verification.

In general we assume that each individual, received signal has unique correlation properties in either the time or the frequency domain. The blind adaptive processor will exploit these properties to detect and extract a particular signal of interest. For example, a CW signal in the time domain is a sinusoid and in the frequency domain is a peak, or spike, which implies that a small number of data points in the frequency domain contain all the information about this type of signal. Similarly, any sufficiently long window of data points of the CW signal in the time domain contains the same kind of information that the spike or the peak does in the frequency domain. In the case of a BURST signal (which we define as a white Gaussian noise (WGN) signal, which is periodically turned ON and OFF) has a flat frequency response, but has a well-defined amplitude profile in the time domain. This idea can be extended for signals other than CW or BURST signals, whose time and frequency responses are more complicated than those of CW or BURST signals. Our only requirement is that the signal properties are such that sufficient information to identify the desired signal is mostly contained within a window of frequency or time samples and can be fully, analytically modeled by first and second order statistics.

Recently developed recursive algorithms [1, 11] and advances in the digital signal processing architecture have motivated us to study this problem and propose a potential solution for commercial applications in the areas of wireless navigation and communication and ad-hoc networks.

Here, we begin the investigation of the blind adaptive algorithm. Consider the input data matrix **X** defined as

$$\mathbf{X} = \left[\mathbf{x}(1)^{\mathbf{H}} \quad \cdots \quad \mathbf{x}(N)^{\mathbf{H}}\right], \tag{4.19}$$

where, $\mathbf{x}(i)$ is a vector of size M, $\forall i \in \{1, 2, \ldots, N\}$, and N denotes the number of data samples not to be confused with the vector size, M, which is equal to the number of antenna elements.

The signal vector, $\mathbf{x}(i)$, represents the superposition of all individual signals, $s_j(i)$, scaled appropriately by the steering vector (also known as the pointing vector [12], *Chap. 6 as a much better detailed discussion on pointing or steering vectors*), $\mathbf{a}_j(\theta)$, and WGN vector, $\mathbf{n}(i)$, which is expressed as

$$\mathbf{x}(i) = \sum_{j=1}^{J} \mathbf{a}_j(\theta_j) s_j(i) + \mathbf{n}(i). \tag{4.20}$$

It is assumed that the magnitude in (dB) and the PSD of the input data from every sensor (or antenna element) is either known or generated.

What one antenna element receives is the superposition of a variety of signals and noise, some of which can be recognized, in closed form in the time domain and others in the frequency domain. Nevertheless, it may not be possible to identify these signals based on a visual inspection of the time domain plot because the degrees of freedom (signals) are lost in the process of superposition. On the other hand, the frequency response of the signal vector, $\mathbf{x}(i)$, may indicate the presence of several signals if the frequency response of the composite signal is other than flat spectrum. Moreover, if we have prior knowledge of what the frequency response of the signal looks like (perhaps we store the PSD of signals with well-defined properties in a database); we may be able to either visually identify or blind detect it. This leads to the second definition of *blind detection*; i.e., to compare the information contained in a narrow section (window of samples for time domain processing and some bandwidth window for frequency domain processing) of the data with the rest of or remaining data information, as indicated by the algorithms of Bromberg [6, 7], and Progri [1, 10].

We begin here the description of the blind adaptive algorithm. An input data vector $\mathbf{x}(n)$ feeds the algorithm at discrete time n. Since, the detection criterion we are presenting is based only on the first and second order statistics, we can compute the autocorrelation matrix, $\mathbf{R}_{\mathbf{xx}}$ corresponding the input vector $\mathbf{x}(n)$. By selecting an appropriate window of sample data points, we form the vector, $\mathbf{x}_{\mathrm{ON}}(n)$, whose autocorrelation matrix, $\mathbf{R}_{\mathbf{xx}}^{\mathrm{ON}}$, contains the information about the signal of interest. If the number of sources is smaller than the number of antenna array elements then there are enough degrees of freedom to select a weight vector, $\mathbf{w}$, such that by applying it to the incoming signal vector, $\mathbf{x}(n)$, adjusts the gain and phases of the individual antenna element gain pattern to maximize the total array gain towards the signal (or source) of interest and minimize the total array gains towards the remaining signals (or sources). When this is achieved then an estimate, $\hat{y}(n)$, of the signal of interest is produced. In order to remove the amplitude uncertainty of the signal, $\hat{y}(n)$, we divide it by its amplitude, which produces, a constant modulus signal, $\hat{d}(n)$, which is the closest estimate of the transmitted signal of interest, $d(n)$. After this we can compute two statistics: the autocorrelation function, $\mathbf{r}_{dd}$, of the estimated signal of interest, $\hat{d}(n)$, and the crosscorrelation vector, $\mathbf{r}_{\mathbf{x}d}$, of the input signal vector, $\mathbf{x}(n)$, with the estimated signal of interest, $\hat{d}(n)$. These statistics provide enough information to get an estimate on the steering vector, $\hat{\mathbf{a}}$, towards the desired RF source of interest. Because the steering vector is a function of the angle of arrival, θ, we can generate a steering vector, $\mathbf{a}(\theta)$, and vary the angle, θ, until $\hat{\mathbf{a}}$ is the best estimate of $\mathbf{a}(\theta)$. This concludes the description of the first stage of blind adaptive algorithm as shown in Fig. 4.6. For the second stage we perform classical adaptive filter processing.

$$\mathbf{x}_{\mathrm{ON}}(:,i) = \begin{cases} \mathbf{x}_{\mathrm{ON}}(N_{\mathrm{S}}:N_{\mathrm{E}},i), & \text{Time processing,} \\ \mathrm{FT}^{-1}\{\mathbf{x}_{\mathrm{ON}}(N_{\mathrm{S}}:N_{\mathrm{E}},i)\}, & \text{Frequency processing,} \end{cases} \tag{4.21}$$

$$\mathbf{X}(:,i) = \mathrm{FT}\{\mathbf{x}(:,i)\}. \tag{4.22}$$

where FT denotes the Fourier transform of $\mathbf{x}(i)$ and the ON time signal vector $\mathbf{x}_{\mathrm{ON}}(:, i)$ is defined as

Second, we define the ON time autocorrelation matrix, $\mathbf{R}_{\mathbf{xx}}^{\mathrm{ON}}$, as

$$\mathbf{R}_{\mathbf{xx}}^{\mathrm{ON}} = \frac{1}{N_{\mathrm{E}} - N_{\mathrm{S}}} \sum_{k=N_{\mathrm{S}}}^{N_{\mathrm{E}}} [\mathbf{x}_{\mathrm{ON}}(k,:)\mathbf{x}_{\mathrm{ON}}(k,:)^{\mathrm{H}}]. \tag{4.23}$$

Third, we define the OFF signal window sample matrix; i.e., either the remainder of the data without the ON section or the whole data samples, time autocorrelation matrix, $\mathbf{R}_{\mathbf{xx}}^{\mathrm{OFF}}$, as

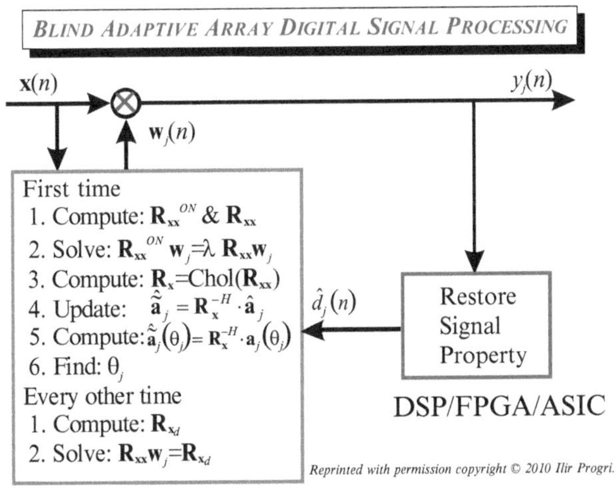

Fig. 4.6 The block diagram of a generic blind adaptive digital signal processing. Reprinted with permission © 2010 Ilir Progri

$$\mathbf{R}_{\mathbf{xx}}^{\mathrm{OFF}} = \mathbf{R}_{\mathbf{xx}} = \frac{1}{N} \sum_{k=1}^{N} \left[\mathbf{x}(k,:)\mathbf{x}(k,:)^{\mathrm{H}}\right]. \tag{4.24}$$

This shows that the OFF signal autocorrelation matrix employs all the input data information. Fourth, the desired signal can be successfully detected by maximizing the following cost function

$$\max_{\mathbf{w}_j} \left(\frac{\hat{\mathbf{w}}_j^{\mathrm{H}} \mathbf{R}_{\mathbf{xx}}^{\mathrm{ON}} \hat{\mathbf{w}}_j}{\hat{\mathbf{w}}_j^{\mathrm{H}} \mathbf{R}_{\mathbf{xx}} \hat{\mathbf{w}}_j} \right) = \lambda, \tag{4.25}$$

which is equivalent to solving the generalized eigen-value problem [6, 10].

$$\mathbf{R}_{\mathbf{xx}}^{\mathrm{OFF}} \hat{\mathbf{w}}_j = \lambda \mathbf{R}_{\mathbf{xx}} \hat{\mathbf{w}}_j. \tag{4.26}$$

Next, suppose that we have an estimate of the steering vector, $\hat{\mathbf{a}}_j$, which is in fact generated [7] and we will see this later, then we form the error vector, $\mathbf{e}_j$

$$\mathbf{e}_j = \hat{\mathbf{a}}_j - g_j \mathbf{a}_j(\theta_j), \tag{4.27}$$

where g_j is some gain parameter assigned to the jth steering vector. For a fixed (or maximum is even better) gain, g_j, and an estimate of the steering vector, $\hat{\mathbf{a}}_j$, we can perform an exhaustive search and find out what the angle of arrival is to an accuracy, which is limited by the signal to noise ratio.

Hence, it remains to find the values of g_j and θ_j that minimize the norm of the error vector, **e**; i.e.,

$$\min_{g_j}\min_{\theta_j}\mu_j(g_j,\theta_j)=\min_{g_j}\min_{\theta_j}\parallel\hat{\mathbf{a}}_j-g_j\mathbf{a}_j(\theta_j)\|^2=\min_{\theta_j}\min_{g_j}\parallel\hat{\mathbf{a}}_j-g_j\mathbf{a}_j(\theta_j)\|^2. \tag{4.28}$$

Taking the first derivative with respect to g_j produces

$$\frac{\partial\mu_j(g_j,\theta_j)}{\partial g_j}=\mathbf{a}_j(\theta_j)^{\mathrm{H}}[\hat{\mathbf{a}}_j-g_j\mathbf{a}_j(\theta_j)]=0. \tag{4.29}$$

Hence, solving for g_j yields

$$g_j=\frac{\mathbf{a}_j(\theta_j)^{\mathrm{H}}\hat{\mathbf{a}}_j}{\mathbf{a}_j(\theta_j)^{\mathrm{H}}\mathbf{a}_j(\theta_j)}, \tag{4.30}$$

which takes care of the first minimization gain argument, g_j.

The minimum value of the norm of the error vector is achieved when we substitute the gain, g_j, given by (4.30), into (4.28)

$$\mu_j(\theta_j)=\parallel\hat{\mathbf{a}}_j\|^2-\frac{|\mathbf{a}_j(\theta_j)^{\mathrm{H}}\hat{\mathbf{a}}_j|^2}{\parallel\mathbf{a}_j(\theta_j)\|^2}=\parallel\hat{\mathbf{a}}_j\|^2\left[1-\frac{|\mathbf{a}_j(\theta_j)^{\mathrm{H}}\hat{\mathbf{a}}_j|^2}{\parallel\mathbf{a}_j(\theta_j)\|^2\parallel\hat{\mathbf{a}}_j\|^2}\right]. \tag{4.31}$$

Minimizing the cost function given by (4.31) is equivalent to maximizing the following cost function

$$\max_{\theta_j}\tilde{\mu}_j(\theta_j)=\frac{|\mathbf{a}_j(\theta_j)^{\mathrm{H}}\hat{\mathbf{a}}_j|^2}{\parallel\mathbf{a}_j(\theta_j)\|^2\parallel\hat{\mathbf{a}}_j\|^2}. \tag{4.32}$$

So far we have presented an analytical approach for determining the weights to extract the jth signal; i.e., when these weights are applied to the input data matrix the resulting output is the extracted signal of interest. Next, we will present an approach for updating the steering vector, which is employed to yield the direction (or angle) of arrival of the extracted signal source (see Fig. 4.6). The antenna elements are utilized to receive the RF analog signal. The analog RF signal from every element is down-converted to the IF and finally to the baseband frequency. The ADC processor samples and digitizes the analog signal to produce the discrete signal vector $\mathbf{x}(n)$. From this point the rest of the processing is done in the DSP, which is what is shown in Fig. 4.6. The algorithm that runs on the DSP contains two stages: the first time (or detection and extraction stage) and the second time (or tracking stage).

During the first time (stage), which is the main theme of this publication, the signal/source detection and extraction is performed and it is, symbolically, accomplished in three main steps

(1) Computation of the *ON* time autocorrelation matrix, $\mathbf{R}_{\mathbf{xx}}^{\mathrm{ON}}$, computation of the OFF time autocorrelation matrix, $\mathbf{R}_{\mathbf{xx}}^{\mathrm{OFF}}$
(2) Solving for the weight vector, $\hat{\mathbf{w}}_j$
(3) Compute the angle of arrival, θ_j

During the second time (stage) the detected signal tracking is performed; nevertheless, this is not the objective of this publication and this has been the theme of our previous publications [12–16].

Once the weight vector is computed, we form the inner product with the weight matrix, $\mathbf{X}$, in post processing or the weight vector, $\mathbf{x}(n)$, in real time to yield output signal, $y_j(n)$, as follows

$$y_j(n) = \mathbf{w}^{\mathrm{H}}\mathbf{X}(1,n)^{\mathrm{H}} \quad \text{or} \quad y_j(n) = \mathbf{w}^{\mathrm{H}}\mathbf{x}(n). \tag{4.33}$$

The signal properties are restored utilizing

$$\hat{d}_j(n) = \frac{y_j(n)}{|y_j(n)|}. \tag{4.34}$$

If we consider for the moment that the mth input signal vector, $\mathbf{x}(m)$, is a superposition of the jth signal of interest and the interfering signal, which can be written mathematically as

$$\mathbf{x}(m) = \mathbf{a}_j(\theta_j)s_j(m) + \mathbf{i}(m), \tag{4.35}$$

where

$$\mathbf{i}(m) = \sum_{k=1,k\neq j}^{J} \mathbf{a}_j(\theta_j)s_j(m) + \mathbf{n}(m). \tag{4.36}$$

Suppose that we know the inverse Cholesky factor of the interference autocorrelation matrix, $\mathbf{R}_{\mathbf{i}}^{-\mathrm{H}}$, then we can multiply both sides of (4.35) by it and the following yields

$$\mathbf{R}_{\mathbf{i}}^{-\mathrm{H}}\mathbf{x}(m) = \mathbf{R}_{\mathbf{i}}^{-\mathrm{H}}\mathbf{a}_j(\theta_j)s_j(m) + \mathbf{R}_{\mathbf{i}}^{-\mathrm{H}}\mathbf{i}(m). \tag{4.37}$$

Next, define $\hat{\hat{\mathbf{a}}}_j$ as

$$\hat{\hat{\mathbf{a}}}_j = \mathbf{R}_{\mathbf{i}}^{-\mathrm{H}}\hat{\mathbf{a}}_j \quad \text{where } \hat{\mathbf{a}}_j = \hat{\mathbf{R}}_{\mathbf{x}s_j}/r_{s_j s_j}. \tag{4.38}$$

Define also $\tilde{\mathbf{a}}_j(\theta_j)$

$$\tilde{\mathbf{a}}_j(\theta_j) = \mathbf{R}_\mathbf{i}^{-\mathrm{H}}\mathbf{a}_j(\theta_j). \tag{4.39}$$

However, when $\mathbf{R}_\mathbf{i}^{-\mathrm{H}}$ is not available we utilize $\mathbf{R}_\mathbf{x}^{-\mathrm{H}}$; i.e., $\hat{\tilde{\mathbf{a}}}_j$ and $\tilde{\mathbf{a}}_j(\theta_j)$ which can be written as

$$\hat{\tilde{\mathbf{a}}}_j = \mathbf{R}_\mathbf{x}^{-\mathrm{H}}\hat{\mathbf{a}}_j \quad \text{and} \quad \tilde{\mathbf{a}}_j(\theta_j) = \mathbf{R}_\mathbf{x}^{-\mathrm{H}}\mathbf{a}_j(\theta_j). \tag{4.40}$$

Hence, taking into consideration (4.40), the angle of arrival can be determined from

$$\max_{\theta_j} \tilde{\mu}_j(\theta_j) = \frac{|\tilde{\mathbf{a}}_j(\theta_j)^{\mathrm{H}}\hat{\tilde{\mathbf{a}}}_j|^2}{\|\,\tilde{\mathbf{a}}_j(\theta_j)\|^2\;\|\,\hat{\tilde{\mathbf{a}}}_j\|^2}. \tag{4.41}$$

The cost function (4.41) was indeed utilized to estimate the angle of arrival instead of (4.42).

In the following section we consider a simulation that demonstrates the effectiveness and the feasibility of the blind adaptive algorithm when signal (or sources) and noise properties satisfy the assumptions made in earlier in this section.

4.6 2001 Simulation Test Setup

In 2001 the results obtained from a simulation of the blind adaptive algorithm, described in the previous section, are presented in this section.

In 2001 we constructed a scenario in which the input data is composed of signals coming from four different signals/sources: QAM, CW, BURST, and FM the center frequency and bandwidth of each one of these signals are less than 4 kHz [10]. This scenario is motivated by a proposed navigation and communication system called MC-CDMA [16].

The angles of arrival for the QAM, CW, BURST, and FM signals with respect to a reference array element are –70°, –15°, 116°, and 100° respectively. Because the lowest frequency we are trying to detect is 4 kHz then a sampling rate of 8 KHz was used and 8,192 samples of the input matrix **X** are required to generate a high fidelity FFT. The steps that we undertook to generate the composite signal are described in order: First, we generate an eight-symbol QAM signal. The angle of the arrival for this signal is –70° with respect to the reference antenna element. Based on the array and the signal locations we determine the original steering vector for the QAM signal. This original steering vector of the QAM signal is applied to the original QAM signal corrupted with multipath error to yield the QAM signal vector for a particular discrete sample. Second, we generate a sine wave (or a CW) signal vector with –15° angle of arrival for the same discrete sample and add it to the QAM signal vector with multipath. Third, we generate a WGN BURST signal with 116° angle of arrival, which is initially turned OFF and it is turned ON when the number of samples reaches 4,096. Fourth, we generate the FM signal vector with 100° angle of arrival and add it to the total input signal vector. Fifth, these signals are also corrupted with WGN, which functions as thermal or front end receiver noise. Finally, the input signal is modulated on a carrier at 1.1 GHz frequency.

The time and frequency domain representation of the input composite signal is shown in Fig. 4.7. A visual inspection of the signal magnitude in the time domain (see Fig. 4.7) gives the idea of a noisy signal environment. However, the frequency domain plot indicates that there may be present at least three signals: (1) one between 0 and 1,000 samples, (2) a spike near 1,575 sample, and (3) another signal between 7,000 and 8,000 samples. Hence, it remains to detect and extract them and to estimate the angles of arrival corresponding to each individual source. Knowing that in general the number of antenna elements should be greater than the number of sources we are trying to extract we selected a six-element array and each antenna element is positioned in the point of a hexagonal, with radius half the wavelength.

The reminder of this section is organized as follows: First, we perform signal detection and extraction in the frequency domain, which includes the signal extraction and estimation of the angle of arrival for the extracted signal source. Second, we perform the signal detection and extraction and direction finding in the time domain. We suspect that only BURST signals can be successfully detected in either

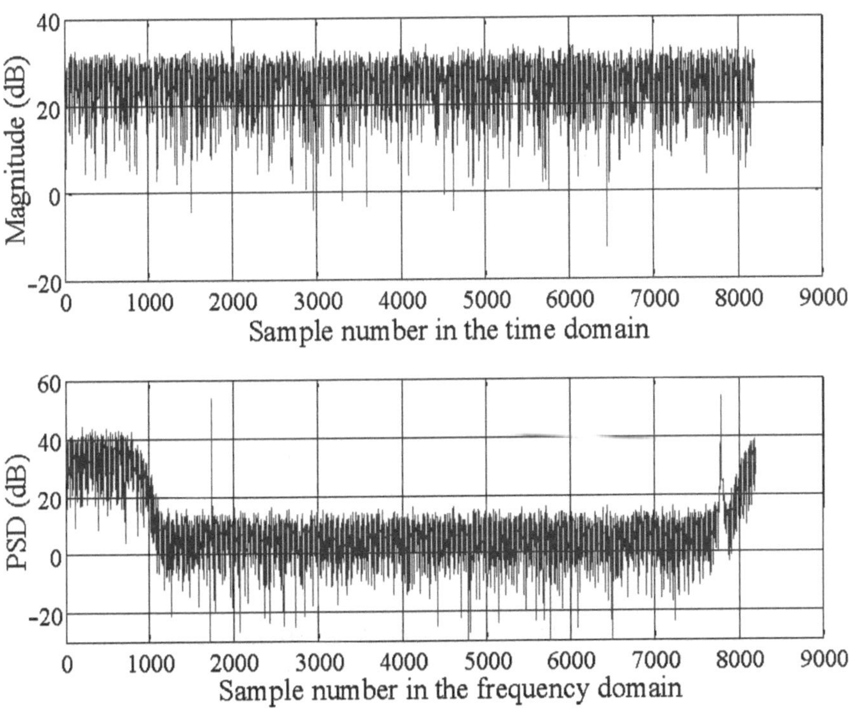

Fig. 4.7 Time and frequency representation of the input data from the first sensor from 2001 data

the time or the frequency domain, because they contain useful correlation information in both domains. Third, we compare the angle of arrival of the BURST extracted signal of the frequency/time domain with the angle of arrival of the BUSRT original signal.

4.6.1 Frequency Domain Processing

Once the input signal is downconverted to the baseband frequency, we take $N = 8{,}192$ samples to calculate the FFT of the input signal. In order to perform the blind extraction we determine values for the start and end samples N_S and N_E, which are taken from the following vectors

$$N_\mathrm{S} = [1\ 1\mathrm{e}3\ 3\mathrm{e}3\ 7\mathrm{e}3] \quad \text{and} \quad N_\mathrm{E} = [1\mathrm{e}3\ 3\mathrm{e}3\ 5\mathrm{e}3\ N]. \tag{4.42}$$

The selection of the start and end samples was made in this example by visually inspecting the profile of the frequency response of the signal. Ultimately, this process must be automated for commercial applications.

The first ON data section corresponds to the input data with samples from $N_\mathrm{S}(1)$ to $N_\mathrm{E}(1)$. For these data samples we compute the inverse FFT and then compute the ON autocorrelation matrix $\mathbf{R}_\mathbf{xx}^\mathrm{ON}$ from (4.23). The OFF autocorrelation matrix $\mathbf{R}_\mathbf{xx}^\mathrm{OFF}$ is computed utilizing all the data as suggested by (4.24). Next, we solve the generalized eigenvalue problem (see (4.25)) and compute the weight vector, $\hat{\mathbf{w}}_1$ (see the first two columns of Table 4.1). These weights are then applied to the input signal matrix, $\mathbf{X}$, the output of which is the signal, $\hat{s}_1$, based on (4.33) and (4.34) and it is compared to the original QAM signal, s_QAM, in forming the detection criterion

$$\frac{\| \hat{s}_1 - s_\mathrm{QAM}\|^2}{N} = \begin{cases} \sigma_n^2 & \text{if } \hat{s}_1 = \hat{s}_\mathrm{QAM} \approx s_\mathrm{QAM}, \\ P_{\hat{s}_1} + P_\mathrm{QAM} + \sigma_n^2 & \text{if } \hat{s}_1 \neq \hat{s}_\mathrm{QAM}, \end{cases} \tag{4.43}$$

as shown in Fig. 4.8. The result of the detection process, (4.43), indicates that the extracted signal is indeed the QAM signal; hence, $\hat{s}_1 = \hat{s}_\mathrm{QAM}$. Next, we estimate the steering vector for this QAM signal with multipath, $\hat{\mathbf{a}}_1$ as suggested by (4.38) and then perform an exhaustive search to maximize the cost function (4.41) employing an angle step size of one degree (because the angle accuracy for this problem was known to be one degree).

Similarly, we followed the same procedure for the three remaining segments; i.e., following the blind adaptive algorithm, and obtained the following output:

(a) Number of emitters is $N_\mathrm{emitters} = 4$, which is correct because there are indeed four different emitters (or sources).
(b) The estimated, complex copy weight matrix is shown in Table 4.1.
(c) The Direction (angle) of arrival (deg) for every emitter is

$$D_\mathrm{oa} = [290^\circ\ (\text{or} - 70^\circ);\ 348^\circ\ (-12^\circ);\ 117^\circ;\ \text{and}\ 100^\circ].$$

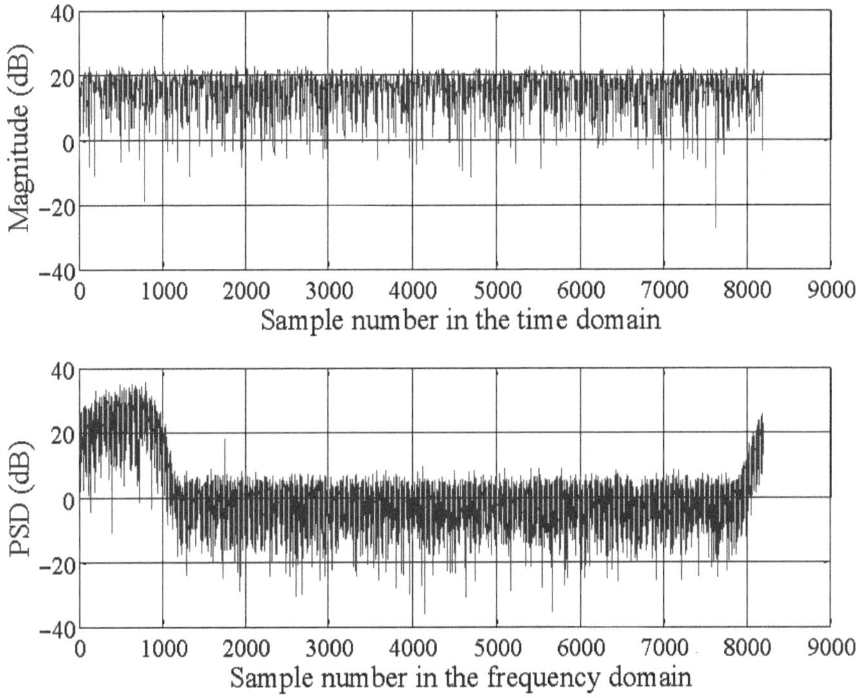

Fig. 4.8 Estimated QAM signal in the time domain and frequency domain from 2001 data

Table 4.1 Resultant weight matrix ($M \times J$) for (a) frequency and (b) time processing from 2001 data

Real1	Imag1	Real2	Imag2	Real3	Imag3	Real4	Imag4
Weight matrix for frequency domain processing							
0.4231	−0.0329	0.6731	−0.0317	0.0488	0.0186	0.5199	0.3022
−0.3403	0.0457	−0.5326	0.2466	0.1670	−0.3213	−0.0218	−0.3417
−0.1556	−0.0337	0.1344	0.1321	−0.6217	0.1764	0.3712	−0.0064
−0.0646	−0.1653	0.2186	0.0171	0.4362	0.2040	0.3436	0.3246
−0.0672	0.3801	−0.1173	−0.2740	−0.3930	−0.0535	−0.0423	−0.2126
−0.0025	0.7043	−0.0624	0.1587	0.2388	0.0487	0.2059	0.2656
Weight matrix for time domain processing							
0.3818	−0.0767	0.3213	0.2185	0.3513	0.1895	0.3100	−0.0769
0.0720	−0.5782	0.4602	−0.3608	0.4016	−0.4269	0.1506	−0.5902
−0.3274	0.2267	−0.3843	−0.0693	−0.3720	−0.0120	−0.3930	0.2337
0.1071	−0.5038	0.4323	−0.2950	0.3870	−0.3648	0.1474	−0.4327
−0.0408	−0.1093	0.0504	−0.0961	0.0461	−0.1007	−0.0813	−0.1322
0.1848	−0.1928	0.2624	−0.0097	0.2467	−0.0520	0.2111	−0.2006

The first detected signal is obtained by applying the first set of weights to the input signal vector, which is shown in Fig. 4.8. The procedure for determining that this was the QAM signal with multipath was explained little earlier in this section. By comparing the frequency response of extracted signal with the frequency response of original QAM signal we conclude that this is a QAM signal analytically summarized by (4.43). Similar result can be obtained if the detection criterion is performed in the time domain because of the uniqueness of the FFT and the adequate sampling rate and number of samples.

This QAM source is located at 290° (or −70°) and its direction of arrival is presented in Fig. 4.9 along with the array beam pattern. Note that a perfect estimation for the angle of arrival for the QAM signal is obtained.

Next, by applying the second set of weights to the input signal vector the second signal source is produced. It appears that this signal is a CW signal because the frequency response, shown in Fig. 4.10, indicates the presence of a single spike or peak at about 55 dB around the 1,575 sample. Nevertheless, we applied the detection criterion (4.43) and confirmed that this was indeed the CW signal. The estimated angle of arrival for the CW source is 348° (−12°) azimuth and is presented in Fig. 4.11 along with the array beam pattern. Note that an angle error of 3° was observed for the CW source.

Next, similarly the third signal source is produced as shown in Fig. 4.12. Note that in this case the detection criterion in either the time domain or frequency representation indicates that there is a BURST signal; i.e., a DC WGN signal, which

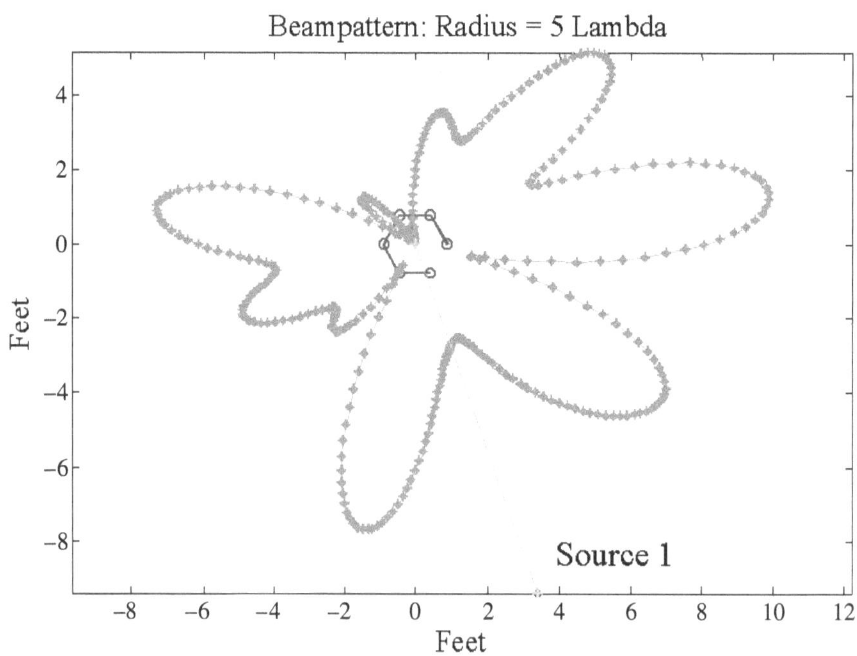

Fig. 4.9 Estimated direction of the QAM signal from 2001 data

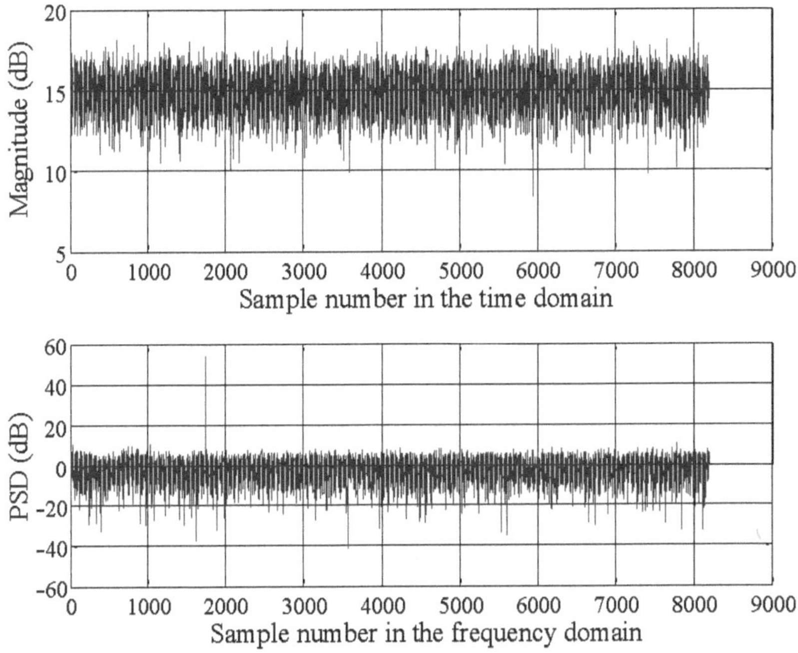

Fig. 4.10 Estimated CW signal in the time and frequency domain from 2001 data

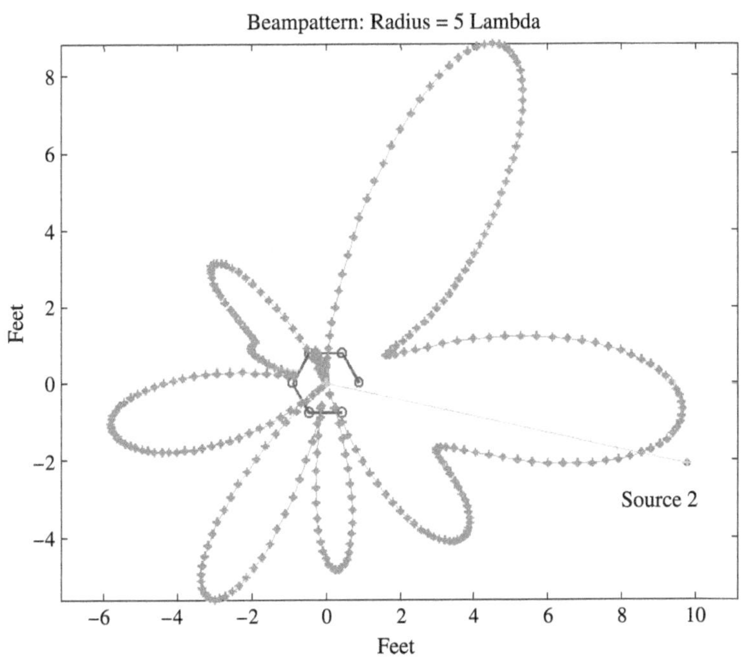

Fig. 4.11 Estimated direction of the CW signal from 2001 data

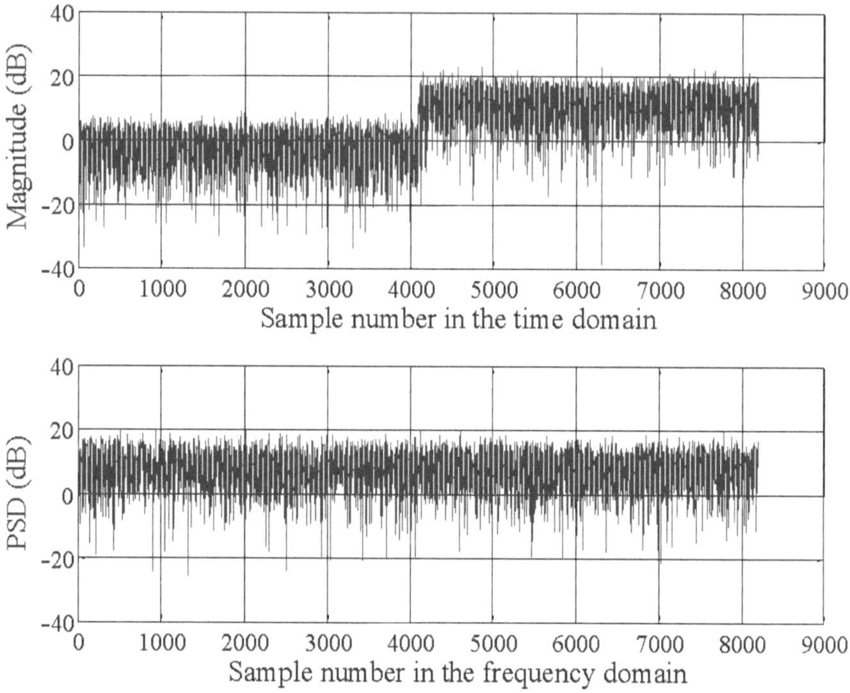

Fig. 4.12 Estimated BURST signal in the time and frequency domain from 2001 data

is OFF and it is turned ON in the middle of the data point in time. The estimated angle of arrival for the BURST signal 117° azimuth and it is presented in Fig. 4.13 along with the array beam pattern. We note here a 1° signal error.

And finally, by applying the fourth (or last) set of weights the last signal source is obtained, which is determined to the extracted FM signal as shown in Fig. 4.14.

The estimated angle of arrival for FM source is 100° azimuth and its angle of arrival is presented in Fig. 4.15 along with the array beam pattern.

This illustration appears to indicate that it is possible to blind detect and extract QAM, CW, BURST, and FM signals in the frequency domain and to determine the angle of arrival to these sources up to 3° accuracy. Hence, these simulation results conform the feasibility of the blind adaptive algorithm when the QAM, CW, BURST, and FM signal (or source) and noise properties satisfy the assumptions made in the "Blind Adaptive Algorithm" section of the chapter.

4.6.2 Time Domain Processing

Although blind processing in the time domain is simpler than processing in the frequency domain, the extraction of signals other than BURST signals is

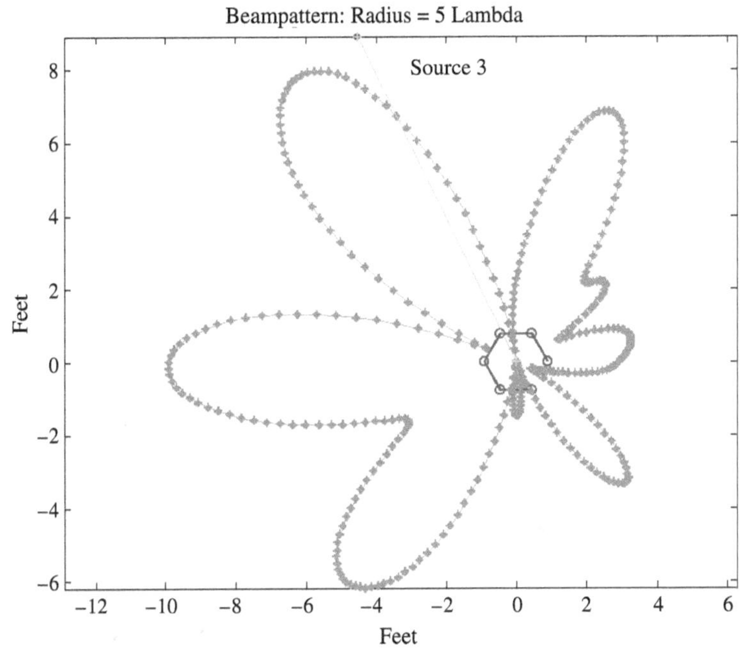

Fig. 4.13 Estimated direction of the BURST signal from 2001 data

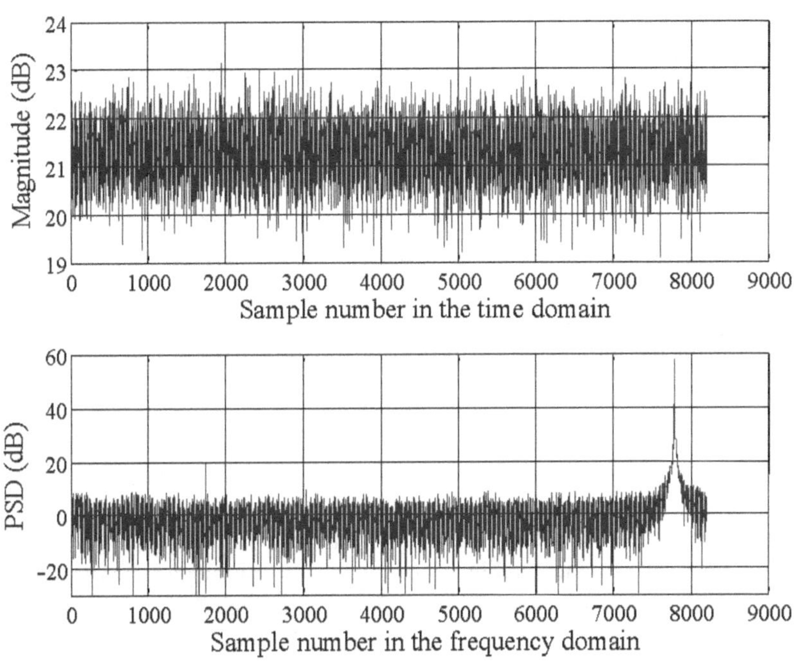

Fig. 4.14 Estimated FM signal in the time and frequency domain from 2001 data

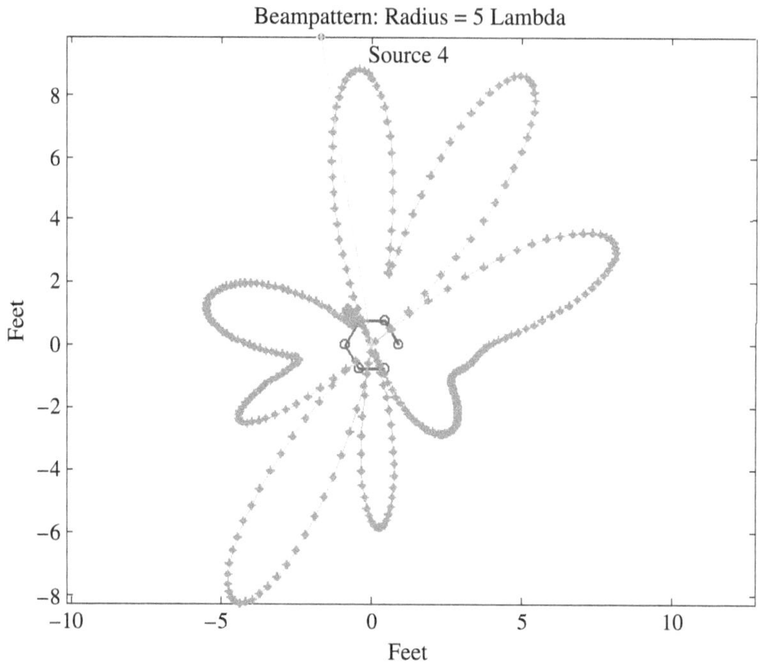

Fig. 4.15 Estimated direction of the FM signal from 2001 data

impossible, because the correlation information in the time domain is almost uniformly distributed. Therefore, we will only concentrate in the extraction the BURST signal.

The numerical results based on the time domain processing (searching only for the BURST signal) are provided below:

(a) Number of emitters is $N_{\text{emitters}} = 1$, which is indeed correct because we were only able to detect and extract one BURST signal from the frequency domain processing.
(b) The estimated, complex weight matrix is shown in Table 4.1.
(c) The Direction (or angle) of arrival for every emitter is

$$D_{\text{oa}} = 116^\circ.$$

The BURST extracted signal is shown in Fig. 4.16, which is very similar to the BURST signal shown in Fig. 4.12.

The angle of arrival for the extracted BURST signal is 116° and its direction (or angle) of arrival is presented in Fig. 4.17, which is very similar to the one presented in Fig. 4.13, along with the array beam pattern utilizing processing in the time domain.

This illustration appears to indicate that it is possible to blind detect and extract BURST signals in the time domain and estimate the direction of arrival to the

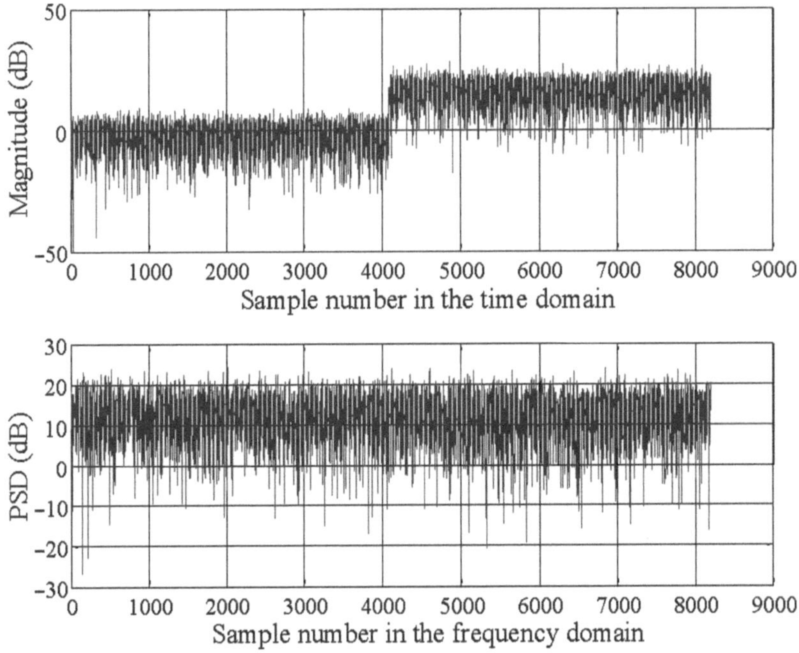

Fig. 4.16 Estimated BURST signal in the time and frequency domain from 2001 data

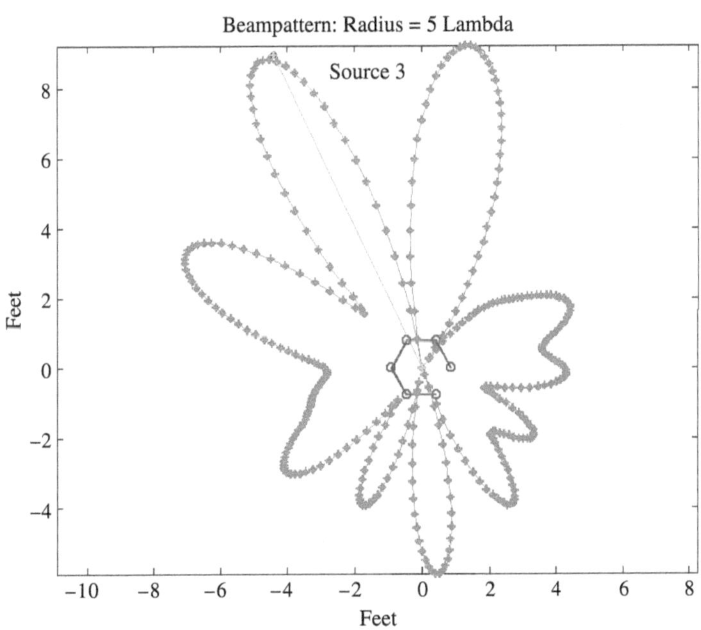

Fig. 4.17 Estimated direction of the BURST signal from 2001 data

corresponding sources with perfect accuracy. Hence, these simulation results conform the feasibility of the blind adaptive algorithm when the BUSRT signal (or source) and noise properties satisfy the assumptions made in the "Blind Adaptive Algorithm" section of the chapter.

4.6.3 2001 Summary and Conclusions

There are several conclusions to be drawn from this work in 2001. First, it is possible to blind detect and extract signals such as QAM, CW, and FM signals, when they appear in the frequency spectrum. Second, it is always possible to blind detect and extract BURST signals in the time and sometimes in the frequency domain. Third, for every extracted signal it is possible to determine the source angle of arrivals within at most 3° of uncertainty.

If these signals are undesired or interfering signals then by blind detecting and extracting them we can get rid of them. If these signals are desired signals then by blind detecting and extracting them we can analyze them and find out what information can be extracted from them.

Although the optimization of this work is underway at the present time, it appears that there are several areas in which improvements to be made for future applications, some of which may be of great benefit to the future GPS. One area would be the detection and extraction of multiple signals of the same source. One would be the full adaptation of the detection algorithm. Another would be the full and multiple stage adaptation for the angle of arrival search. One would be the azimuth and elevation angle of arrival search as opposed to only azimuth.

4.7 2002 Simulation Test Setup

In 2002 we modified the test setup with the sole purpose to achieve direction finding with 0° precision the results obtained from a simulation of the blind adaptive algorithm, described in the beginning of this chapter, are presented in this section.

In 2002 we constructed a scenario in which the input data is composed of signals coming from four different signals/sources: QAM, CW, BURST, and FM the center frequency and bandwidth of each one of these signals are less than 1 kHz [17]. This scenario is motivated by a proposed navigation and communication system called MC-CDMA [16].

The angles of arrival for the CW, QAM, BURST, and FM signals with respect to a reference array element are 123°, 23°, −50°, and −90° respectively. Because the lowest frequency we are trying to detect is 1 kHz, then a sampling rate of 2 kHz was used and 2,084 samples of the input matrix **X** are required to generate a high fidelity FFT. The steps that we undertook to generate the composite signal are described in order: First, we generate a sine wave (or a CW) signal vector with 123° angle of arrival for the same discrete sample. Second, we generate an eight-symbol QAM signal with multipath and add it to the CW signal vector. The angle of the arrival for this signal is 23° with respect to the reference antenna element. Based on the array and the signal locations we determine the original steering vector for the QAM signal. This original steering vector of the QAM signal is applied to the original QAM signal corrupted with multipath error to yield the QAM signal vector for a particular discrete sample. Third, we generate a WGN BURST signal with −50° angle of arrival, which is initially turned OFF and it is turned ON when the number of samples reaches 1,024. Fourth, we generate the FM signal vector with –90° angle of arrival and add it to the total input signal vector. Fifth, these signals are also corrupted with WGN, which functions as thermal or front end receiver noise. Finally, the input signal is modulated on a carrier at 1.1 GHz frequency.

The time and frequency domain representation of the input composite signal is shown in Fig. 4.18. A visual inspection of the signal magnitude in the time domain (see Fig. 4.18) gives the idea of a noisy signal environment. However, the frequency domain plot indicates that there may be present at least three signals: (1) one between 0 and 500 samples, (2) a spike near 30 sample, and (3) another signal between 1,500 and 2,030 samples. Hence, it remains to detect and extract them and to estimate the angles of arrival corresponding to each individual source. Knowing that in general the number of antenna elements should be greater than the number of

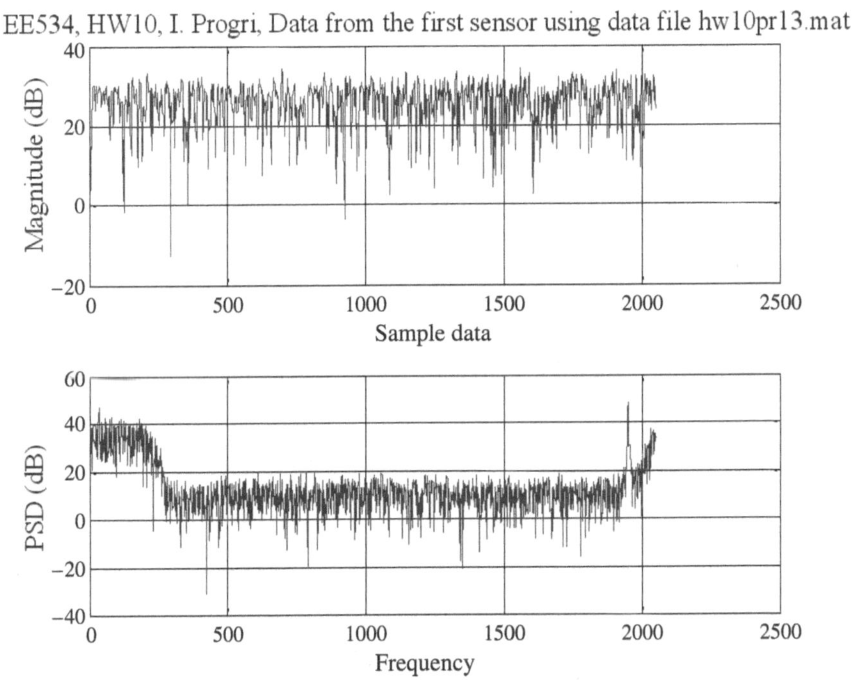

Fig. 4.18 Time and frequency representation of the 2002 input data from the first sensor

sources we are trying to extract we selected a eight-element array and each antenna element is positioned in the point of a hexagonal, with radius half the wavelength.

The reminder of this section is organized as follows: First, we perform signal detection and extraction in the frequency domain, which includes the signal extraction and estimation of the angle of arrival for the extracted signal source. Second, we perform the signal detection and extraction and direction finding in the time domain. We suspect that only BURST signals can be successfully detected in either the time or the frequency domain, because they contain useful correlation information in both domains. Third, we compare the angle of arrival of the BURST extracted signal of the frequency/time domain with the angle of arrival of the BUSRT original signal.

4.7.1 Frequency Domain Processing

Once the input signal is downconverted to the baseband frequency, we take $N = 2{,}048$ samples to calculate the FFT of the input signal. In order to perform the blind extraction we determine values for the start and end samples N_S and N_E, which are taken from the following vectors

$$N_{\mathrm{S}} = [1\ 50\ 90\ 1{,}900] \quad \text{and} \quad N_{\mathrm{E}} = [70\ 300\ 1{,}300\ 2{,}048]. \tag{4.44}$$

The selection of the start and end samples was made in this example by visually inspecting the profile of the frequency response of the signal. Ultimately, this process must be automated for commercial applications.

The first ON data section corresponds to the input data with samples from $N_{\mathrm{S}}(1)$ to $N_{\mathrm{E}}(1)$. For these data samples we compute the inverse FFT and then compute the ON autocorrelation matrix $\mathbf{R}_{\mathbf{xx}}^{\mathrm{ON}}$ from (4.23). The OFF autocorrelation matrix $\mathbf{R}_{\mathbf{xx}}^{\mathrm{OFF}}$ is computed utilizing all the data as suggested by (4.24). Next, we solve the generalized eigenvalue problem (see (4.25)) and compute the weight vector, $\hat{\mathbf{w}}_1$ (see the first two columns of Table 4.2). These weights are then applied to the input signal matrix, $\mathbf{X}$, the output of which is the signal, $\hat{s}_1$, based on (4.32) and (4.33) and it is compared to the original CW signal, s_{CW}, in forming the detection criterion

Table 4.2 Resultant weight matrix ($M \times J$) from 2002 frequency domain processing from 2002 data

0.0447 + 0.0291i	0.1678−0.0365i	−0.0505 + 0.1704i	0.3254 + 0.1836i
−0.3324 + 0.0822i	0.2998−0.0552i	0.0244 + 0.2184i	0.0278 + 0.0768i
−0.2522 + 0.1265i	−0.0617 + 0.2658i	0.1474 + 0.2824i	−0.0778 + 0.1314i
−0.1403 + 0.2278i	0.0254−0.4110i	0.1158−0.5196i	−0.2018−0.0280i
0.5188−0.0672i	−0.3245−0.09090i	0.3515−0.1879i	0.0189 + 0.2848i
0.0098 + 0.6268i	0.2476−0.6092i	0.1959 + 0.0871i	0.4996−0.3071i
−0.1218−0.1004i	−0.0229 + 0.0089i	0.3322 + 0.1733i	0.3040 + 0.3872i
0.0728 + 0.1787i	−0.2132 + 0.2036i	0.1394 + 0.4131i	0.3407−0.0715i

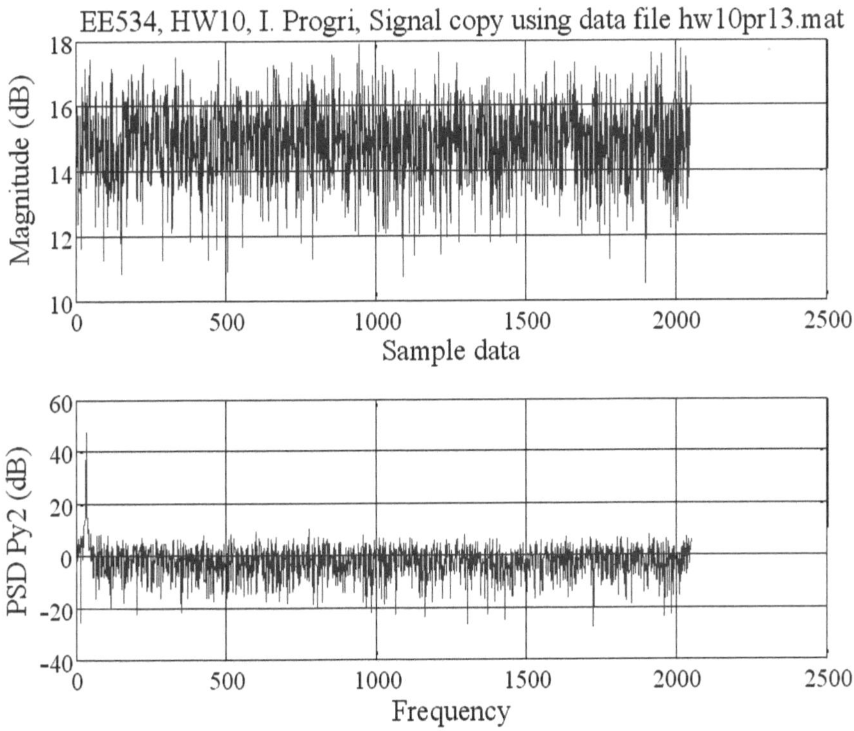

Fig. 4.19 Estimated CW signal in the time domain and frequency domain from the 2002 data

$$\frac{\|\hat{s}_1 - s_{\mathrm{CW}}\|^2}{N} = \begin{cases} \sigma_n^2 & \text{if } \hat{s}_1 = \hat{s}_{\mathrm{CW}} \approx s_{\mathrm{CW}}, \\ P_{\hat{s}_1} + P_{\mathrm{CW}} + \sigma_n^2 & \text{if } \hat{s}_1 \neq \hat{s}_{\mathrm{CW}}, \end{cases} \tag{4.45}$$

as shown in Fig. 4.19. The result of the detection process, (4.44), indicates that the extracted signal is indeed the CW signal; hence, $\hat{s}_1 = \hat{s}_{\mathrm{CW}}$. Next, we estimate the steering vector for this CW signal with multipath, $\hat{\mathbf{a}}_1$, as suggested by (4.37) and then perform an exhaustive search to maximize the cost function (4.40) employing an angle step size of one degree (because the angle accuracy for this problem was known to be one degree).

Similarly, we followed the same procedure for the three remaining segments; i.e., following the blind adaptive algorithm, and obtained the following output:

(a) Number of emitters is $N_{\mathrm{emitters}} = 4$, which is correct because there are indeed four different emitters (or sources).
(b) The estimated, complex copy weight matrix is shown in Table 4.2.
(c) The direction (angle) of arrival (deg) for every emitter is.

$$D_{\mathrm{oa}} = [123^\circ;\ 23^\circ;\ 310^\circ\,(-50^\circ);\ \text{and}\, 270^\circ\,(\text{or} - 90^\circ)].$$

The first detected signal is obtained by applying the first set of weights to the input signal vector, which is shown in Fig. 4.19. The procedure for determining that this was the CW signal was explained little earlier in this section indicates the presence of a single spike or peak at about 55 dB around the 30 sample. By comparing the frequency response of extracted signal with the frequency response of original CW signal we conclude that this is a CW signal analytically summarized by (4.44). Similar result can be obtained if the detection criterion is performed in the time domain because of the uniqueness of the FFT and the adequate sampling rate and number of samples.

This CW source is located at 123° and its direction of arrival is presented in Fig. 4.20 along with the array beam pattern. Note that a perfect estimation for the angle of arrival for the CW signal is obtained.

Next, by applying the second set of weights to the input signal vector the second signal source is produced. It appears that this signal is a QAM signal with multipath because the frequency response, shown in Fig. 4.21. Nevertheless, we applied the detection criterion (4.44) and confirmed that this was indeed the CW signal. The estimated angle of arrival for the QAM source is 23° azimuth and is presented in Fig. 4.22 along with the array beam pattern. Note that an angle error of 0° was observed for the QAM source with multipath.

Next, similarly the third signal source is produced as shown in Fig. 4.23. Note that in this case the detection criterion in either the time domain or frequency representation indicates that there is a BURST signal; i.e., a DC WGN signal, which

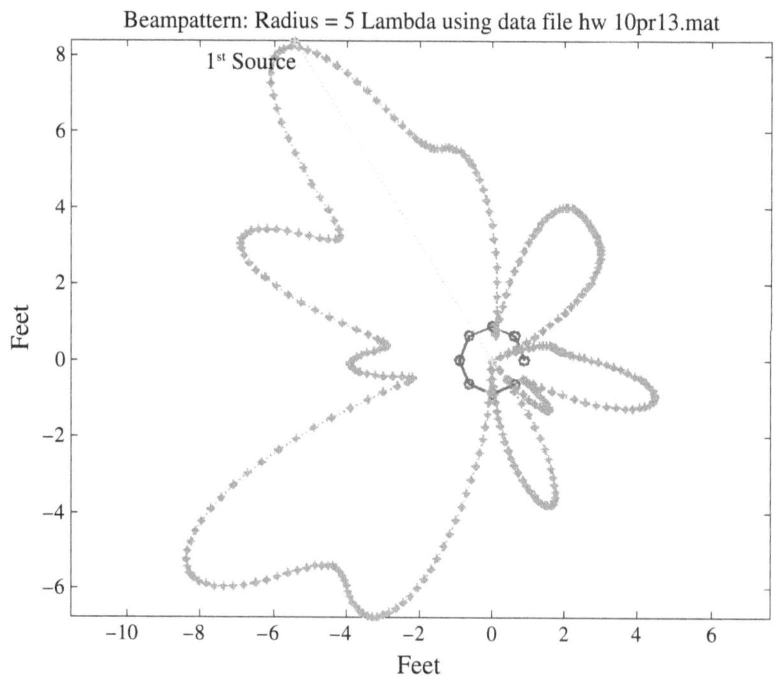

Fig. 4.20 Estimated direction of the CW signal from 2002 input data

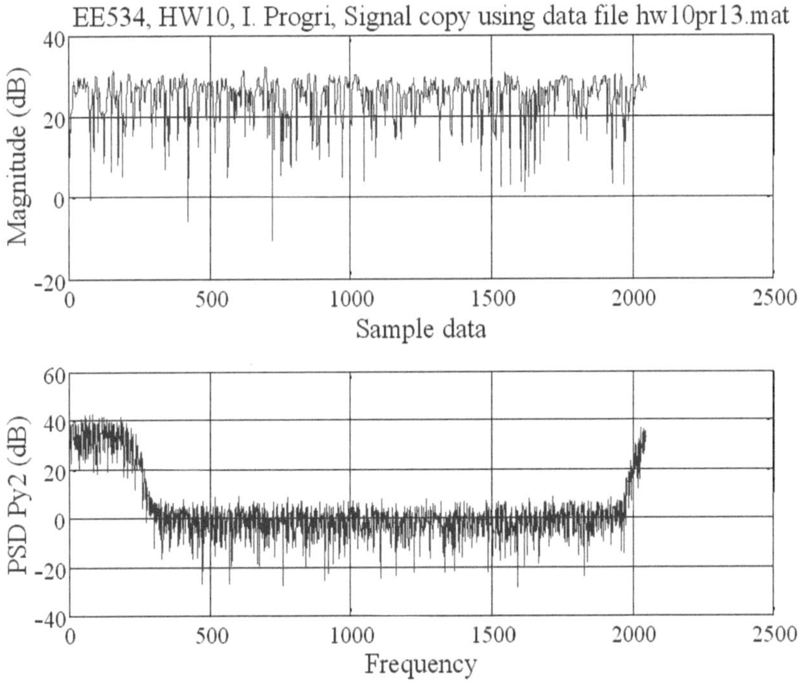

Fig. 4.21 Estimated QAM signal with multipath in the time and frequency domain from 2002 data

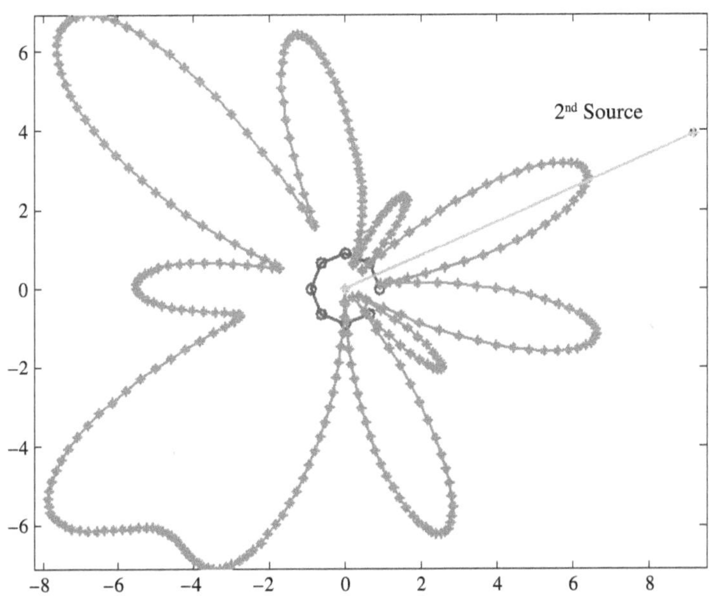

Fig. 4.22 Estimated direction of arrival of the QAM signal with multipath from 2002 data

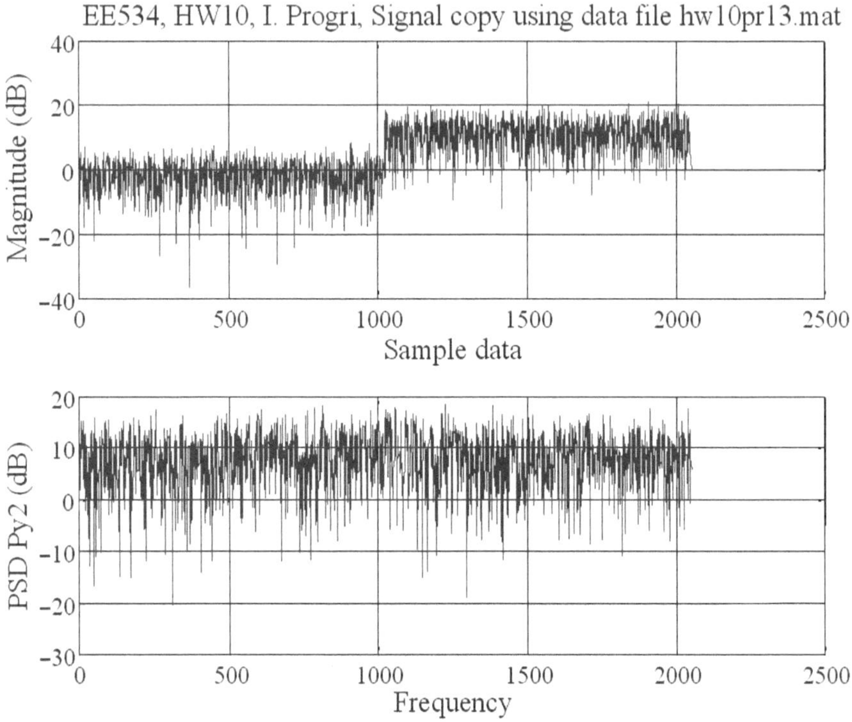

Fig. 4.23 Estimated BURST signal in the time and frequency domain from 2002 data

is OFF and it is turned ON in the middle of the data point in time. The estimated angle of arrival for the BURST signal 310° (−50°) azimuth and it is presented in Fig. 4.24 along with the array beam pattern. We note here a 0° signal error.

And finally, by applying the fourth (or last) set of weights the last signal source is obtained, which is determined to the extracted FM signal as shown in Fig. 4.25.

The estimated angle of arrival for FM source is 270° (or −90°) azimuth and its angle of arrival is presented in Fig. 4.26 along with the array beam pattern.

This illustration appears to indicate that it is possible to blind detect and extract QAM, CW, BURST, and FM signals in the frequency domain and to determine the angle of arrival to these sources up to 0° accuracy. Hence, these simulation results conform the feasibility of the blind adaptive algorithm when the CW, QAM, BURST, and FM signal (or source) and noise properties satisfy the assumptions made in the "Blind Adaptive Algorithm" section of the chapter.

4.7.2 Time Domain Processing

Although blind processing in the time domain is simpler than processing in the frequency domain, the extraction of signals other than BURST signals is

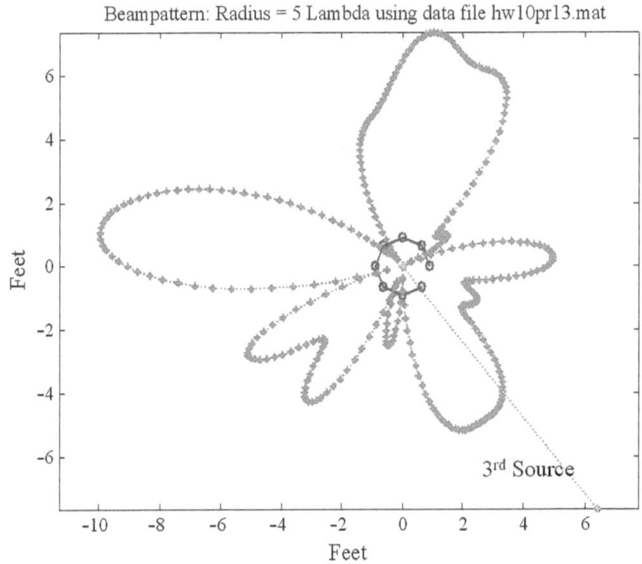

Fig. 4.24 Estimated direction of the BURST signal from 2002 data

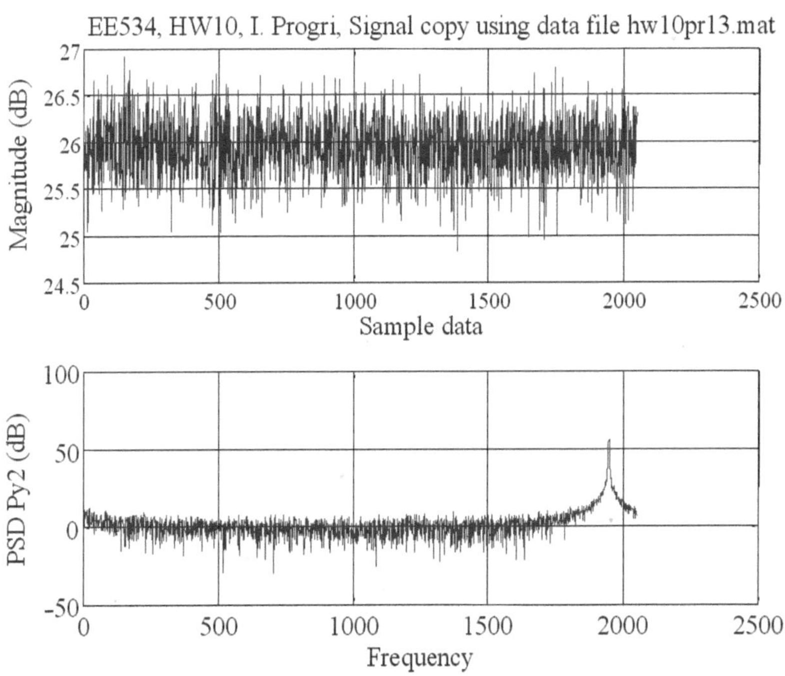

Fig. 4.25 Estimated FM signal in the time and frequency domain from 2002 data

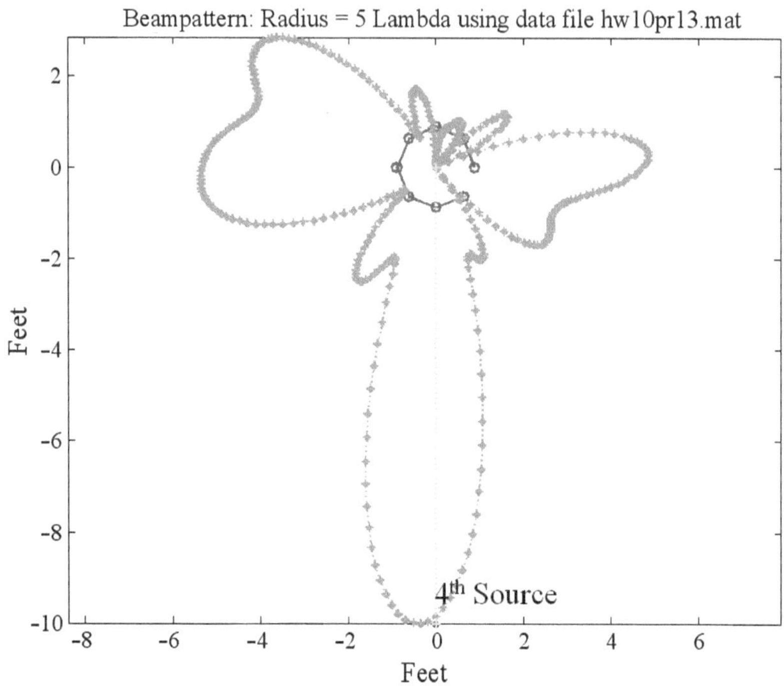

Fig. 4.26 Estimated direction of the FM signal from 2002 data

impossible, because the correlation information in the time domain is almost uniformly distributed. Therefore, we will only concentrate in the extraction the BURST signal.

The numerical results based on the time domain processing (searching only for the BURST signal) are provided below:

(a) Number of emitters is $N_{\text{emitters}} = 1$, which is indeed correct because we were only able to detect and extract one BURST signal from the frequency domain processing.
(b) The estimated, complex weight matrix.
(c) The Direction (or angle) of arrival for every emitter is

$$D_{\text{oa}} = 310^{\circ}(-50^{\circ}).$$

The BURST extracted signal is shown in Fig. 4.27, which is very similar to the BURST signal shown in Fig. 4.28.

The angle of arrival for the extracted BURST signal is 310° (−50°) and its direction (or angle) of arrival is presented in Fig. 4.27, which is very similar to the one presented in Fig. 4.28, along with the array beam pattern utilizing processing in the time domain.

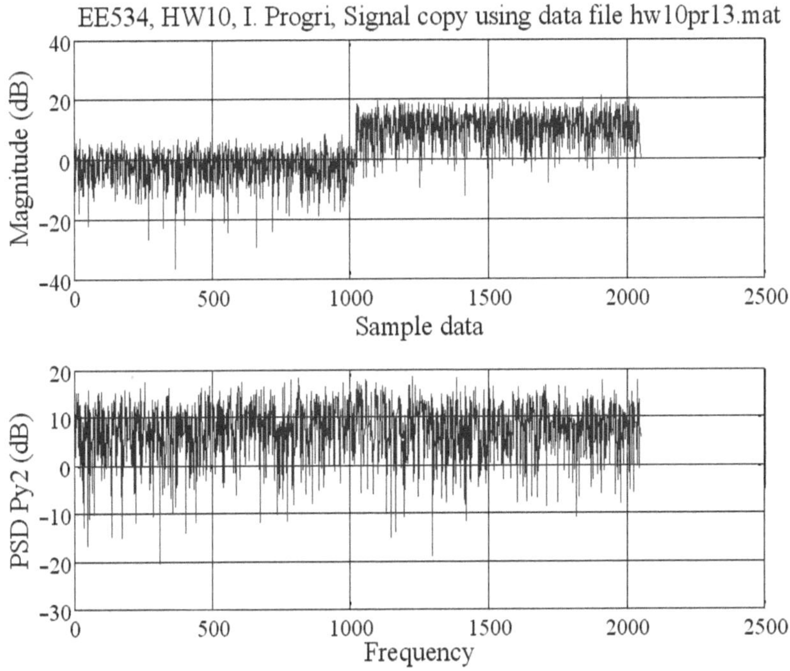

Fig. 4.27 Estimated BURST signal in the time and frequency domain from 2002 data

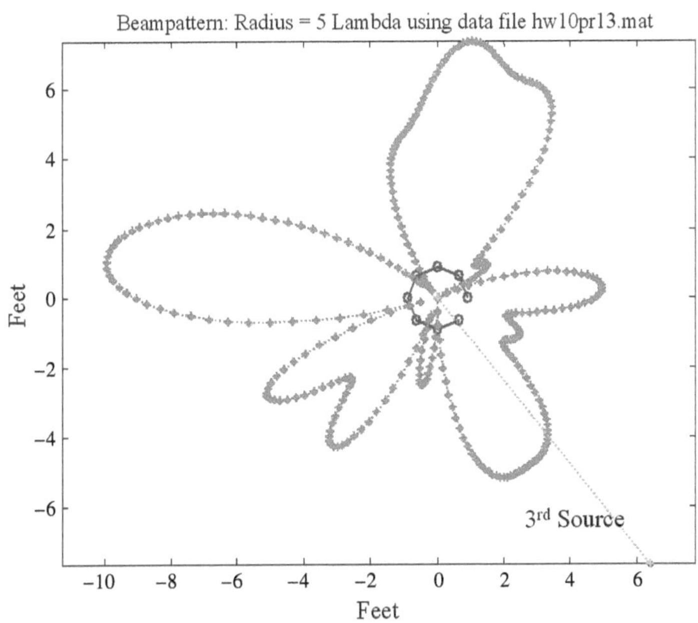

Fig. 4.28 Estimated direction of the BURST signal from 2002 data

This illustration appears to indicate that it is possible to blind detect and extract BURST signals in the time domain and estimate the direction of arrival to the corresponding sources with perfect accuracy. Hence, these simulation results conform the feasibility of the blind adaptive algorithm when the BUSRT signal (or source) and noise properties satisfy the assumptions made in the "Blind Adaptive Algorithm" section of the chapter.

4.7.3 2002 Summary and Conclusions

There are several conclusions to be drawn from this work in 2002. First, it is possible to blind detect and extract signals such as CW, QAM, and FM signals, when they appear in the frequency spectrum. Second, it is always possible to blind detect and extract BURST signals in the time and sometimes in the frequency domain. Third, in contrast to the 2001 summary and conclusions, for every extracted signal it is possible to determine the source angle of arrivals within *at most* 0° *of uncertainty* 100% *of the time*.

If these signals are undesired or interfering signals then by blind detecting and extracting them we can get rid of them. If these signals are desired signals then by blind detecting and extracting them we can analyze them and find out what information can be extracted from them.

Although the optimization of this work is underway at the present time, it appears that there are several areas in which improvements to be made for future applications, some of which may be of great benefit to the future GPS. One area would be the detection and extraction of multiple signals of the same source. One would be the full adaptation of the detection algorithm. Another would be the full and multiple stage adaptation for the angle of arrival search. One would be the azimuth and elevation angle of arrival search as opposed to only azimuth.

4.8 2010+ Simulation Test Setup Requirements for Future Work

2010+ Simulation Test Setup Requirements for Future Work would be based on two major avenues: (1) 2010+ DoD Simulation Test Setup Requirements for DoD Future Work as discussed in Sect. 4.8.1 and (2) 2010+ Simulation Test Setup Requirements for non DoD Future Work as presented in Sect. 4.8.2.

4.8.1 2010+ DoD Simulation Test Setup Requirements for DoD Future Work

Now that we have develop the complete system requirements in Chap. 1, an extensive description of the system in Chap. 2, a detailed description of some of the RF signals of interests in the frequency band of 100 MHz until 66 GHz in Chap. 3 and provided additional material in Chap. 4, what should be some of the more reasonable and more complex simulation test setups that should be considered in the future.

1. The work that we begun in Chaps. 2 and 3 should be expanded with more realistic simulation scenarios; i.e., we need to show that we have all the available signal description, waveforms, software development, configuration, equipment etc. to receive, decode, process, display etc. all the signal discussed in configuration descriptions of Chap. 2.
2. We need to determine if there are any common variances among all the typical scenarios; i.e., we should arrive to at least 39 typical scenarios as illustrated in Chap. 2 GRFS system descriptions.
3. We need to determine the amount of simulation parameters, data, time, and recourses that are needed to successfully perform blind adaptive array GRFS system processing and provide information (location or direction of arrival for all RF sources of interest) to a program manager, commander or the personnel who is in charge to receive this information for further verification or validation via image, video processing or other means.
4. Propose the necessary modifications for future or further developments or integration of the blind adaptive algorithm with GIS maps, geospatial database or other geospatial tools.
5. Be flexible to add other requirements as suggested by the DoD client(s).

So if we find a DoD client (or several DoD clients; i.e., or a market within the DoD or other US government agencies) who is serious about this work or who plans to develop new R & D system prototypes along these lines then we can certainly perform the tasks proposed in the 2010+ Simulation Test Setup Requirements because the volume, the scope, and the resources required to do what we are proposing is way beyond the scope of this book. The second edition (or other future editions) of this book will benefit as a means to illustrate results, the description, the mathematics etc.

4.8.2 2010+ Simulation Test Setup Requirements for Non DoD Future Work

What if we cannot find a DoD client or if it turns out that we marketing efforts of this edition of the book will not produce a successful DoD contract? What happens then? In that case the assumption would be that we can still perform future work and publish the material in future conferences by seeking private donors, sponsors, or other means of non-DoD funding. In that context future work will depend on what is being negotiated and on what might be made available for future as part of information that is obtained from future conference papers, journal articles, other publically available proposals etc. As of today some of this work could be included as follows:

1. Include number of sources that exceeds the number of antenna elements.
2. Include angle discrimination simulation setup step that is comparable to the Cramér-Rao Lower Bound and Cramér-Rao Lowest Possible Bound on Angle Accuracy instead of 1° that was arbitrary chosen in for the 2001 and 2002 test setup.
3. Perform the necessary simulation updates for more realistic waveforms in the frequency band of 100 MHz to 66 GHz.
4. Provide the necessary system description of the adaptive arrays prototypes and associated costs.
5. Be flexible to add other requirements as suggested by the client(s).

4.9 Summary and Conclusions

The work that we have discussed in this chapter can be augmented should DoD or non-DoD funding becomes available in the direction that have proposed in the 2010+ DoD Simulation Test Setup Requirements for DoD Future Work in Sect. 4.8.1 and 2010+ Simulation Test Setup Requirements for non DoD Future Work in Sect. 4.8.2.

Should funding become available we are included to believe that blind adaptive array signal processing is possible and for every extracted signal it is possible to determine the source angle of arrivals (in azimuth, elevation, and location) within *at most* 0° *of uncertainty* 100% *of the time or (the Cramér-Rao Lower Bound and Cramér-Rao Lowest Possible Bound on Angle Accuracy*).

References

1. Progri, I.F., Michalson, W.R., and Bromberg, M.C., A study of a blind adaptive algorithm in the time and frequency domain, in *Proc. ION-NTM 2002*, San Diego, pp. 439–447, Jan. 2002.
2. Manolakis, D.G., Ingle, V.K., and Kogon, S.M., *Statistical and Adaptive Signal Processing: Spectral Estimation, Signal Modeling, Adaptive Filtering, and Array Processing*, Boston: MC Graw Hill, 2000.
3. Codara, L.C., *Smart Antennas*, Boca Raton: CRC Press, 2004.
4. Xu, Z., and Tsatsanis, M.C., Blind adaptive algorithms for minimum variance CDMA receivers, *IEEE Trans. Commun.*, vol. 49, pp. 180–194, Jan. 2001.
5. Martone, M., Blind adaptive detection of DS/CDMA signals on time-varying multipath channels with antenna arrays using high-order statistics, *IEEE Trans. Commun.*, vol. 48, pp. 1590–1600, Sep. 2000.
6. Bromberg, M., and Brown, D., The use of programmable DSPs in antenna array processing in *The Application of Programmable DSPs in Mobile Communications*, New York: Wiley, Nov. 2001.

7. Bromberg, M., and Agee, B., Direction finding for unstructured emitters in the presence of structured interferers, in *Proc. ICASSP*, May 1998.
8. Gromov, K., Akos, D., Pullen, S., Enge, P. Parkinson, B., and Pervan, B., Interference direction finding for aviation applications of GPS, in *Proc. 12th Inter. Tech. Mtg. Sat. Div. ION (ION GPS 1999)*, Nashville, pp. 115–124, Sep. 1999.
9. Gromov, K., Akos, D., Pullen, S., Enge, P., and Parkinson, B., GIDL: generalized interference detection and localization system, in *Proc. 13th Inter. Tech. Mtg. Sat. Div. ION (ION GPS 2000)*, Salt Lake City, pp. 447–457, Sep. 2000.
10. Progri, I.F., HW#9, Homework report on EE534 course, Worcester Polytechnic Institute, Worcester, Apr. 2001.
11. Progri, I.F., Michalson, W.R., and Bromberg, M.C., A comparison between the recursive Cholesky and MGSO algorithms, in *Proc. 2002 Nat. Tech. Mtg. ION*, San Diego, pp. 655–665, Jan. 2002.
12. Progri, I.F., and Michalson, W.R., Adaptive spatial and temporal selective attenuator in the presence of mutual coupling and channel errors, in *Proc. 13th Inter. Tech. Mtg. Sat. Div. ION (ION GPS 2000)*, Salt Lake City, pp. 462–470, Sep. 2000.
13. Progri, I.F., and W.R. Michalson, An improved adaptive spatial temporal selective attenuator, in *Proc. ION-GPS*, Salt Lake City, Sep. 2001.
14. Progri, I.F., Michalson, W.R., Bogdanov, G., and Ramanna, V.C., An investigation of a GPS adaptive temporal selective attenuator, *NAVIGATION J. Inst. Nav.*, vol. 49, nr. 3, pp. 137–147, fall 2002.
15. Michalson, W.R., and Progri, I.F., An investigation of the adaptive spatial and temporal selective attenuator, in *Proc. 14th Inter. Tech. Mtg. Sat. Div. ION (ION-GPS 2001)*, Salt Lake City, pp. 1985–1996, Sep. 2001.
16. Progri, I., A MC-CDMA indoor geolocation system, in *Proc. PIMRC 2005*, Berlin, vol. 4, pp. 2535–2542, 9–14, Sep. 2005.
17. Progri, I.F., HW#10 Solution, Homework solution report on EE534 course, Worcester Polytechnic Institute, Worcester, Apr. 2002.
18. Blostein, S.D., and Leib, H., Multiple antenna systems: their role and impact in future wireless access, *IEEE Com. Mag.*, vol. 41, nr. 7, pp. 94–101, Jul. 2003.
19. Polydoros, A., Rautio, J., Razzano, G., Bogucka, H., Ragazzi, D., Dallas, P.I., Mammela, A., Benedix, M., Lobeira, M., and Agarossi, L., WIND-FLEX: developing a novel testbed for exploring flexible radio concepts in an indoor environment, *IEEE Com. Mag.*, vol. 41, nr. 7, pp. 116–122, Jul. 2003.
20. Palicot, J., and Roland, C., A new concept for wireless reconfigurable receivers, *IEEE Com. Mag.*, vol. 41, nr. 7, pp. 124–132, Jul. 2003.
21. Srikanteswara, S., Palat, C., Reed, J.H., and Athanas, P., An overview of configurable computing machines for software radio handsets, *IEEE Com. Mag.*, vol. 41, nr. 7, pp. 134–141, Jul. 2003.
22. Nanda, S., Walton, R., Ketchum, J., Wallace, M., and Howard, S., A high-performance MIMO OFDM wireless LAN, *IEEE Com. Mag.*, vol. 43, nr. 2, pp. 101–109, Feb. 2005.
23. Zhang, X., Tang, J., Chen, H.-H., Ci, S., and Guizani, M., Cross-layer-based modeling for quality of service guarantees in mobile wireless networks, *IEEE Com. Mag.*, vol. 44, nr. 1, pp. 100–106, Jan. 2006.
24. Falletti, E., Laddomada, M., Mondin, M., and Sellone, F., Integrated services from high-altitude platforms: a flexible communication system, *IEEE Com. Mag.*, vol. 44, nr. 2, pp. 124–133, Feb. 2006.
25. Zhang, Y., Cosmas, J., Song, Y., and Bard, M., Future transmitter/receiver diversity schemes in broadcast wireless networks, *IEEE Com. Mag.*, vol. 44, nr. 10, pp. 120–127, Oct. 2006.
26. Dohler, M., Mclaughlin, S., Laurenson, D., Beach, M., Tan, C.M., and Aghvami, A.H., Implementable wireless access for B3G networks – part 1: MIMO channel measurements, analysis, and modeling, *IEEE Com. Mag.*, vol. 45, nr. 3, pp. 85–92, Mar. 2007.

27. Mishra, S.M., Broodersen, R.W., Brink, S.t., and Mahadevappa, R., Detect and avoid: an ultra-wideband/WiMAX coexistence mechanism, *IEEE Com. Mag.*, vol. 45, nr. 6, pp. 68–75, Jun. 2007.
28. Groe, J., Polar transmitters for wireless communications, *IEEE Com. Mag.*, vol. 45, nr. 9, pp. 58–63, Sep. 2007.
29. Li, J., Li, R., and Kato, J., Future trust management framework for mobile Ad Hoc networks, *IEEE Com. Mag.*, vol. 46, nr. 4, pp. 108–114, Apr. 2008.
30. Greenspan, A., Klerer, M., Tomcik, J., Canchi, R., and Wilson, J., IEEE 802.20: mobile broadband wireless access for the twenty-first century, *IEEE Com. Mag.*, vol. 46, nr. 7, pp. 56–63, Jul. 2008.
31. Kim, S., Kim, H.W., Kang, K., and Ahn, D.S., Performance enhancement in future mobile satellite broadcasting services, *IEEE Com. Mag.*, vol. 46, nr. 7, pp. 118–124, Jul. 2008.
32. She, J., and Ho, P.-H., Cooperative coded video multicast for IPTV services under EPON-WiMAX integration, *IEEE Com. Mag.*, vol. 46, nr. 8, pp. 104–110, Aug. 2008.
33. Hyeon, S., Kim, J., and Choi, S., Evaluation and standardization of the smart antenna system for software defined radio, *IEEE Com. Mag.*, vol. 46, nr. 9, pp. 68–74, Sep. 2008.
34. Hashemi, H., Chu, T.-S., and Roderick, J., Integrated true-time-delay-based ultra-wideband array processing, *IEEE Com. Mag.*, vol. 46, nr. 9, pp. 162–172, Sep. 2008.
35. Singh, H., Oh, J., Kweon, C.Y., Qin, X., Shao, H.-R., and Ngo, C., A 60 GHz wireless network for enabling uncompressed video communication, *IEEE Com. Mag.*, vol. 46, nr. 12, pp. 71–78, Dec. 2008.
36. Andrews, J., Shakkattai, S., Heath, R., Jindal, N., Haenggi, M., Berry, R., Guo, D., Neely, M., Weber, S., Jafer, S., and Yener, A., Rethinking information theory for mobile Ad Hoc networks, *IEEE Com. Mag.*, vol. 46, nr. 12, pp. 94–101, Dec. 2008.
37. Stojmenovic, I., Simulations in wireless sensor and Ad Hoc networks: matching and advancing models, metrics, and solutions, *IEEE Com. Mag.*, vol. 46, nr. 12, pp. 102–107, Dec. 2008.
38. Lee, J., Lou, H.-L., Toumpakaris, D., Jang, E.M., and Cioffi, J.M., Transceiver design for MIMO wireless systems incorporating hybrid ARQ, *IEEE Com. Mag.*, vol. 47, nr. 1, pp. 32–40, Jan. 2009.
39. Sung, Y., Lim, Y., Tong, L., and Veen, A.-J.v.d., Signal processing advances for 3G WCDMA: from Rake receivers to blind techniques, *IEEE Com. Mag.*, vol. 47, nr. 1, pp. 48–54, Jan. 2009.
40. Stevenson, C.R., Chouinard, G., Lei, Z., Hu, W., Shellhammer, S.J., and Caldwell, W., IEEE 802.22: the first cognitive radio wireless regional area network standard, *IEEE Com. Mag.*, vol. 47, nr. 1, pp. 130–138, Jan. 2009.
41. Mudumbai, R., Brown III, D.R., Madhow, U., and Poor, V.H., Distributed transmit beamforming: challenges and recent progress, *IEEE Com. Mag.*, vol. 47, nr. 2, pp. 102–110, Feb. 2009.
42. Astely, D., Dahlman, E., Furuskar, A., Jading, Y., Lindstrom, M., and Parkvall, S., LTE: the evolution of mobile broadband, *IEEE Com. Mag.*, vol. 47, nr. 4, pp. 44–51, Apr. 2009.
43. Jungnickel, V., Schellmann, M., Thiele, L., Wirth, T., Haustein, T., Koch, O., Zirwas, W., and Schulz, E., Interference-aware scheduling in the multiuser MIMO-OFDM downlink, *IEEE Com. Mag.*, vol. 47, nr. 6, pp. 56–66, Jun. 2009.
44. Li, Q., Lin, X.E., Zhang, J., and Roh, W., Advancement of MIMO technology in WiMAX: from IEEE 802.16d/e/j to 802/16m, *IEEE Com. Mag.*, vol. 47, nr. 6, pp. 100–107, Jun. 2009.
45. Zhu, J., and Yin, H., Enabling collocated coexistence in IEEE 802.16 networks via perceived concurrency, *IEEE Com. Mag.*, vol. 47, nr. 6, pp. 108–114, Jun. 2009.
46. Cai, L.X., Shen, X., and Mark, J.W., Efficient MAC protocol for ultra-wideband networks, *IEEE Com. Mag.*, vol. 47, nr. 6, pp. 179–185, Jun. 2009.
47. Berger, S., Kuhn, M., Wittneben, A., Unger, T., and Klein, A., Recent advances in amplify-and-forward two-hop relaying, *IEEE Com. Mag.*, vol. 47, nr. 7, pp. 50–56, Jul. 2009.

48. Issariyakul, T., Pillutla, L.S., and Krishnamurthy, V., Tuning radio resource in an overlay cognitive radio network for TCP: green isn't good, *IEEE Com. Mag.*, vol. 47, nr. 7, pp. 57–63, Jul. 2009.
49. Li, Y., Distributed coding for cooperative wireless networks: an overview and recent advances, *IEEE Com. Mag.*, vol. 47, nr. 8, pp. 71–77, Aug. 2009.
50. Akan, O.B., Isik, M.T., and Baykal, B., Wireless passive sensor networks, *IEEE Com. Mag.*, vol. 47, nr. 8, pp. 92–99, Aug. 2009.
51. Melvin, W.L., A STAP overview, *IEEE A&E Sys. Mag.*, vol. 19, nr. 1, part. 2, pp. 19–35, Jan. 2004.
52. Sadler, B.M., Fundamentals of energy-constrained sensor network systems, *IEEE A&E Sys. Mag.*, vol. 20, nr. 8, part. 2, pp. 17–35, Aug. 2005.
53. Cantrell, B., Rao, J., Tavik, G., Dorsey, M., and Krishevsky, V., Wideband array antenna concept, *IEEE A&E Sys. Mag.*, vol. 21, nr. 1, pp. 9–12, Jan. 2006.
54. Kalinin, A.V., Anechoic chamber wideband antenna measurements, *IEEE A&E Sys. Mag.*, vol. 21, nr. 1, pp. 21–24, Jan. 2006.
55. Nickel, U., Overview of generalized monopulse estimation, *IEEE A&E Sys. Mag.*, vol. 21, nr. 6, part. 2, pp. 27–56, Jun. 2006.
56. Richards, M.A., A beginner's guide to interferometric SAR concepts and signal processing, *IEEE A&E Sys. Mag.*, vol. 22, nr. 9, part. 2, pp. 5–29, Sep. 2007.
57. Huang, Y., WiMAX dynamic beamforming antenna, *IEEE A&E Sys. Mag.*, vol. 23, nr. 8, pp. 26–31, Aug. 2008.
58. Gardikis, G., and Kourtis, A., Using DVB-S2 adaptive coding and modulation for provision of Satellite triple play services, *IEEE Com. Mag.*, vol. 46, nr. 12, pp. 128–135, Dec. 2008.

Chapter 5
Recursive Algorithms for Adaptive Array Systems

5.1 Introduction

The signal processing *heart* of several applications related to filter design and implementation can be modeled as a linear system of equations. In order to solve this linear system of equations, an estimate of the autocorrelation matrix and crosscorrelation vector for given input measurement vectors is required. The autocorrelation matrix and the crosscorrelation vector are then later utilized to yield the desired solution vector or state vector; such are the cases of the Kalman filter update and the multiplier computation for the sample matrix inversion adaptive criterion [1–41].

Based on the properties of the autocorrelation matrix and the application nature, among the other computational techniques, the recursive Cholesky and Modified Graham-Schmidt Orthogonalization (MGSO) can yield a fast, efficient, and robust computation of the desired solution, which is desirable because it reduces the system cost. The dense, recursive Cholesky's and MGSO methods are developed, explained, and compared in this paper. After this comparison is completed, the recursive Cholesky's results twice as fast as the recursive MGSO algorithm and 5/3 as fast as the recursive Sherman–Morrison Formula. Moreover, the recursive Cholesky's and MGSO reduce the eigenvalue ratio by the square root, which is very desirable for most applications.

Previously, we have discussed a recursive solution to the vector normal equation utilizing the recursive Cholesky and MGSO algorithms [1, 2]. Previously, we have also discussed a blind adaptive approach for detection and extraction of signals of interest in the presence of noise and interference without relying on preamble or training sequences. The heart of the blind adaptive algorithm is based upon solving the recursive generalized eigenvalue problem (RGEP), the solution of which is discussed in this chapter based on the recursive Cholesky or *QR* factors and the Householder and *QL* algorithm with implicit shifts. Even though the solution to the RGEP is slightly more efficient than the solution to the *direct* generalized eigenvalue when all the eigenvalues and eigenvectors are required, the improvement is more noticeable when only one eigenvalue and the corresponding eigenvector are required. Therefore, the solution that is proposed here serves well for

I. Progri, *Geolocation of RF Signals*,
DOI 10.1007/978-1-4419-7952-0_5, © Springer Science+Business Media, LLC 2011

recursive generalized eigenvalue problems that require only one eigenvalue and the corresponding eigenvector.

5.2 Recursive Algorithms' Main Description

The Cholesky's method (a.k.a. the direct, dense Cholesky's method in contrast to its recursive, dense counterpart[1]) was first used by Cholesky in France before 1916 in connection with symmetrical systems [3]. In 1938, Th. Banacheiwicz reformulated the Cholesky's method in matrix form in Poland [3]. In 1941, Crout rediscovered and adapted this method to machine computations in the United States [3]. Turing and Zurmühl studied the Cholesky's method again in England in 1948 and in Germany in 1949, respectively [3]. In 1952, Salvadori and Baron provided the algorithm for computing the direct Cholesky factors in the United States [3].

In 1971, Wilkinson and Reinsch adapted the direct Cholesky method for computations in linear systems and linear programming [4] based on the work performed by Fox in 1954 [5].

In the early 1990s and later on, the direct Cholesky's method has been found in almost any good textbook in numerical linear algebra [6–8] due to its fast and easy computation and numerical stability. Moreover, the Cholesky factors are computed as either lower triangular or upper triangular factors [6–8], which makes this method very attractive for different applications.

For a special category of problems, in which the symmetrical matrix is formed recursively, it is very beneficial to compute the Cholesky factors recursively, thus eliminating the computation time to form the matrix. This idea is found as early as with Dongara et al. [9] and Gill et al. [6] and fully incorporated into the MATLAB® software package. Nevertheless, for the purpose of our investigation, a complete and independent derivation of the recursive Cholesky's algorithm is found in Progri's unpublished work [10, 11] and is fully developed and explained in this publication. The first derivation of the fast recursive Cholesky (see [10]) treats the calculation of the lower triangular Cholesky factors recursively for a complex positive-definite matrix (PDM) which satisfies a recursive relation. The second derivation of the fast recursive Cholesky (see [11]) treats the calculation of the upper triangular Cholesky factors recursively for the same matrix discussed in [10]. While both methods produce identical results, depending on the nature of the problem or application, there might be benefit favoring one factorization technique from the other. The reader is also informed in this publication that the recursive Cholesky's method is faster than the Sherman–Morrison formula, which is another important aspect of the Cholesky's almost unknown benefits in the community.

---

[1]The Cholesky's and MGSO methods can be applied to sparse systems as well; nevertheless, the spare Cholesky's and MGSO methods are not considered in this publication.

On the other hand, an early development of the *QR* decomposition for quadratic forms is found in [12]. Nevertheless, Francis [12], Given [13], and Kubalnovskaya [14] are the first who developed the *QR* algorithm, which is conceptually related to the *LR* algorithm developed by Rutishauser [14], later to Barth [17] and Martin [18, 20]. However, a full treatment can be also found as early as with Wilkinson [21–26].

Due to its robustness, the earliest, recursive *QR* algorithm was incorporated into the LINPACK manual in the early 1970s with Dongara et al. [9], which is widely used today in the MATLAB® software package and in any good books on signal processing [27]. Spilker also suggests standard techniques such as that the *QR* and Cholesky's decomposition can be applied to simplify the Kalman computation [28]. Nevertheless, the full and independent derivation of the recursive *QR* algorithm is also found in Progri's work [29] and is polished in this publication.

The generalized eigenvalue problem (GEP) is not new. In the early 1950s, Given [13] presented a method for solving the characteristic values for real, symmetric matrices. In the mid 1950s, Rutishauser [14] discussed the solution to the eigenvalue problem employing the left and right (*LR*) transformation. Kubalnovskaya [15], in the late 1950s, provided also additional algorithms for solution of the complete eigenvalue problem. Another formal treatment of the eigenvalue problem has been found with Hadley in the 1960s [16]. Francis discussed the analogy and uniqueness of the *QR* transformation with respect to the *LR* transformation in early 1960s [12]. Among all, Wilkinson is the leading researcher and the prolific mathematician of the time, who provided the most valuable treatise of the eigenvalue problem in the mid 1960s [21–25]. Then, other works by Martin, Barth, and Wilkinson provide the centerpiece for solving the eigenvalue problem for tridiagonal matrices using the Householder transformation and for reducing a symmetric matrix to a tridiagonal matrix [17–20]. Moler and Stewart also provide a solution to the GEP taken from a more formal perspective [21]. At this point, the research is considered mature and we have among us the first computer programming guides (or packages) on eigenvalue solution package (EISPACK) [22] and linear algebra package (LINPACK) [9]. The design of linear algebra libraries for high performance computers which particularly emphasize on the development of scalable algorithms for multiple instructions multiple data (MIMD) including a brief description of EISPACK, LINPACK, LAPACK, and ScaLAPACK, which is a distributed memory version of LAPACK, is extensively discussed in [34].

On one hand, the eigenvalue problem was considered impractical for most real-time applications due to its high complexity and high computational power requirement. On the other hand, the GEP adds extra complexity and computational burden to the problem. However, for a special class of matrices, such as Hermitian symmetric and semipositive-definite matrices (SPDM), the solution for finding eigenvalues and eigenvectors is somewhat less complex and less computationally intensive than for the general class of square matrices [6–8].

In the last decade of the twentieth century and the beginning of the twenty-first century or so, there has been a growing interest in a special class of the GEP [34–41]. Although there are several applications that require knowledge of all or most of the eigenvalues or both eigenvalues and eigenvectors, for many applications knowing most of the eigenvalues is not necessary. Often, applications only require either the smallest or the largest eigenvalue and the corresponding eigenvector due to the min/max nature of the problem [1, 35]. For example, the solution of a blind adaptive algorithm requires computing the largest eigenvalue and its corresponding eigenvector from the GEP [1, 27]. Hence, if we were to approach this problem, we would only seek to compute the largest eigenvalue and its corresponding eigenvector, leaving the remaining eigenvalues and eigenvectors out of the computation.

Nevertheless, in the treatment presented here we shall consider three approaches: (1) first, compute all eigenvectors or both eigenvalues and eigenvectors; (2) second, compute all the eigenvalues; and (3) third, compute the largest eigenvalue and its corresponding eigenvector which, in fact, is the approach of interest. This methodology will enable us to make comparisons among the three approaches.

Moreover, this chapter is intended to provide a review on the solution to the RGEP and will suggest a methodology that should be followed to yield a fast and robust algorithm for solving the RGEP pertaining to applications discussed in [1, 27].

There are also several works that treat the error analysis or the numerical robustness of these techniques because correct computation is just as important as the accuracy of the method themselves. One of the early classical works that treats the error analysis is presented by Bunch [31]. Refined error analysis of Cholesky's factorization is discussed by Meinguet [32]. Bunch [31] expresses the error as the function of the order of the matrix and Meinguet [32] expresses the error as the function of the largest or smallest eigenvalue or the eigenvalue spectral decomposition.

This chapter is organized as follows. First, we start out with the Cholesky's method, where we treat the Cholesky factors of a complex Hermitian matrix and the recursive Cholesky Method in Sect. 5.3. Second, we discuss the MGSO method and its recursive counterpart in Sect. 5.4. Third, we perform an assessment of both methods and discuss their applications to adaptive signal processing and applications related to adaptive arrays and navigation filter design and implementation in Sect. 5.5. Fourth, we introduce the concept of the GEP in Sect. 5.6. Fifth, we introduce the solution to the RGEP which includes the applications of interests in Sect. 5.7 [1, 27]. Sixth, we perform an assessment for both algorithms in Sect. 5.8. Seventh, we conclude the chapter with conclusions, future work, references, and an appendix in Sect. 5.9, Sect. 5.10.

5.3 The Cholesky's Method for Complex Hermitian Matrices

A complete discussion of the direct and recursive Cholesky's method for a Complex Hermitian Matrix is presented here based on the unpublished work [10]. First, we provide a simple algorithm for obtaining the lower triangular Cholesky factors for a complex Hermitian matrix, in accordance with [10]. Second, we develop the complete algorithm for yielding the recursive Cholesky factors for particular, complex, and Hermitian symmetric matrices [10]. Third, we discuss the solution of the recursive linear system employing the recursive Cholesky factors [10].

The reader is reminded that the upper triangular Cholesky factors can be obtained following the algorithm in Progri's unpublished work [11].

5.3.1 The Direct Cholesky's Method

This is the first step in developing the recursive Cholesky's algorithm. While this is treated in many textbooks, we take the time to explain it for the complex Hermitian matrices, which are less known to the readers and which are the case of interest.

We assume that the matrix **A** is positive-definite and Hermitian symmetric (PDH) because we will see that Cholesky factors, which are explained later, do not exist for nonpositive-definite matrices. Mathematically, the properties of matrix **A** can be expressed as

$$A = A^{\mathrm{H}} = (A^*)^T = \left(A^T\right)^* \quad \text{and} \quad \forall v \neq 0,\ v^{\mathrm{H}}Av > 0. \tag{5.1}$$

We assume the matrix **A** to be decomposed into a lower triangle matrix **L** and an upper triangle matrix whose Hermitian is in fact **L**. The matrix **L** and its Hermitian are also known as the Cholesky *factors* of the matrix **A**, i.e., we want to seek the following matrix relation

$$L \cdot L^{\mathrm{H}} = A. \tag{5.2}$$

Recall from [7] that the components of **L** can be determined from

$$l_{ii} = \sqrt{a_{ii} - \sum_{k=1}^{i-1} l_{ik} l_{ik}^*}, \quad \forall i = j \tag{5.3}$$

and

$$l_{ji} = \frac{1}{l_{ii}}\left(a_{ij}^* - \sum_{k=1}^{i-1} l_{ik}^* l_{jk}\right), \quad \forall i < j. \tag{5.4}$$

It is important to mention here that the Cholesky *factors* of a complex PDH matrix *exist* (we just showed that) and are *unique* because the *LU* factorization of a square matrix is unique [7].

From relation (5.3), note that l_{ii} must be real; thus, **A** must be a Hermitian symmetric matrix and that l_{ii} must be greater than 0, which implies that **A** must be a positive-definite matrix.

The algorithm for obtaining the Cholesky factors of **A** (see Fig. 5.1 for the graphical representation of what follows) is as follows

For $i = 1 : M$
    For $j = i : M$
        $\sigma = a_{ij}^*(k)$
        For $t = i - 1 : M$
            $\sigma = \sigma - l_{it}^*(k) l_{jt}(k)$
        End
        If $j = i$
            $l_{ii}(k) = \sqrt{\sigma}$
        Else
            $l_{ij}(k) = \frac{\sigma}{l_{ii}(k)}$
        End
    End
End

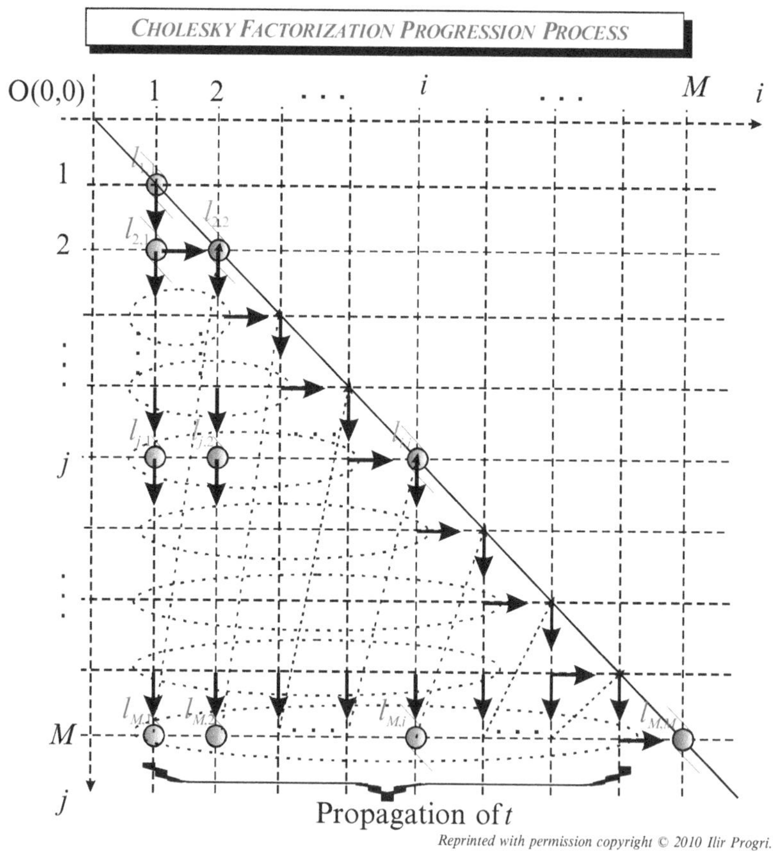

Fig. 5.1 Graphical interpretation for obtaining the Cholesky factorization progression process by the columns given by (5.3) and (5.4). Reprint with permission copyright 2010 Ilir Progri

Also, the Cholesky's method is a fast way to determine the rank of the matrix because its computation complexity is $O(M^3/6)$ (M is the rank of matrix **A**) and this occurs because this method takes advantage of the matrix Hermitian symmetry, compared to $O(M^3/3)$ corresponding to the regular LU decomposition.

Note: The reader may also develop a very similar algorithm that will perform Cholesky factors of **A** doing the substitutions by the rows which will correspond to a very similar graphical representation as the one shown in Fig. 5.1.

For many applications related to adaptive signal processing and navigation filter implementation, the autocorrelation matrices are PDH; therefore, the Cholesky's method can be safely and efficiently exploited to yield fast and efficient computation.

Having discussed the direct Cholesky's method, we analyze, next, the recursive Cholesky's algorithm for particular, complex PDH matrices.

5.3.2 The Recursive Cholesky's Algorithm

While the direct Cholesky's method is faster than the direct LU decomposition $O(M^3/3)$, for large M this (the direct Cholesky's method) algorithm is still expensive. The reader is reminded that the computation complexity of $O(M^3/6)$ was achieved by exploiting the mathematical properties of the correlation matrix. However, someone may wonder: Is it possible to reduce the computation complexity by exploring the physical characteristics of the correlation matrix? The derivation that follows answers this question positively.

Assume that the PDH matrix **A** is formed recursively, the analytical expression of which obeys the physical principle (or law) defined, at the kth cycle, by

$$A(k) = (1 - \alpha)A(k - 1) + \alpha x(k) \cdot x^{\mathrm{H}}(k). \tag{5.5}$$

We will see in the assessment section that this expression for the correlation matrix is a very suitable model for many applications of interest.

In the above expression, **x** is a complex (or real) vector of input data of size M (which is also the rank of matrix **A**) and α is a real, adjusting parameter. For an important observation of the parameter α, the reader should refer to Appendix A.

Expressions of $a_{ii}(k)$ and $a_{ij}^*(k)$ in light of (5.5) are determined from

$$a_{ii}(k) = (1 - \alpha)a_{ii}(k - 1) + \alpha|x_i(k)|^2, \quad \forall i = j, \tag{5.6}$$

$$a_{ij}(k) = (1 - \alpha)a_{ij}(k - 1) + \alpha x_i(k)x_j^*(k), \quad \forall i \neq j, \tag{5.7}$$

$$a_{ij}^*(k) = (1 - \alpha)a_{ij}^*(k - 1) + \alpha x_i^*(k)x_j(k), \quad \forall j \neq i. \tag{5.8}$$

Since calculating the direct Cholesky factors for every cycle (or epoch) is still computationally expensive, $O(M^3/6)$, we would like to obtain the Cholesky factors of $\mathbf{A}(k)$ (see (5.5)), assuming that we previously know the Cholesky factors of $\mathbf{A}(k-1)$ and the vector of input data $\mathbf{x}(k)$. Thus, (5.5) can be rewritten as

$$\begin{aligned}\mathbf{L}(k)\cdot\mathbf{L}^{\mathrm{H}}(k) &= (1-\alpha)\mathbf{L}(k-1)\cdot\mathbf{L}^{\mathrm{H}}(k-1)+\alpha\mathbf{x}(k)\cdot\mathbf{x}^{\mathrm{H}}(k)\\ &= L(k-1)\left[(1-\alpha)\mathbf{I}+\alpha\mathbf{y}(k)\cdot\mathbf{y}^{\mathrm{H}}(k)\right]\mathbf{L}^{\mathrm{H}}(k-1),\end{aligned} \tag{5.9}$$

where

$$\mathbf{x}(k)=\mathbf{L}(k-1)\cdot\mathbf{y}(k) \quad \text{and} \quad \mathbf{x}^{\mathrm{H}}(k)=\mathbf{y}^{\mathrm{H}}(k)\cdot\mathbf{L}^{\mathrm{H}}(k-1). \tag{5.10}$$

If we are, however, able to obtain the Cholesky factors of

$$(1-\alpha)\mathbf{I}+\alpha\mathbf{y}(k)\cdot\mathbf{y}^{\mathrm{H}}(k)=\tilde{\mathbf{L}}(k)\cdot\tilde{\mathbf{L}}^{\mathrm{H}}(k), \tag{5.11}$$

then (5.9) can be modified as

$$\mathbf{L}(k)\cdot\mathbf{L}^{\mathrm{H}}(k)=\mathbf{L}(k-1)\cdot\tilde{\mathbf{L}}(k)\cdot\tilde{\mathbf{L}}^{\mathrm{H}}(k)\cdot\mathbf{L}^{\mathrm{H}}(k-1), \tag{5.12}$$

hence, the following must be satisfied

$$\mathbf{L}(k)=\mathbf{L}(k-1)\cdot\tilde{\mathbf{L}}(k). \tag{5.13}$$

We just showed that, in principle, the Cholesky factors of $\mathbf{A}(k)$ could be obtained from the Cholesky factors of $\mathbf{A}(k-1)$ and the vector of input data $\mathbf{x}(k)$. On the other hand, these *factors* are unique due to the uniqueness of the Cholesky's decomposition, which confirms the uniqueness of (5.13). Therefore, the remainder of this section deals with the approach of obtaining recursively and efficiently the Cholesky factors of $\mathbf{A}(k)$ by satisfying the property (5.13). In order to accomplish this goal, first consider the factorization of

$$\tilde{\mathbf{A}}(k)=(1-\alpha)\mathbf{I}+\alpha\mathbf{y}(k)\cdot\mathbf{y}^{\mathrm{H}}(k). \tag{5.14}$$

Clearly, the element-by-element decomposition of (5.14) produces

$$\tilde{a}_{ii}(k)=(1-\alpha)+\alpha|y_i(k)|^2, \quad \forall i=j, \tag{5.15}$$

$$\tilde{a}_{ij}(k)=\alpha y_i(k)y_j^*(k), \quad \forall i\neq j, \tag{5.16}$$

Where, on the other hand, exploiting the Cholesky's decomposition (5.11) yields

$$\tilde{a}_{ii}(k)=\tilde{l}_{ii}(k)\tilde{l}_{ii}^*(k)+\sum_{t=1}^{i-1}\left[\tilde{l}_{it}(k)\tilde{l}_{it}^*(k)\right], \quad \forall i=j, \tag{5.17}$$

$$\tilde{a}_{ij}(k)=\tilde{l}_{ii}(k)\tilde{l}_{ij}^*(k)+\sum_{t=1}^{i-1}\left[\tilde{l}_{it}(k)\tilde{l}_{jt}^*(k)\right], \quad \forall i\neq j. \tag{5.18}$$

Equating both sides of (5.16) with (5.17) and (5.16) with (5.18), respectively, yields

$$\tilde{l}_{ii}(k) = \sqrt{(1-\alpha) + \alpha|y_i(k)|^2 - \sum_{t=1}^{i-1}\left[\tilde{l}_{it}(k)\tilde{l}_{it}^*(k)\right]}, \quad \forall i = j, \tag{5.19}$$

$$\tilde{l}_{ji}(k) = \frac{\alpha y_i^*(k)y_j(k) - \sum_{t=1}^{i-1}\left[\tilde{l}_{it}^*(k)\tilde{l}_{jt}(k)\right]}{\tilde{l}_{ii}(k)}, \quad \forall j > i. \tag{5.20}$$

Intuitively, we realize that

$$\tilde{l}_{ij}(k) = \beta_j(k)y_i^*(k) = y_i^*(k)\beta_j(k), \quad \forall j \le i, \tag{5.21}$$

$$\tilde{l}_{ji}(k) = \tilde{l}_{ij}^*(k) = \beta_i^*(k)y_j(k) = y_j(k)\beta_i^*(k), \quad \forall j \ge i, \tag{5.22}$$

where $\beta_j(k)$ is a coefficient that is defined later in this document. The proof of (5.21) and (5.22) is given with the help of the total mathematical induction. First, we note that

$$\tilde{l}_{ii}(k) = (1-\alpha) + \alpha|y_1(k)|^2, \quad \forall i = 1, \tag{5.23}$$

$$\tilde{l}_{i1}(k) = \alpha\frac{y_i^*(k)y_1(k)}{\tilde{l}_{11}(k)}, \quad \forall i > 1. \tag{5.24}$$

If we define, in accordance with (5.21) or (5.22), the coefficient

$$\beta_1(k) = \alpha\frac{y_1(k)}{\tilde{l}_{11}(k)}, \quad \forall i > 1, \tag{5.25}$$

then we can rewrite (5.24) as

$$\tilde{l}_{i1}(k) = \beta_1(k)y_i^*(k) = y_i^*(k)\beta_1(k), \quad \forall i > 1. \tag{5.26}$$

Now assuming that (5.21) is true, we apply the result of (5.26) into (5.19) and (5.20) and the following is obtained

$$\tilde{l}_{ii}(k) = \sqrt{(1-\alpha) + \left(\alpha - \sum_{t=1}^{i-1}|\beta_t(k)|^2\right)|y_i(k)|^2}, \quad \forall i = j, \tag{5.27}$$

$$\tilde{l}_{ji}(k) = \frac{\left(\alpha - \sum_{t=1}^{i-1}|\beta_t(k)|^2\right)y_i^*(k)y_j(k)}{\tilde{l}_{ii}(k)} = \beta_i^*(k)y_j(k), \quad \forall j > i, \tag{5.28}$$

with the new coefficient

$$\beta_i(k) = \frac{\alpha - \sum_{t=1}^{i-1} |\beta_t(k)|^2}{\tilde{l}_{ii}(k)} y_i(k), \tag{5.29}$$

and therefore completing the proof. If we define

$$\alpha^i(k) = \alpha - \sum_{t=1}^{i-1} |\beta_t(k)|^2, \tag{5.30}$$

then a recursive relation for the $\alpha^i(k)$ is determined from

$$\alpha^{i+1}(k) = \alpha^i(k) - |\beta_i(k)|^2. \tag{5.31}$$

Employing (5.30) into the expression of $\beta_i(k)$, given by (5.29), produces

$$\beta_i(k) = \frac{\alpha^i(k)}{\tilde{l}_{ii}(k)} y_i(k), \tag{5.32}$$

and the expression for the recursive relation of $\alpha^i(k)$ becomes

$$\alpha^{i+1}(k) = \alpha^i(k) - \frac{|\alpha^i(k)|^2}{|\tilde{l}_{ii}(k)|^2} |y_i(k)|^2 = \frac{\alpha^i(k)}{|\tilde{l}_{ii}(k)|^2}. \tag{5.33}$$

Finally, we present the equations for obtaining the diagonal and the off-diagonal elements of $\mathbf{L}(k)$, by writing out in component (5.13), which produces

$$l_{ii}(k) = l_{ii}(k-1)\tilde{l}_{ii}(k) = l_{ii}(k-1)\sqrt{[1-\alpha^i(k)] + \alpha^i(k)|y_i(k)|^2}, \quad \forall i = j, \tag{5.34}$$

$$\begin{aligned} l_{ij}(k) &= l_{ij}(k-1)\tilde{l}_{ij}(k) + \sum_{t=j+1}^{n} [\tilde{l}_{it}(k-1)\tilde{l}_{tj}(k)] \\ &= l_{ij}(k-1)\sqrt{[1-\alpha^j(k)] + \alpha^j(k)|y_j(k)|^2} + \sum_{t=j+1}^{n} [l_{it}(k-1)\beta_j^*(k)y_t(k)] \\ &= l_{ij}(k-1)\sqrt{[1-\alpha^j(k)] + \alpha^j(k)|y_j(k)|^2} + \beta_j^*(k)x_i^j(k), \quad \forall i > j, \end{aligned} \tag{5.35}$$

where

$$x_i^j(k) = \sum_{t=j+1}^{n} [l_{it}(k-1)y_t(k)]. \tag{5.36}$$

Intuitively, we also realize that

$$x_i^0(k) = \sum_{t=1}^{n} [l_{it}(k-1)y_t(k)] = x_i(k), \tag{5.37}$$

and recursively

$$x_i^{j+1}(k) = x_i^j(k) - l_{ij}(k-1)y_j(k). \tag{5.38}$$

Also, recall that

$$y_j(k) = \frac{x_j(k) - \sum_{t=1}^{j}[l_{jt}(k-1)y_t(k)]}{l_{jj}(k-1)} = \frac{x_i^j(k)}{l_{jj}(k-1)}. \tag{5.39}$$

Substituting expression (5.39) into (5.34) and (5.35) finally produces the desired expressions of the diagonal and off-diagonal elements of $\mathbf{L}(k)$, which are

$$l_{ii}(k) = l_{ii}(k-1)\sqrt{[1-\alpha^i(k)] + \alpha^i(k)\frac{\left|x_i^j(k)\right|^2}{|l_{ii}(k-1)|^2}}, \quad \forall i = j, \tag{5.40}$$

$$l_{ij}(k) = l_{ij}(k-1)\sqrt{[1-\alpha^j(k)] + \alpha^j(k)\frac{\left|x_j^j(k)\right|^2}{\left|l_{jj}(k-1)\right|^2}} + \beta_j^*(k)x_i^j(k), \quad \forall i > j. \tag{5.41}$$

Clearly, the recursive algorithm for obtaining the diagonal and the off-diagonal elements of $\mathbf{L}(k)$ according to (5.40) and (5.41) is in the order on $O(M^2/2)$, where M is the rank of the complex Hermitian matrix $\mathbf{A}(k)$, which is $O(M/3)$ times faster than the direct Cholesky's decomposition. Nevertheless, this is only the smallest portion of the computational benefit. The largest portion of the benefit is discussed in the assessment section.

The above recursive relations can be implemented efficiently in the following algorithm for obtaining the recursive Cholesky factors by the columns (see Fig. 5.2)

For $i = 1 : M$

$\gamma(k) = \sqrt{[1-\alpha^i(k)] + \alpha^i(k)\frac{\left|x_i^j(k)\right|^2}{|l_{ii}(k-1)|^2}}$

$l_{ii}(k) = l_{ii}(k-1)\gamma(k)$

$\beta_i(k) = \frac{\alpha^i(k)}{l_{ii}(k)}x_i^i(k)$

For $j = i+1 : M$

$x_j^i(k) = x_j^i(k) - \frac{l_{ji}(k-1)}{l_{ii}(k-1)}x_i^i(k)$

$l_{ji}(k) = l_{ji}(k-1)\gamma(k) + \beta_i^*(k)x_j^i(k)$

End

$\alpha^i(k) = \frac{\alpha^i(k)}{\gamma^2(k)}$

End

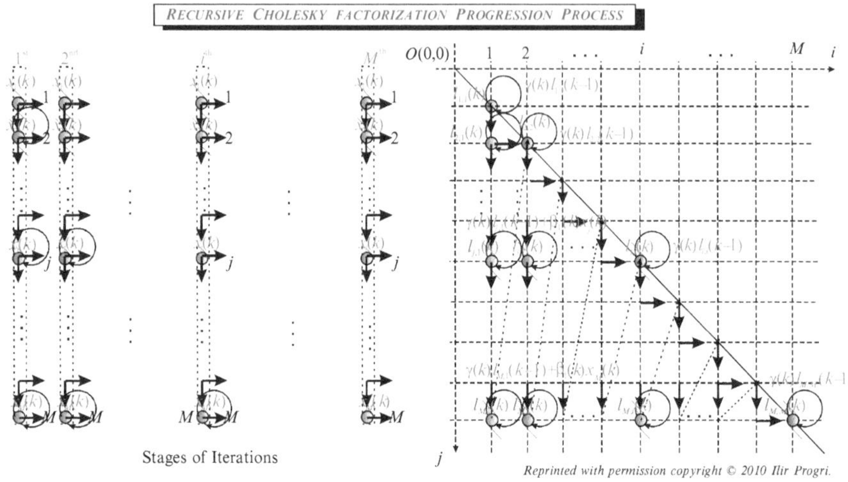

Fig. 5.2 Graphical interpretation for obtaining the recursive Cholesky factorization progression process by the columns given by (5.37–5.41). Reprint with permission copyright 2010 Ilir Progri

Figure 5.2 should also aide in the understanding and intuitive progression of the recursive Cholesky's factorization which was so brilliantly introduced before as part of (5.37) through (5.41). By all means, Fig. 5.2 is not meant to substitute the algorithm; it is only meant to provide a visual interpretation of how recursive Cholesky progresses as far the elements of $l_{ii}(k)$ and $l_{ji}(k)$ and the $x_j^i(k)$ get computed. Without having to burden the reader with too much detail, the reader should be able to fill in the rest of the computations of $\gamma(k)$, $\beta_i(k)$, and $\alpha^i(k)$. *Again, a very similar algorithm can be obtained using the recursive Cholesky factorization by the rows.*

This then completes the discussion of the recursive Cholesky's algorithm and we continue next with the discussion on the recursive solution of a linear system.

5.3.3 The Recursive Solution of a Complex Linear System

The ultimate goal of any method that we use is the computation of the desired vector, which is of the classical form

$$\mathbf{z}(k) = \mathbf{A}^{-1}(k) \cdot \mathbf{b}(k). \tag{5.42}$$

Instead of solving the above linear system, we will use an alternate method, which is symbolically known as

$$\mathbf{z}(k) = \mathbf{A}(k)/\mathbf{b}(k) \quad \text{or} \quad \mathbf{z}(k) = \mathbf{A}(k)\backslash\mathbf{b}(k), \tag{5.43}$$

because it is less expensive than the solution obtained from inverting the direct matrix (5.42), result which to some extent is known. Nevertheless, we refresh the memory of our welcomed readers.

Suppose that the updated Cholesky factors of $\mathbf{A}(k)$ are available then the vector components of $\mathbf{z}(k)$ are computed utilizing the forward and the backward substitution, which is described in details in the reminder of this subsection employed by (5.43) instead of (5.42).

In order to accomplish this goal, first, after obtaining the Cholesky factors of $\mathbf{A}$, the linear system of equation $\mathbf{A}(k) \cdot \mathbf{z}(k) = \mathbf{b}(k)$ can be written as

$$\mathbf{L}(k) \cdot \mathbf{L}^{\mathrm{H}}(k) \cdot \mathbf{z}(k) = \mathbf{b}(k). \tag{5.44}$$

Next, we define

$$\mathbf{v}(k) = \mathbf{L}^{\mathrm{H}}(k) \cdot \mathbf{z}(k), \tag{5.45}$$

then solve

$$\mathbf{L}(k) \cdot \mathbf{v}(k) = \mathbf{b}(k). \tag{5.46}$$

Writing out (5.45) in components yields, $\sum_{j=1}^{i-1} l_{ij}(k)v_j(k) + l_{ii}(k)v_i(k) = b_i(k)$, from where we obtain (5.46).

Notice that in order to compute the ith value $v_i(k)$, we need to know first all the previous components $\{v_1(k) \cdots v_{i-1}(k)\}$. This recursive technique is given the name of *forward substitution*, as we move forward in substituting the values of v_i.

The above recursive relations are implemented efficiently in the following algorithm

For $i = 1 : M$
  $\sigma = 0$
  For $j = 1 : i-1$
    $\sigma = \sigma + l_{ij}(k)v_j(k)$
  End
  $v_i(k) = \frac{1}{l_{ii}(k)}\left[b_i(k) - \sigma\right]$
End

Next, it remains to compute the values of $\mathbf{z}(k)$ given by (5.46). Similar to the *forward substitution* technique, we write out in component equation (5.46) as, $\sum_{j=i+1}^{M} l_{ji}^*(k)z_j(k) + l_{ii}(k)z_i(k) = v_i(k)$, from where we obtain

$$z_i(k) = \frac{1}{l_{ii}(k)}\left[v_i(k) - \sum_{j=i+1}^{M} l_{ji}^*(k)z_j(k)\right]. \tag{5.47}$$

In contrast to the forward substitution technique, we move backwards, which means that in order to compute the ith value $z_i(k)$, we need to know first the last components $\{z_{i+1}(k)\cdots z_M(k)\}$ and the technique is known with the name of *backward substitution.*

The above recursive relations are implemented efficiently in the following algorithm

For $i = 1 : M$
  $\sigma = 0$
  For $j = i + 1 : M$
    $\sigma = \sigma + l^*_{ji}(k)z_j(k)$
  End
  $z_i(k) = \frac{1}{l_{ii}(k)}[v_i(k) - \sigma]$
End

So far we have discussed the direct classical Cholesky method and the modified recursive Cholesky method for a recursive, complex PDH matrices which satisfy (5.5). For these matrices, we have analytically justified the benefit of employing the recursive Cholesky factorization.

The computation complexity exploiting the direct Cholesky factorization is faster than the algorithm presented in Johnson's patent [30], because we never invert the Cholesky's factor as the inversion requires another $O(M^3/6)$. Based on this argument, the recursive Cholesky's is more than an order of magnitude faster than the Johnson's algorithm in patent [30].

Moreover, the recursive Cholesky's reduces the eigenvalue ration by the square root and is numerically very sound [11, 32].

This discussion continues on the Assessment of Both Algorithms section, where we present a couple of very interesting applications that are related to adaptive signal and filter processing. Next, we begin the investigation of the other competing algorithm, which is the MGSO method.

5.4 The MGSO Method for Complex PDH Matrices

The MGSO is another competing algorithm to Cholesky's, both in computation time and numerical robustness. Hence, there is value for a complete investigation of this method for complex PDH matrices as discussed earlier.

This section includes first a simple algorithm for obtaining the QR factors for a complex PDH matrix, based on the initial work [29]. Next, we develop the complete algorithm for yielding the QR factors recursively for recursive, complex PDH matrices, which satisfy (5.5), [29]. Third, we discuss the solution of the recursive linear system employing the recursive QR factors.

5.4.1 The Direct MGSO Algorithm

Assume that the PDH matrix $\mathbf{A}$ (see (5.1)) can be expressed in the form of

$$\mathbf{A} = \mathbf{Q} \cdot \mathbf{R}, \tag{5.48}$$

where $\mathbf{R}$ is an upper triangular matrix and $\mathbf{Q}$ is orthogonal that means

$$\mathbf{Q}^{\mathrm{H}} \cdot \mathbf{Q} = \mathbf{I}. \tag{5.49}$$

While there are several algorithms for obtaining the QR factors, the standard algorithm is based on the Householder transformation [7]; hence, we will only give the general idea behind this transformation. First, we arrange the matrix $\mathbf{Q}_1$, which is going to zero all elements in the first column of the matrix $\mathbf{A}$ below the first element. Similarly, $\mathbf{Q}_2$ zeros all elements of the second column, below the second element, and this transformation continues until $\mathbf{Q}_{n-1}$ leaving the last element of the last column; thus, the matrix $\mathbf{A}$ is transformed into an upper triangular matrix

$$\mathbf{R} = \mathbf{Q}_{n-1} \cdots \mathbf{Q}_1 \cdot \mathbf{A}. \tag{5.50}$$

Also, knowing that the Householder matrices are orthogonal produces

$$\mathbf{Q} = (\mathbf{Q}_{n-1} \cdots \mathbf{Q}_1)^{-1} = \mathbf{Q}_1 \cdots \mathbf{Q}_{n-1}. \tag{5.51}$$

When solving the linear system, it is going to become obvious that the $\mathbf{Q}$ matrix is not required explicitly; hence, we store it in the form given by (5.51). Moreover, pivoting is not usually necessary unless the matrix is close to singular, which is not the case anyway.

Define the input data in matrix $\mathbf{X}(k-1)$

$$\mathbf{X}(k-1) = \left[\mathbf{x}_1^{\mathrm{H}}(k-1) \quad \cdots \quad \mathbf{x}_{N-1}^{\mathrm{H}}(k-1)\right], \tag{5.52}$$

where, $\mathbf{x}_i(k-1)$ is a vector of size M, $\forall i \in \{1, 2, \cdots, N-1\}$.

Also, the training signal sample vector $\mathbf{s}(k-1)$ is determined from

$$\mathbf{s}(k-1) = \left[s_1^*(k-1) \quad \cdots \quad s_{N-1}^*(k-1)\right]^T. \tag{5.53}$$

We form the correlation matrix $\mathbf{A}(k-1)$ (of size $M \times M$) in accordance with

$$\mathbf{A}(k-1) = \alpha \mathbf{X}(k-1)^{\mathrm{H}} \cdot \mathbf{X}(k-1), \tag{5.54}$$

and the crosscorrelation vector $\mathbf{b}(k-1)$ (of size $M \times 1$) as

$$\mathbf{b}(k-1) = \alpha \mathbf{X}(k-1)^{\mathrm{H}} \cdot \mathbf{s}(k-1). \tag{5.55}$$

Assuming that the QR decomposition of $\mathbf{X}(k-1)$ is possible, i.e.,

$$\mathbf{X}(k-1) = \mathbf{Q}(k-1) \cdot \mathbf{R}(k-1), \tag{5.56}$$

where $\mathbf{R}(k-1)$ is an upper triangular matrix and $\mathbf{Q}(k-1)$ is a matrix which satisfies the following

$$\mathbf{Q}(k-1)^{\mathrm{H}} \cdot \mathbf{Q}(k-1) = \mathbf{I}, \tag{5.57}$$

then the following yields

$$\begin{aligned}\mathbf{A}(k-1) &= \alpha\mathbf{R}(k-1)^{\mathrm{H}} \cdot \mathbf{Q}(k-1)^{\mathrm{H}} \cdot \mathbf{Q}(k-1) \cdot \mathbf{R}(k-1) \\ &= \alpha\mathbf{R}(k-1)^{\mathrm{H}} \cdot \mathbf{R}(k-1),\end{aligned} \tag{5.58}$$

which is the scaled upper triangular Cholesky factor. This fact is indeed the great motivation that we should never form the matrix $\mathbf{A}(k-1)$, but instead try to find the QR factors of the $\mathbf{X}(k-1)$ matrix and use them to find the solution for $\mathbf{z}(k-1)$, which we expect to be in similar form as discussed in the Cholesky's decomposition. Nevertheless, we should mention here that the computation complexity of the MGSO is equal to $O(M^3/6 + 3M^2/6)$, which is at least twice as slow as the direct Cholesky's factorization.

Although we suspect that the recursive MGSO would be slower than the recursive Cholesky's, the following section discusses the recursive MGSO method anyway.

5.4.2 The Recursive MGSO Algorithm

The discussion of the recursive MGSO algorithm reveals the necessary steps of computing the QR factors recursively for complex PDH matrices of interest.

Assume that we want to add the new vector of data $\mathbf{x}(k)$ into the existing set of data. The new correlation matrix, $\mathbf{A}(k)$, can be obtained as

$$\mathbf{A}(k) = (1-\alpha)\mathbf{X}(k-1) + \alpha\mathbf{x}(k) \cdot \mathbf{x}^{\mathrm{H}}(k). \tag{5.59}$$

Similarly, the new crosscorrelation vector reads

$$\mathbf{b}(k) = (1-\alpha)\mathbf{X}(k-1) \cdot \mathbf{s}(k-1) + \alpha\mathbf{x}(k)s(k). \tag{5.60}$$

Note that the real factor α is fully discussed in Appendix A.

Assume that $\mathbf{Q}(k)$ and $\mathbf{R}(k)$ are the new factors of $\sqrt{\alpha}\mathbf{X}(k) = [\sqrt{1-\alpha}\mathbf{X}(k-1)$ $\sqrt{\alpha}\mathbf{x}^{\mathrm{H}}(k)]^T$ such that

$$\mathbf{X}(k) = \mathbf{Q}(k) \cdot \mathbf{R}(k) \quad \text{and} \quad \mathbf{Q}^{\mathrm{H}}(k) \cdot \mathbf{Q}(k) = \mathbf{I}. \tag{5.61}$$

To make things easier, we can write

$$\mathbf{X}(k) = [\,\beta\mathbf{X}(k-1) \quad \mathbf{x}^{\mathrm{H}}(k)\,]^{T}, \tag{5.62}$$

where $\beta = \sqrt{(1-\alpha)/\alpha}$.

If we were to apply (5.61) to obtain the factors for matrix $\mathbf{A}(k)$, then we end up in the same algorithm as explained in the recursive Cholesky decomposition. Nevertheless, we seek an alternative algorithm to obtain the solution for the weigh vector.

The new solution for the weight vector can be written as

$$\mathbf{z}(k) = [\mathbf{R}^{\mathrm{H}}(k) \cdot \mathbf{R}(k)]^{-1} \cdot \mathbf{R}^{\mathrm{H}}(k) \cdot \mathbf{Q}(k) \cdot \mathbf{s}(k) = \mathbf{R}(k)/\mathbf{v}(k), \tag{5.63}$$

where

$$\mathbf{v}(k) = \mathbf{Q}^{\mathrm{H}}(k) \cdot \mathbf{s}(k). \tag{5.64}$$

Question: Is it possible to recursively obtain updates for $\mathbf{Q}(k)$, $\mathbf{R}(k)$, and $\mathbf{v}(k)$, given $\mathbf{R}(k)$, and $\mathbf{v}(k)$, $\mathbf{x}(k)$, and $\mathbf{s}(k)$?

Fortunately, we already know that the answer to this question should be yes. We can rewrite (5.61) and (5.64) in the form of

$$\mathbf{Q}^{\mathrm{H}}(k)[\,\mathbf{X}(k) \quad \mathbf{s}(k)\,] = \mathbf{Q}^{\mathrm{H}}(k)\begin{bmatrix} \beta\mathbf{X}(k-1) & \beta\mathbf{s}(k-1) \\ \mathbf{x}^{\mathrm{H}}(k) & s^{*}(k) \end{bmatrix} = [\,\mathbf{R}(k) \quad \mathbf{v}(k)\,]. \tag{5.65}$$

Substituting the matrix, $\mathbf{X}(k-1) = \mathbf{Q}(k-1) \cdot \mathbf{R}(k-1)$, and the vector, $\mathbf{s}(k-1) = \mathbf{Q}(k-1) \cdot \mathbf{v}(k-1)$, into (5.65) produces

$$\begin{aligned} &\mathbf{Q}^{\mathrm{H}}(k)\begin{bmatrix} \beta\mathbf{Q}(k-1) \cdot \mathbf{R}(k-1) & \beta\mathbf{Q}(k-1) \cdot \mathbf{v}(k-1) \\ \mathbf{x}^{\mathrm{H}}(k) & s^{*}(k) \end{bmatrix} \\ &= \mathbf{Q}^{\mathrm{H}}(k)\begin{bmatrix} \mathbf{Q}(k-1) & 0 \\ 0 & 1 \end{bmatrix}\begin{bmatrix} \beta\mathbf{R}(k-1) & \beta\mathbf{v}(k-1) \\ \mathbf{x}^{\mathrm{H}}(k) & s^{*}(k) \end{bmatrix} = [\,\mathbf{R}(k) \quad \mathbf{v}(k)\,]. \end{aligned} \tag{5.66}$$

Denote with $\mathbf{Q}_1$ the matrix

$$\mathbf{Q}_1 = \begin{bmatrix} \mathbf{Q}(k-1) & 0 \\ 0 & 1 \end{bmatrix}, \quad \mathbf{Q}_1^{\mathrm{H}} \cdot \mathbf{Q}_1 = \mathbf{I}. \tag{5.67}$$

Denote also the factors of the matrix

$$\begin{bmatrix} \beta\mathbf{R}(k-1) & \beta\mathbf{v}(k-1) \\ \mathbf{x}^{\mathrm{H}}(k) & s^{*}(k) \end{bmatrix} = \mathbf{Q}_2 \cdot [\,\mathbf{R}_2 \quad \mathbf{v}_2\,], \quad \mathbf{Q}_2^{\mathrm{H}} \cdot \mathbf{Q}_2 = \mathbf{I}. \tag{5.68}$$

Based on the uniqueness of the QR decomposition, it yields

$$\mathbf{Q}(k) = \mathbf{Q}_1 \cdot \mathbf{Q}_2 \quad \text{and} \quad [\,\mathbf{R}_2 \quad \mathbf{v}_2\,] = [\,\mathbf{R}(k) \quad \mathbf{v}(k)\,]. \tag{5.69}$$

We can write the matrix $\mathbf{Q}_2$ in the form of

$$\mathbf{Q}_2 = [\,\mathbf{Q}_3 \quad \mathbf{q}_3\,]^T \quad \text{and} \quad \mathbf{Q}(k) = [\,\mathbf{Q}(k-1)\cdot\mathbf{Q}_3 \quad \mathbf{q}_3\,]^T. \tag{5.70}$$

Therefore, the update equation for $\mathbf{v}(k)$ is the following

$$\begin{aligned}\mathbf{v}(k) &= \mathbf{Q}^{\mathrm{H}}(k)\cdot\mathbf{s}(k) = \left[\mathbf{Q}_2^{\mathrm{H}}\cdot\mathbf{Q}^{\mathrm{H}} \quad \mathbf{q}_3^{\mathrm{H}}\right][\,\beta\mathbf{s}(k-1) \quad s^*(k)\,]^T \\ &= \beta\mathbf{Q}_3^{\mathrm{H}}\cdot\mathbf{v}(k-1) + \mathbf{q}_3^{\mathrm{H}}s^*(k).\end{aligned} \tag{5.71}$$

Next, we provide the recursive MGSO algorithm as follows

For $i = 1 : M$
    $\sigma = 0$
    For $j = 1 : i$
        $\sigma = \sigma + |X_{ji}|^2$
    End
    $\sigma = \sigma + |X_{Ni}|^2$
    $r_{ii} = \sqrt{\sigma}$
    $\sigma = 0$
    For $j = 1 : i$
        $q_j = \frac{X_{ji}}{r_{ii}}$
        $\sigma = \sigma + q_j^* v_j$
    End
    $q_N = \frac{X_{Ni}}{r_{ii}}$
    $v_i = \sigma + q_N^* s_N$
    For $j = i + 1 : M$
        $\sigma = 0$
        For $t = 1 : i$
            $\sigma = \sigma + q_t^* X_{tj}$
        End
        $r_{ij} = \sigma + q_N^* X_{Nj}$
        For $t = 1 : i$
            $X_{tj} = X_{tj} - r_{tj} q_t$
        End
        $X_{Nj} = X_{Nj} - r_{tj} q_N$
    End
End

Thus far we have completed the derivation of the recursive MGSO algorithm. Not surprisingly, the computation complexity of this algorithm is also in the order of $O(M^2)$ which is twice as slow as the recursive Cholesky's.

5.4.3 *The Solution of a Complex, Recursive Linear System*

The solution for the weight vector is given by

$$\mathbf{z}(k) = \mathbf{A}(k)^{-1} \cdot \mathbf{b}(k) = [\mathbf{X}^{\mathrm{H}}(k) \cdot \mathbf{X}(k)]^{-1} \cdot \mathbf{X}^{\mathrm{H}}(k) \cdot \mathbf{s}(k). \tag{5.72}$$

Substituting (5.56) and (5.57) into (5.72) yields

$$\mathbf{z}(k) = [\mathbf{R}^{\mathrm{H}}(k) \cdot \mathbf{R}(k)]^{-1} \cdot \mathbf{R}^{\mathrm{H}}(k) \cdot \mathbf{Q}^{\mathrm{H}}(k) \cdot \mathbf{s}(k) = \mathbf{R}(k)/\mathbf{v}(k), \tag{5.73}$$

where

$$\mathbf{v}(k) = \mathbf{Q}^{\mathrm{H}}(k) \cdot \mathbf{s}(k). \tag{5.74}$$

In order to obtain the solution for the vector, $\mathbf{z}(k)$, we perform *backward substitution*, which is in the order of $O(M^2/2)$, and thus the total computation complexity is of the order of $O(3M^2/2)$, which is the same as the recursive Cholesky's. Nevertheless, this may not be the case of interest because updating the solution vector every iteration may be expensive.

Note that the recursive MGSO is, however, faster than the direct orthogonalization algorithm presented in [33] by an order of magnitude, which makes the recursive MGSO very appealing and desirable.

Two real application scenarios are discussed next.

5.5 Assessment of Both Recursive Algorithms

Thus far we have explained and studied both the direct and recursive methods; now, it remains to analyze a couple of important application cases – *simple block processing* and *recursive block processing* – which are also based on the information of Appendix A.

5.5.1 *Simple Block Processing*

Steps required to complete this scenario are described in the following procedure. First, compute recursively the autocorrelation matrix factors for given input data vectors, N (e.g., iterations, cycles, or epochs). Second, update the weights once at the end of the processing of the input data. Third, we reset (or zero out) the autocorrelation matrix factors.

If we were to perform the recursive Cholesky's decomposition than the computation complexity would be of the order of $O(NM^2/2 + M^2)$.[2]

If we were, however, able to perform the recursive MGSO algorithm than the computation complexity would be of the order of $O(NM^2 + M^2/2)$.

For many applications, the block size N is much larger than the rank of matrix M; thus, roughly speaking although for every cycle both recursive methods yield identical results for this case the recursive Cholesky's is twice as fast as the recursive MGSO.

5.5.2 *Recursive Block Processing*

In order to complete this scenario the following procedure should be performed. First, compute recursively the autocorrelation matrix factors for given input data vectors, N (e.g., iterations, cycles, or epochs). Second, update the weights once at the end of the input data processing. Third, apply a forgetting factor to the old autocorrelation matrix factors.

Note that the *recursive block processing* is different from the *simple block processing* only as regards the third step. Due to this, the *recursive block processing* is supposed to be slightly faster than the *simple block processing*; nevertheless, even for this scenario, the recursive Cholesky's computes the final solution twice as fast as the recursive MGSO.

Appendix B treats the recursive Cholesky's vs. the Sherman–Morrison Formula, a result which is almost unknown to most scientists and engineers.

5.6 Generalized Eigenvalue Problem

There are several applications in engineering, physics, and applied mathematics that can be modeled with the help of the GEP. There are also a large number of publications that deal with the GEP. The work by Tisseur and Meerbergen [40] is an excellent review of most of the engineering applications related to the GEP or quadratic eigenvalue problem (QEP). For example, some of the applications that can be reduced to a GEP or QEP are: (1) second-order differential equations; (2) vibration analysis of structural systems – modal superposition method; (3) vibro-acoustics; (4) fluid mechanics; (5) constrained least square problems; (6) (adaptive) signal processing; and (7) (multiple input multiple output) MIMO systems [40]. The applications that we are most interested in are the three latter ones due to ties with the communication engineering and the science of navigation and indoor geolocation.

[2]Throughout this chapter we have used interchangeably N for M and vice versa. The redefinition is only done on section by section bases not within the same section.

Here, we formulate and explain the GEP. Denote with **A** and **B** two matrices of the same size ($N \times N$). In this case the GEP is defined as, [7]

$$\mathbf{A} \cdot \mathbf{x} = \lambda \mathbf{B} \cdot \mathbf{x}. \tag{5.75}$$

For the applications of interest, the **B** matrix is an autocorrelation matrix of a complex input vector, which contains desired data plus noise; hence, **B** is nonsingular, Hermitian symmetric, and positive-definite and (5.75) can be handled as [7]

$$(\mathbf{B}^{-1} \cdot \mathbf{A}) \cdot \mathbf{x} = \lambda \mathbf{x}. \tag{5.76}$$

On the other hand, the matrix **A** is also an autocorrelation matrix of some sort, which means that **A** is Hermitian symmetric; however, **A** may be singular.

There are three steps to solve the standard GEP (see [1]): (1) reduce the GEP to a regular symmetric eigenvalue problem; (2) reduce then the symmetric (or Hermitian symmetric) matrix to a tridiagonal matrix using any of the transformations (e.g, Householder transformation); and (3) solve the tridiagonal system employing any of the algorithms (e.g., *QL* with implicit shifts). At the end of this process, the symmetric matrix is reduced to a lower diagonal matrix whose diagonal elements are in fact the eigenvalues and another matrix which contains all the eigenvectors (see Fig. 5.3).

In the case where we are interested in finding only the largest eigenvalue and its corresponding eigenvector, there are also three steps for solving the direct GEP (see Fig. 5.4): (1) reduce the GEP to a regular symmetric eigenvalue problem; (2) find the largest eigenvalue problem; and (3) find the eigenvector corresponding to the largest eigenvalue. At the end of this process, the largest eigenvalue and its corresponding eigenvector are computed (see Fig. 5.4).

The reader is reminded that the *first step* is common for both algorithms; therefore, we start out with recovering a symmetry property to the GEP. Because the matrix $\mathbf{B}^{-1} \cdot \mathbf{A}$ is not symmetric, we need to recover a symmetry property from (5.76) to enable us to use the Cholesky and MGSO algorithms developed earlier [2]. This can be accomplished by, first, obtaining the Cholesky factors of $\mathbf{B} = \mathbf{L_B} \cdot \mathbf{L_B^H}$ [35] and, second, multiplying both sides of (5.76) by $\mathbf{L_B^{-1}}$, which yields [7]

$$\mathbf{C} \cdot (\mathbf{L_B^H} \cdot \mathbf{x}) = \lambda (\mathbf{L_B^H} \cdot \mathbf{x}), \tag{5.77}$$

where the Hermitian symmetric matrix **C** is determined from

$$\mathbf{C} = \mathbf{L_B^{-1}} \cdot \mathbf{A} \cdot \mathbf{L_B^H}. \tag{5.78}$$

Now that we have recovered the symmetry property of (5.76), the eigenvalues of **C** are the same as those of (5.75) because we have applied a linear transformation;

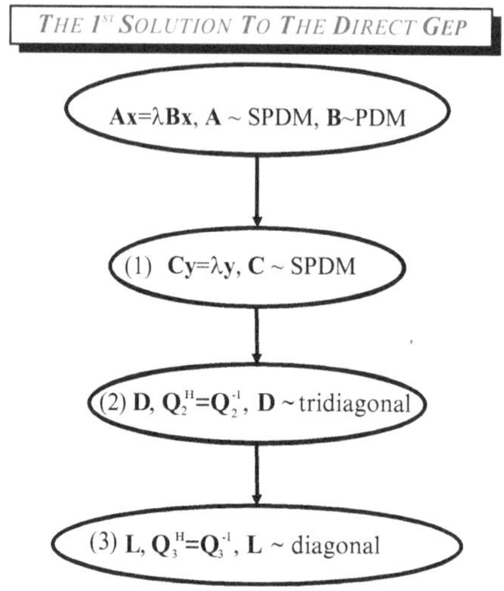

Fig. 5.3 Three steps for solving all the eigenvalues and all the eigenvectors of the direct GEP given by (5.75). Reprint with permission copyright 2010 Ilir Progri

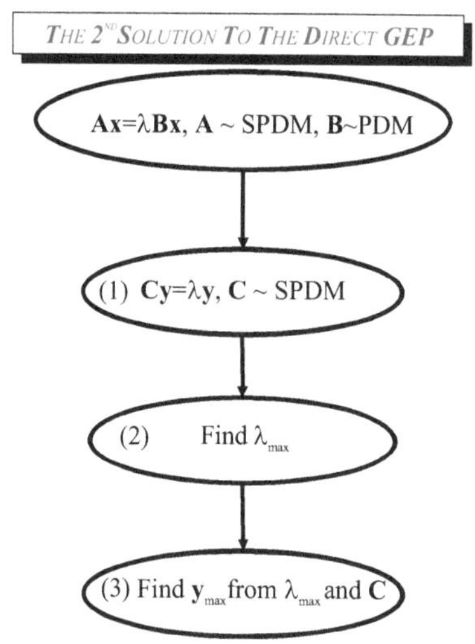

Fig. 5.4 Three steps for solving the largest eigenvalue and the corresponding eigenvector of the direct GEP given by (5.75). Reprint with permission copyright 2010 Ilir Progri

hence, it suffices to solve (5.77). Although $\mathbf{A}$ is not positive-definite, we can still obtain its Cholesky factors with deficient rank

$$\mathbf{A} = \mathbf{L_A} \cdot \mathbf{L_A^H}. \tag{5.79}$$

Substituting the result of (5.79) into (5.78) yields

$$\mathbf{C} = (\mathbf{L}_{\mathbf{B}}^{-1} \cdot \mathbf{L}_{\mathbf{A}}) \cdot (\mathbf{L}_{\mathbf{A}}^{\mathrm{H}} \cdot \mathbf{L}_{\mathbf{B}}^{\mathrm{H}}) = \mathbf{L}_{\mathbf{C}} \cdot \mathbf{L}_{\mathbf{C}}^{\mathrm{H}}, \tag{5.80}$$

where

$$\mathbf{C} = \mathbf{L}_{\mathbf{B}}^{-1} \cdot \mathbf{L}_{\mathbf{A}}. \tag{5.81}$$

At this point, we have concluded the first step in the process of solving the GEP and the new GEP is in the form of

$$\mathbf{C} \cdot \mathbf{y} = \lambda \mathbf{y},\ \mathbf{C} \sim SPDM, \quad and \quad \mathbf{y} = \mathbf{L}_{\mathbf{B}}^{\mathrm{H}} \cdot \mathbf{x}. \tag{5.82}$$

Next, we pursue the *second step* according to Fig. 5.3. Once the factors of the **C** matrix are computed, we can obtain the solution to the regular eigenvalue problem using any of the transformations such as the Householder transformation, Given's rotation, or reduction to Hessenberg form. Due to its popularity and efficiency, we present the Householder transformation.

The basic idea behind the Householder transformation is a Householder matrix **P** such that

$$\mathbf{P} = \mathbf{I} - \frac{\mathbf{u} \cdot \mathbf{u}^{\mathrm{H}}}{s}, \tag{5.83}$$

where **I** is the identity matrix and s is a scalar defined as

$$s \equiv 0.5|\mathbf{u}|^2. \tag{5.84}$$

It can easily be shown that the matrix **P** is an orthogonal transformation because, first

$$\mathbf{P}^{\mathrm{H}} = \mathbf{I} - \frac{(\mathbf{u} \cdot \mathbf{u}^{\mathrm{H}})^{\mathrm{H}}}{s} = \mathbf{I} - \frac{\mathbf{u} \cdot \mathbf{u}^{\mathrm{H}}}{s} = \mathbf{P}, \tag{5.85}$$

and second

$$\begin{aligned} \mathbf{P}^2 = \left(\mathbf{I} - \frac{\mathbf{u} \cdot \mathbf{u}^{\mathrm{H}}}{s}\right) \cdot \left(\mathbf{I} - \frac{\mathbf{u} \cdot \mathbf{u}^{\mathrm{H}}}{s}\right) &= \mathbf{I} - 2\frac{\mathbf{u} \cdot \mathbf{u}^{\mathrm{H}}}{s} + \frac{\mathbf{u} \cdot \mathbf{u}^{\mathrm{H}}}{s}\frac{\mathbf{u} \cdot \mathbf{u}^{\mathrm{H}}}{s} \\ &= \mathbf{I} - 2\frac{\mathbf{u} \cdot \mathbf{u}^{\mathrm{H}}}{s} + 2\frac{\mathbf{u} \cdot \mathbf{u}^{\mathrm{H}}}{s} = \mathbf{I}. \end{aligned} \tag{5.86}$$

Therefore

$$\mathbf{P} \cdot \mathbf{P}^{\mathrm{H}} = \mathbf{I} \rightarrow \mathbf{P}^{\mathrm{H}} = \mathbf{P}^{-1}, \tag{5.87}$$

which completes the orthogonality proof of the matrix **P**.

To reduce a Hermitian symmetric matrix into a tridiagonal form, we choose a vector $\mathbf{x}$ to be the first $N-1$ elements of the first column of $\mathbf{C}$. The second Householder transformation zeros out the bottom $N-2$ rows of the second column and leaves the first column unchanged. In order to complete the transformation, we have to left and right multiply by matrix $\mathbf{P}$. The new matrix $\mathbf{D}$ is of the form

$$\mathbf{D} = \mathbf{P} \cdot \mathbf{C} \cdot \mathbf{P}. \tag{5.88}$$

Actually carrying out the multiplication, we obtain

$$\mathbf{C} \cdot \mathbf{P} = \mathbf{C} \cdot \left(\mathbf{I} - \frac{\mathbf{u} \cdot \mathbf{u}^{\mathrm{H}}}{s} \right) = \mathbf{C} - \mathbf{C} \cdot \frac{\mathbf{u} \cdot \mathbf{u}^{\mathrm{H}}}{s} = \mathbf{C} - \mathbf{p} \cdot \mathbf{u}^{\mathrm{H}}, \tag{5.89}$$

where

$$\mathbf{p} \equiv \frac{\mathbf{C} \cdot \mathbf{u}}{s}. \tag{5.90}$$

Next, we compute the product

$$\begin{aligned} \mathbf{D} = \mathbf{P} \cdot \mathbf{C} \cdot \mathbf{P} &= \mathbf{P} \cdot \mathbf{C} - \mathbf{P} \cdot \mathbf{C} \cdot \frac{\mathbf{u} \cdot \mathbf{u}^{\mathrm{H}}}{s} = \mathbf{P} \cdot \mathbf{C} - \mathbf{P} \cdot \mathbf{p} \cdot \mathbf{u}^{\mathrm{H}} \\ &= \mathbf{C} - \mathbf{u} \cdot \mathbf{p}^{\mathrm{H}} - \mathbf{p} \cdot \mathbf{u}^{\mathrm{H}} + \frac{\mathbf{u} \cdot \mathbf{u}^{\mathrm{H}}}{s} \mathbf{p} \cdot \mathbf{u}^{\mathrm{H}} \\ &= \mathbf{C} - \mathbf{u} \cdot \mathbf{p}^{\mathrm{H}} - \mathbf{p} \cdot \mathbf{u}^{\mathrm{H}} + \mu \mathbf{u} \cdot \mathbf{u}^{\mathrm{H}}, \end{aligned} \tag{5.91}$$

where

$$\mu = \frac{\mathbf{u}^{\mathrm{H}} \cdot \mathbf{p}}{s}. \tag{5.92}$$

This transformation continues until $\mathbf{D}$ is turned into a tridiagonal matrix.

If the eigenvectors of the matrix $\mathbf{D}$ are found, the eigenvectors of the original matrix $\mathbf{C}$ can be computed by applying the accumulated transformation

$$\mathbf{Q} = \mathbf{P}_1 \cdot \mathbf{P}_2 \cdots \mathbf{P}_{N-2}, \tag{5.93}$$

to the eigenvectors of the tridiagonal matrix, $\mathbf{D}$.

Therefore, we form the matrix $\mathbf{Q}$ once the $\mathbf{P}$'s are computed [7] as follows

$$\mathbf{Q}_{N-2} = \mathbf{P}_{N-2}, \quad \mathbf{Q}_j = \mathbf{P}_j \cdot \mathbf{Q}_{j+1}, \quad j = \{n-3, \ldots, 1\}, \ \mathbf{Q} = \mathbf{Q}_1. \tag{5.94}$$

For large N, the operation count of the Householder transformation is $2N^3/3$ for eigenvalues only and $4N^3/3$ for both eigenvalues and eigenvectors. This concludes the development of the *second step* of the algorithm shown in Fig. 5.3.

Because the computation of the whole eigenvalues is computationally very inefficient, we should seek alternate algorithms for computing the largest eigenvalue and the corresponding eigenvector. Therefore, we present the *second step* of the algorithm (see Fig. 5.4).

Because the matrix **C** is either PDM or SPDM, the largest eigenvalue of **C** can be approximated as [32]

$$\lambda_{\max} \cong \alpha \|\mathbf{C}\|, \tag{5.95}$$

where $\|\mathbf{C}\|$ denotes the first norm of the matrix **C** and it is determined from

$$\|\mathbf{C}\| = \max(\text{sum}(\text{abs}(\mathbf{C}))), \tag{5.96}$$

and α is a real parameter in the vicinity of

$$0.85 \leq \alpha \leq 0.95. \tag{5.97}$$

Figures 5.5 and 5.6 illustrate the absolute and relative error, respectively, between the estimate of the largest eigenvalue given by (5.95) and the true largest eigenvalue for a PDM and SPDM **C** for $N = \{1, 11, \ldots, 120\}$.

Note that this approximation is better for PDMs as opposed to SPDMs. This reduces the computation of the largest eigenvalue in the order of N^2.

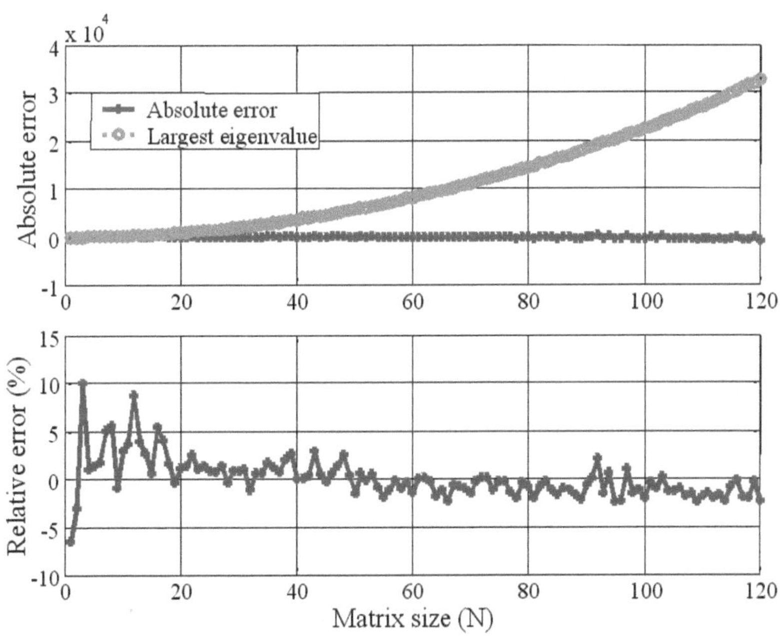

Fig. 5.5 Absolute and relative error for estimating the largest eigenvalue for positive-definite matrices, $\alpha = 0.935$. Reprint with permission copyright 2010 Ilir Progri

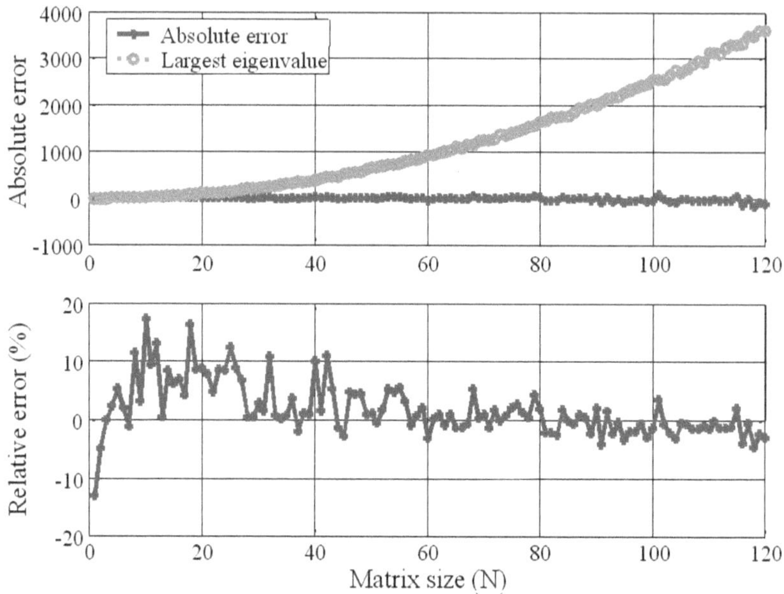

Fig. 5.6 Absolute and relative error for estimating the largest eigenvalue for semipositive-definite matrices, $\alpha = 0.87$. Reprint with permission copyright 2010 Ilir Progri

At this point, we have completed the *second step* of both algorithms and we shall continue with the *third (final) step*.

The *third step* of algorithm shown in Fig. 5.3 is introduced first. Once the matrix **C** is transformed into a tridiagonal matrix, **D**, we apply the *QL* algorithm with implicit shifts to get all the eigenvalues and eigenvectors of the matrix **D** [7]. The *QL* algorithm consists in a sequence of transformations

$$\mathbf{D}_k = \mathbf{Q}_k \cdot \mathbf{L}_k \quad and \quad \mathbf{Q}_k^{\mathrm{H}} = \mathbf{Q}_k^{-1}, \tag{5.98}$$

$$\mathbf{D}_k = \mathbf{L}_k \cdot \mathbf{Q}_k = \mathbf{Q}_k^{\mathrm{H}} \cdot \mathbf{L}_k \cdot \mathbf{Q}_k. \tag{5.99}$$

These transformations may as well be Householder transformations; therefore, it is more efficient to employ plane rotations, $\mathbf{P}_{i,i+1}$, to annihilate the $d_{i,i+1}$. Due to the symmetry of matrix **D**, the subdiagonal element $d_{i,i+1}$ will be annihilated too. Hence, each $\mathbf{Q}_k$ is a product of plane rotations [7]

$$\mathbf{Q}_k^H = \mathbf{P}_1^{(k)} \cdot \mathbf{P}_2^{(k)} \cdots \mathbf{P}_{N-1}^{(k)}. \tag{5.100}$$

If the matrix **D** has an eigenvalue, λ_i, with multiplicity p (a.k.a. a superdiagonal element), then there must be at least $p - 1$ zeros on the sub- or superdiagonal. This leads to splitting the matrix into submatrices and each submatrix can be diagonalized separately [7]. A superdiagonal element converges to zero at the rate of

$$d_{ij}^{(k)} \sim \left(\frac{\lambda_i}{\lambda_j}\right)^k. \tag{5.101}$$

The convergence rate can be very slow if λ_i is close to λ_j. A well-known technique called *shifting* can help accelerate the rate of convergence: For a given real constant s, the matrix $\mathbf{D} - s\mathbf{1}$ has eigenvalues $\lambda_i - s$; therefore, the rate of convergence for the new superdiagonal elements is of the order of

$$\tilde{d}_{ij}^{(k)} \sim \left(\frac{\lambda_i - s}{\lambda_j - s}\right)^k, \tag{5.102}$$

and, in general, the convergence rate is determined by the ratio

$$\frac{\lambda_i - s}{\lambda_j - s}. \tag{5.103}$$

A proper choice of the parameter $s \equiv s^{(k)}$ at each kth iteration can maximize the rate of convergence. Nevertheless, when the elements of the matrix, **D**, differ a lot, for example an order of magnitude, then subtracting a large constant s can lead to loss of accuracy for small eigenvalues. The *QL* algorithm with *implicit shifts* can eliminate this anomaly. The implicit *QL* is mathematically equivalent to the shifted *QL*; however, it does not require that the constant s be actually subtracted from the diagonal elements of **D** [7].

The total number of operations for eigenvalues only is about $30N^2$ and for both the eigenvalues and eigenvectors is about $3N^3$ operations [7].

The *third step* of the algorithm shown in Fig. 5.4 is the last development presented in this section. For this algorithm, it remains to compute the eigenvector corresponding to the largest eigenvalue. One method for solving the eigenvector is the method of *inverse iteration* [7]. The basic idea behind this algorithm is explained below.

Let **z** be the solution to the linear system

$$(\mathbf{C} - \tau 1) \cdot \mathbf{z} = \mathbf{b}, \tag{5.104}$$

where **b** is a random vector and τ is a real number closer to the eigenvalue λ of **C**. Then the solution to **z** will be close to the eigenvector corresponding to λ, replacing **b** by **z** and solving for the new **z**, which is going to be closer to the true eigenvector. Suppose that we can write vectors **b** and **z** as linear combinations of the eigenvectors $\mathbf{y}_j$ of **C**

$$\mathbf{z} = \sum_j \alpha_j \mathbf{y}_j \quad \text{and} \quad \mathbf{b} = \sum_j \beta_j \mathbf{y}_j, \tag{5.105}$$

hence, (5.104) can be written as

$$\sum_j \alpha_j(\lambda_j - \tau)\mathbf{y}_j = \sum_j \beta_j \mathbf{y}_j, \tag{5.106}$$

from where we get

$$\alpha_j \equiv \frac{\beta_j}{\lambda_j - \tau} \tag{5.107}$$

and

$$\mathbf{z} \equiv \sum_j \frac{\beta_j \mathbf{y}_j}{\lambda_j - \tau}. \tag{5.108}$$

The algorithm with *inverse iterations* works as follows: Suppose that τ_k and $\mathbf{b}_k$ are our current guesses. We then solve the normal equations at the kth iteration

$$\mathbf{z} \equiv \sum_j \frac{\beta_j \mathbf{y}_j}{\lambda_j - \tau_k}. \tag{5.109}$$

Normalize $\mathbf{b}_k$ so that $\mathbf{b}_k^{\mathrm{H}} \cdot \mathbf{b}_k = 1$. Since $\mathbf{z}$ is an improved approximation of $\mathbf{y}$, the new $\mathbf{b}_{k+1}$ is normalized such that

$$\mathbf{b}_{k+1} = \frac{\mathbf{z}}{|\mathbf{z}|}, \tag{5.110}$$

which leads into an improved eigenvalue solution

$$\tau_{k+1} = \tau_k + \frac{1}{\mathbf{b}_k^{\mathrm{H}} \cdot \mathbf{z}}. \tag{5.111}$$

While this method is in the order of N^3 for symmetric matrices, it is many times less efficient than the *QL* method with implicit shifts as soon as more than 25% of eigenvectors are required [7].

Having discussed all the three steps of both algorithms, we conclude the following:

1. Both algorithms have similar computational performance on *the first step*.
2. The second algorithm is an order of magnitude faster that the first one on *the second step*.
3. The second algorithm is expected to be somewhat faster than the first algorithm on *the third step*.

Although this presentation uses real operations, complex arithmetic is also very important. For more details on complex arithmetic multiplication, the reader may refer to Appendix C.

5.7 Recursive Generalized Eigenvalue Problem

So far we have considered the GEP assuming that we have to form the matrices **A** and **B**. For a discussion on the special structure of matrices **A** and **B**, the reader should refer to Appendix D. Now, we will assume that the complex matrices **A** and **B** are formed recursively in accordance with [1]

$$\mathbf{A}(k) = (1-\alpha)\mathbf{A}(k-1) + \alpha\mathbf{x}_{\mathbf{A}}(k)\cdot\mathbf{x}_{\mathbf{A}}^{\mathrm{H}}(k), \tag{5.112}$$

$$\mathbf{B}(k) = (1-\alpha)\mathbf{B}(k-1) + \alpha\mathbf{x}_{\mathbf{B}}(k)\cdot\mathbf{x}_{\mathbf{B}}^{\mathrm{H}}(k). \tag{5.113}$$

We will still require that the matrix $\mathbf{A}(k)$ be Hermitian complex SPD and the matrix $\mathbf{B}(k)$ be Hermitian complex PDM.

Based on the development of Progri et al [1, 2], we can obtain the Cholesky factors for $\mathbf{A}(k)$ and $\mathbf{B}(k)$ recursively; therefore, we can write

$$\mathbf{L}_{\mathbf{A}}(k)\cdot\mathbf{L}_{\mathbf{A}}^{\mathrm{H}}(k) = (1-\alpha)\mathbf{L}_{\mathbf{A}}(k-1)\cdot\mathbf{L}_{\mathbf{A}}^{\mathrm{H}}(k-1) + \alpha\mathbf{x}_{\mathbf{A}}(k)\cdot\mathbf{x}_{\mathbf{A}}^{\mathrm{H}}(k), \tag{5.114}$$

$$\mathbf{L}_{\mathbf{B}}(k)\cdot\mathbf{L}_{\mathbf{B}}^{\mathrm{H}}(k) = (1-\alpha)\mathbf{L}_{\mathbf{B}}(k-1)\cdot\mathbf{L}_{\mathbf{B}}^{\mathrm{H}}(k-1) + \alpha\mathbf{x}_{\mathbf{B}}(k)\cdot\mathbf{x}_{\mathbf{B}}^{\mathrm{H}}(k). \tag{5.115}$$

Since the matrix **B** is PDM, its Cholesky factors exist and numerically they are very stable [2, 32] and therefore

$$\mathbf{L}_{\mathbf{B}}^{-1}(k) = f(\mathbf{L}_{\mathbf{B}}(k)). \tag{5.116}$$

This concludes the first step for both algorithms shown in Figs. 5.3 and 5.4. The first step of the solution to the recursive GEP presented here saves $N^3/3$ operations compared to the first step of the GEP presented in the previous section because we do not form the matrices **A** and **B**, but we compute the Cholesky factors of **A** and **B** recursively [2].

At this point, we proceed exactly as in the previous section in two approaches shown in Figs. 5.3 and 5.4. We have already discussed the first step of the first approach (see Fig. 5.3) in the first part of this section. The last two steps of the first approach are (1) the Householder reduction and (2) the *QL* algorithm with implicit shifts. Also, we have already discussed the first step of the second approach (see Fig. 5.4). The last two steps of the second approach are (1) we compute the largest eigenvalue and (2) use the inverse iteration algorithm to compute the eigenvector corresponding to the largest eigenvalue (Table 5.1).

If only eigenvalues are required, then algorithm (1) (see Fig. 5.3) will require $2N^3/3 + 30N^2$ operations and for both the eigenvalue and eigenvectors, algorithm (2) will require $13N^3/3$ operations. On the other hand, algorithm (2) (see Fig. 5.4) will require $N^3 + N^2$ operations.

Table 5.1 An assessment of the direct GEP vs. RGEP when both eigenvalues and eigenvectors are required (see Fig. 5.3)

Step	Direct GEVP	Recursive GEVP
Form A and B	MN^2	0
Form L_A and L_B	$N^3/3$	MN^2
Invert L_B	$N^3/6$	$N^3/6$
Form L_C	$N^3/6$	$N^3/6$
Form C	$N^3/6$	$N^3/6$
Householder	$4N^3/3$	$4N^3/3$
QL	$3N^3$	$3N^3$
Total	$31N^3/6+ MN^2$	$29N^3/6+ MN^2$

It is important to emphasize here that the improvement of both algorithms is gained only as a result of utilizing the information from the recursive property. This improvement, however, is going to affect only the *first step* of both algorithms equally leaving the *second* and the *third steps* unchanged.

The following section discusses the assessment of both the direct and the recursive algorithms in the context of applications pertaining to (adaptive) signal processing, communication engineering, indoor geolocation, and science of navigation.

5.8 Assessment of Both Algorithms

Suppose that we want to process a block of M data samples, i.e., at the end of M samples, we want to have the eigenvalues and eigenvectors of a RGEP given by (5.75) with matrices **A** and **B** formed recursively (see (5.112) and (5.113)). Table 5.1 provides the steps required and the number of operation counts for both the direct and the recursive GEP when both eigenvalues and eigenvectors are required.

Note that when employing algorithm (1) (see Fig. 5.3), there is very little improvement for the case when the eigenvalues are required because most of the burden falls in the Householder reduction and the *QL* algorithm with implicit shifts.

Table 5.2 provides the steps required and the operation counts for both the direct and the recursive GEP when only the eigenvalues are required and again the processing is carried out based on algorithm (1) (see Fig. 5.3).

Note that when only eigenvalues are required (see Figs. 5.3 and Table 5.2), then the difference in the number of complex operation counts between the direct and the recursive GEP is more noticeable than the same difference using the requirements of Table 5.1.

Table 5.3 provides the steps required and the operation counts for both the direct and the recursive GEP when only the largest eigenvalue and its corresponding eigenvector are required utilizing the second algorithm (see Fig. 5.4).

Note also that when both one eigenvalue and its corresponding eigenvector are required (see Fig. 5.4 and Table 5.3), then the difference in the number of complex

Table 5.2 An assessment of the direct GEP vs. RGEP algorithms when only eigenvalues are required (see Fig. 5.3)

Step	Direct GEVP	Recursive GEVP
Form A and B	MN^2	0
Form L_A and L_B	$N^3/3$	MN^2
Invert L_B	$N^3/6$	$N^3/6$
Form L_C	$N^3/6$	$N^3/6$
Form C	$N^3/6$	$N^3/6$
Householder	$2N^3/3$	$2N^3/3$
QL	$30N^2$	$30N^2$
Total	$3N^3/2 + (M + 30)N^2$	$7N^3/6 + (M + 30)N^2$

Table 5.3 An assessment of the direct GEP vs. RGEP when only one eigenvalue and its corresponding eigenvector are required (see Fig. 5.4)

STEP	Direct GEVP	Recursive GEVP
Form A and B	MN^2	0
Form L_A and L_B	$N^3/3$	MN^2
Invert L_B	$N^3/6$	$N^3/6$
Form L_C	$N^3/6$	$N^3/6$
Form C	$N^3/6$	$N^3/6$
Eigenvalue	N^2	N^2
Eigenvector	N^3	N^3
Total	$11N^3/6 + (M + 1)N^2$	$3N^3/2 + (M + 1)N^2$

operation counts between the direct and the recursive GEP is more noticeable than the same difference using the requirements of Table 5.1.

5.9 Conclusions

With this work we revisited two fundamental methods in linear algebra for solving complex linear systems of equation of the form $\mathbf{A}(k) \cdot \mathbf{z}(k) = \mathbf{A}(k)$: the Cholesky's decomposition and the MGSO algorithm. During this investigation, we mentioned the historical background, the most significant contributions, and our efforts for achieving the Cholesky's and MGSO algorithms in their best form for adaptive signal processing and navigation filter design and implementation. It appears that for either simple or recursive block processing, the recursive Cholesky's decomposition is twice as fast as the recursive MGSO. Moreover, for every cycle (or iteration), the recursive Cholesky's is 5/3 times (or 167%) faster than the result obtained from applying the Sherman–Morrison Formula – conclusion that is almost unknown to most engineers and mathematicians. Moreover, the recursive Cholesky's and MGSO reduce the eigenvalue ration by the square root, which is very desirable for most applications.

The GEP is being revisited in the context of recursive Cholesky and MGSO algorithms for a special class of PDM or SPDM matrices related to adaptive signal processing and indoor communication and geolocation applications.

When all the eigenvalues and the corresponding eigenvectors are required, then there is computationally little improvement between the solution to the recursive GEP and direct GEP (see Table 5.1).

When only eigenvectors are required, then there is more noticeable improvement between the solution to the recursive GEP and the direct GEP (see Table 5.2).

The most favorable scenario encountered in adaptive signal processing and indoor communication and geolocation applications would be when only the largest eigenvalue and the corresponding eigenvector are required (see Table 5.3). Under this condition, the $N^3/3$ improvement in computation complexity is desirable and noticeable.

For $M = 4N$, the recursive solution to the second approach is 57% faster than the direct solution to the first approach and this is the greatest improvements that we can gain.

Finally, while the solution to the GEP may seem trivial mathematically, it can be very tricky in implementation; therefore, caution must be exercised in implementing this algorithm.

5.10 Future Research

Following are some suggestions for future research. Perhaps the most intriguing area of research is the parallel algorithms for implementing the recursive Cholesky, MGSO, and recursive GEP based on works by Gallivan et al. [42]. Moreover, latest versions of MATLAB already include scripts for parallel algorithms or for breaking sequential algorithms scripts into parallel algorithms scripts. Should microprocessor architecture be in place that would have already enabled the implementation of parallel algorithm scripts, then the parallel algorithms should be investigated for speed and storage. I would not be surprised to hear that the parallel structures might be able to produce an order of magnitude improvement in speed and memory storage.

Another area of research is to actually implement these algorithms in real-time and with real live data and see how they in fact perform.

Acknowledgments The authors would like to thank Dr. James Farrell, *Vigil Inc.*, for presenting portions of this chapter at the annual meeting of the Institute of Navigation, June 24–26, 2002, Albuquerque, New Mexico.

Appendix A

In this appendix we observe some useful properties of the real factor α. Let A be the expected value of matrix $\mathbf{A}(k)$; hence

$$A = E[\mathbf{A}(k)]. \tag{5.117}$$

First, it can be easily observed that taking the expected value of both sides of (5.5) reduces it into the following equality

$$A = E[\mathbf{A}(k)] = E[(1-\alpha)\mathbf{A}(k) + \alpha\mathbf{x}(k)\cdot\mathbf{x}^{\mathrm{H}}(k)] = (1-\alpha)\mathbf{A} + \alpha A = A. \quad (5.118)$$

Second, the parameter α must be smaller than one, hence we can write

$$\mathbf{A}(k) = (1-\alpha)\mathbf{A}(k-1) + \alpha\mathbf{x}(k)\cdot\mathbf{x}^{\mathrm{H}}(k), \quad (5.119)$$

$$\mathbf{A}(k) = (1-\alpha)\left[\mathbf{A}(k-1) + \frac{\alpha}{1-\alpha}\mathbf{x}(k)\cdot\mathbf{x}^{\mathrm{H}}(k)\right]. \quad (5.120)$$

Moreover, the parameter $\alpha/(1-\alpha)$ should also be smaller than one, otherwise the matrix $\mathbf{A}(k)$ will blow up; hence, $\alpha < 0.5$.

There are at least two important cases in principle for forming the matrix $\mathbf{A}(k)$ recursively: one using only a block of new data and, on the other case, using both old and new data. For the first scenario, the matrix $\mathbf{A}(k)$ is formed as

$$\mathbf{A}(k) = \frac{N-1}{N}\left[\mathbf{A}(k-1) + \frac{1}{N-1}\mathbf{x}(k)\cdot\mathbf{x}^{\mathrm{H}}(k)\right], \quad (5.121)$$

$$\mathbf{A}(k) = \frac{\sum_{i=1}^{N}\left[\mathbf{x}(k)\cdot\mathbf{x}^{\mathrm{H}}(k)\right]}{N}, \quad (5.122)$$

where

$$\alpha = 1/N \quad \text{and} \quad 1-\alpha = (N-1)/N. \quad (5.123)$$

For the second scenario, the matrix $\mathbf{A}(k)$ is formed as

$$\mathbf{A}(k) = (1-\alpha)\left[\mathbf{A}(k-1) + \frac{\alpha}{1-\alpha}\mathbf{x}_0(k)\cdot\mathbf{x}_0^{\mathrm{H}}(k)\right], \quad (5.124)$$

$$\mathbf{A}(k) = (1-\alpha)\sum_{i=0}^{\infty}\alpha^i\left[\mathbf{x}_i(k)\cdot\mathbf{x}_i^{\mathrm{H}}(k)\right] \quad (5.125)$$

where $0 < \alpha < 0.5$ and α is also known as the *forgetting factor* of the matrix $\mathbf{A}(k)$.

Note that for the first scenario the data are weighted equally as opposed to the second scenario for which the data are weighted by $(1-\alpha)\alpha^i$, which is different for different values of α.

Note also that both scenarios satisfy (5.118), which is trivial anyway.

Appendix B

In this appendix we perform a comparison between the Sherman–Morrison Formula and the recursive Cholesky's and MGSO.

A detailed discussion of the Sherman–Morrison Formula is found in *Numerical Recipes* 7; therefore, we will provide here the result for the complex PDH matrix defined in [7]. If we apply the Sherman–Morrison Formula for this matrix, then

$$\mathbf{A}(k) = (1-\alpha)\left[\mathbf{A}^{-1}(k-1) + \frac{\beta}{1+\beta\lambda}\mathbf{y}(k)\cdot\mathbf{y}^{\mathrm{H}}(k)\right], \tag{5.126}$$

where

$$\beta = \frac{\alpha}{1-\alpha}, \tag{5.127}$$

$$\lambda = \mathbf{x}^{\mathrm{H}}(k)\cdot\mathbf{y}(k). \tag{5.128}$$

The total number of symmetry we can compute with the inverse of only the upper (or lower) triangular matrix; hence, the total number of operations is $O(3M^2/2)$. Including the $O(M^2)$ operations to compute the final solution yields $O(5M^2/2)$ for the number of operations, which is 3/5 times slower than the recursive Cholesky's.

Appendix C (A Review on Complex Arithmetic)

The traditional way of performing the complex arithmetic may not always be the fastest and the best way of computation. While the addition and subtraction are performed in obvious way, complex multiplication can be done in several ways. First, we remind the reader that the traditional complex multiplication is performed in four multiplications, one addition, and one subtraction as

$$(a+ib)(c+id) = (ac-bd) + i(ad+bc). \tag{5.129}$$

On the other hand, if we rearrange the terms of the real and imaginary part of (5.129), the complex multiplication can be done in four different ways. The first and second (for the second approach see [7]) approaches can be performed in three multiplications, three subtractions, and two additions as follows:

$$(a+ib)(c+id) = [(a-b)c + (c-d)b] + i[(d+c)a - (a-b)c], \tag{5.130}$$

$$(a+ib)(c+id) = (ac-bd) + i[(a+b)(c+d) - ac - bd]. \tag{5.131}$$

The third and fourth approaches can be performed in three multiplications, two subtractions, and three additions as

$$(a + ib)(c + id) = [(c + d)a - (a + b)d] + i[(c - d)b + (a + b)d], \tag{5.132}$$

$$(a + ib)(c + id) = [(a + b)(c - d) - ad - bc] + i(ad + bc). \tag{5.133}$$

Although the total number of operations using either one of the nontraditional approaches is higher by two, it may be faster to perform the complex multiplication using either one of the four approaches than the traditional one.

The rest of the complex arithmetic is found in ([7], Sect. 5.4).

Appendix D (A Review on Toeplitz Matrices)

Perhaps one of the most common cases of adaptive filtering and spectral estimation is the case when the matrix $\mathbf{A}(k)$ is Toeplitz with diagonally dominant elements as well as being PDM [7]. In this case we already know that the solution of $\mathbf{A}(k) \cdot \mathbf{z}(k) = \mathbf{b}(k)$ is on the order of $O(M^2)$, so it appears almost as fast as the recursive Cholesky and MGSO, but it may not be as numerically stable as the other two because it does not include pivoting. Some basic statements that should help are as follows:

1. If the matrix $\mathbf{A}(k)$ is Toeplitz, the Cholesky factors (see (5.2)) are in general not Toeplitz.
2. If the matrix $\mathbf{A}(k)$ is Toeplitz, the inverse of the matrix $\mathbf{A}(k)$ (see (5.42)) is in general not Toeplitz.
3. If the matrices $\mathbf{A}(k)$ and $\mathbf{B}(k)$ are Toeplitz, the matrix $\mathbf{C}(k)$ from the GEP (see (5.78)) is in general not Toeplitz.

This example and others like this should provide a plausible explanation as to why the recursive algorithms are so efficient, optimized for speed and storage, and also are the major source of linear algebra libraries such as those provided in [34].

References

1. Progri, I.F., Michalson, W.R., and Bromberg, M.C., "A study of a blind adaptive algorithm in the time and frequency domain," in *Proc. ION-NTM 2002*, San Diego, CA, pp. 439–447, Jan. 2002.
2. Progri, I.F., Michalson, W.R., and Bromberg, M.C., "A comparison between the recursive Cholesky and MGSO algorithms," in *Proc. ION-NTM 2002*, San Diego, CA, pp. 655–665, Jan. 2002.
3. Salvadori, M.G., and Baron, M.L., *Numerical Methods in Engineering*, 1st ed., Englewood Cliffs, NJ: Prentice-Hall, 1952.

4. Wilkinson, J.H., and Reinsch, C., *Linear Algebra* (Handbook for Automatic Computation), New York: Springer, 1971.
5. Fox, L., "Practical solution of linear equations and inversion of matrices," *Nat. Bur. Standards Appl. Math. Ser.* vol. 39, pp. 1–54, 1954.
6. Gill, P.E., Murray, W., and Wright, M.H., *Numerical Linear Algebra and Optimization*, vol. 1, Redwood City, CA: Addison-Wesley, 1991.
7. Press, W.H., Teukolskey, S.A., Vetterling, W.T., and Flannary, B.P., *Numerical Recipes in C: The Art of Scientific Computing*, 2nd ed. New York, NY: Cambridge University Press, 1992.
8. Golub, G.H., and Van Loan, C., *Matrix Computations*, 3rd ed. Baltimore: Johns Hopkins University Press, 1996.
9. Dongarra, J.J., Bunch, J.R., Moler, C.B., and Stewart, G.W., *LINPACK Users' Guide*. Philadelphia: SIAM, 1979.
10. Progri, I., "A dense recursive linear solution algorithm," Project report on MA510 course, Worcester Polytechnic Institute, Worcester, MA, fall 1999.
11. Progri, I., "UU^T recursive factorization," Project report on MA514 course, Worcester Polytechnic Institute, Worcester, MA, Spr. 2000.
12. Francis, J.C.F., "The QR transformation – a unitary analogue of the LR transformation," *Comput. J.*, vol. 4, pp. 265–271 and pp. 332–345, 1961/62.
13. Given, J.W., "Numerical computation of the characteristic values of a real symmetric matrix," Oak Ridge National Laboratory, ORNL-1574, 1954.
14. Rutishauser, H., "Solution of the eigenvalue problems with the LR-transformation," *Nat. Bur. Standards Appl. Math. Ser.*, vol 49, pp. 47–81, 1958.
15. Kubalnovskaya, V.N., "On some algorithms for the solution of the complete eigenvalue problem," *Ž. Vyisl. Mat.*, vol. 1, pp. 555–570, 1960.
16. Hadley, G., *Linear Algebra*, 3rd ed., chp. 7, "Characteristic value problems and quadratic forms," Reading, MA: Addison-Wesley, pp. 236–284, 1961.
17. Barth, W., Martin, R.S., and Wilkinson, J.H., "Calculation of the eigenvalues of a symmetric tridiagonal matrix by the method of bisection," *Numer. Math.*, vol. 9, pp. 386–393, 1967.
18. Wilkinson, J.H., "Global convergence of tridiagonal QR algorithm with origin shifts," *Linear Algebra Appl.*, vol. 1, pp. 409–420, 1968.
19. Martin, R.S., and Wilkinson, J.H., "Reduction of the symmetric eigenproblem $Ax = \lambda Bx$ and related problem to standard form," *Numer. Math.*, vol. 11, pp. 99–110, 1968.
20. Martin, R.S., Reinch, C., and Wilkinson, J.H., "Householder's tridiagonalization of a symmetric matrix," *Numer. Math.*, vol. 11, pp. 181–195, 1968.
21. Moler, C.B., and Stewart, G.W., "An algorithm for generalized matrix eigenvalue problems," *SIAM J. Numer. Anal.*, vol. 10, no. 2, pp 241–256, Apr. 1973.
22. Smith, B.T., Boyle, J.M., Dongarra, J.J., Garbow, B.S., Ikebe, Y., Klema, V.C., Moler, C.B., *Matrix Eigensystem Routines – EISPACK Guide*, 2nd ed., vol. 6 of Lecture Notes in Computer Science, New York: Springer, 1976.
23. Wilkinson, J.H., *The Algebraic Eigenvalue Problem*, London: Oxford University Press, 1965.
24. Wilkinson, J.H., "Convergence of the LR, QR and related algorithms," *Comput. J.*, vol. 8, pp. 77–84, 1965.
25. Wilkinson, J.H., "The QR algorithm for real symmetric matrices with multiple eigenvalues," *Comput. J.*, vol. 8, pp. 85–87, 1965.
26. Wilkinson, J.H. and Reinsch, C., *Linear Algebra*, New York, NY: Springer, 1971.
27. Manolakis, D.G., Ingle, V.K., and Kogan, S.M., *Statistical and Adaptive Signal Processing*, Boston, MA: McGraw Hill, 2000.
28. Parkinson, B.W., Spilker, J.J., Jr., Axelrad, P., and Enge, P., *Global Positioning System: Theory and Applications*, vol. 1, chp. 7, "Fundamentals of signal tracking theory," vol. 163, Washington, DC: AIAA, pp. 245–327, 1996.
29. Progri, I., "HW#8," Homework report on EE534 course, Worcester Polytechnic Institute, Worcester, MA, Apr. 02, 2001.

30. Johnson, R.K., U.S. Patent No. 5,694,416, "Direct sequence spread spectrum receiver and antenna array for the simultaneous formation of a beam on a signal source and a null on an interfering jammer," SN 08/ 393,716, Feb. 24, 1995.
31. Bunch, J.R., "Analysis of sparse elimination," *SIAM J. Numer. Anal.*, vol. 11, no. 5, pp. 847–873, 1974.
32. Meinguet, J., "Refined error analysis of Cholesky's factorization," *SAIM*, vol. 20, pp. 1243–1250, 1983.
33. Bottomley, G.K., U.S. Patent No. 5,615,209, "Method and apparatus for direct signal orthogonalization," SN 08/507,714, Jul. 26, 1995.
34. Dongarra, J.J., and Walker, D.W., "Software libraries for linear algebra computations on high performance computers," *SIAM Rev.*, vol. 37, no. 2, pp. 151–180, 1995.
35. Fernando, K.V., "On computing an eigenvector of a tridiagonal matrix. Part 1: Basic results," *SIAM J. Matrix Anal. Appl.*, vol. 18, no. 4, pp. 1013–1034, 1997.
36. Zhang, T., Law, K.H., and Golub, G.H., "On the homotopy method for perturbed symmetric generalized eigenvalue problems," *SIAM J. Sci. Comput.*, vol. 19, no. 6, pp. 1625–1645, 1998.
37. Drmac, Z., "A tangent algorithm for computing the generalized singular value decomposition," *SIAM J. Numer. Anal.*, vol. 35, no. 5, pp. 1804–1832, 1998.
38. Schandrasekaran, S., "An efficient and stable algorithm for the symmetric-definite generalized eigenvalue problem," *SIAM J. Matrix Anal. Appl.*, vol. 22, no. 2, pp. 392–412, 2000.
39. Arav, M., Hershkowitz, D., Mehrmann, V., and Schneider, H., "The recursive inverse eigenvalue problem," *SIAM J. Matrix Anal. Appl.*, vol. 22, no. 2, pp. 392–412, 2000.
40. Tisseur, F., and Meerbergen, K., "The quadratic eigenvalue problem," *SIAM Rev.*, vol. 43, no. 2, pp. 235–286, 2001.
41. Davies, P.I., Higham, N.J., and Tisseur, F., "Analysis of the Cholesky method with iterative refinement for solving the symmetric definite generalized eigenvalue problem," *SIAM J. Matrix Anal. Appl.*, vol. 23, no. 2, pp. 472–493, 2001.
42. Gallivan, K.A., Plemmons, R.J., and Sameh, A.H., "Parallel algorithms for dense linear algebra computations," *SIAM Rev.*, vol. 43, no. 1, pp. 54–135, 1990.

Chapter 6
Adaptive Array Beamforming for Interference Mitigation for GRFS Systems

6.1 Introduction

Adaptive array beamforming for interference mitigation for GRFS systems is the last chapter of the book on *Geolocation of RF Signals: Principles and Simulations*, which presents the best practices, innovative techniques, research and development processes, analysis and simulation tools in the art and science of adaptive array beamforming for interference mitigation (or suppression) for GRFS systems in the last 20 years or so [1–101] (Fernández-Prades, Closas, and Arribas, 2009, Implementation of digital beamforming in GNSS receivers, personal communications; Lee and Prasanna, 1995, High throughput-rate parallel algorithms for space–time adaptive processing (STAP). Department of EE Systems, USC, Los Angeles, Private Communications; Trinkle and Gray, 2001, Adaptive antenna arrays for GPS interference localisation. University of Adelaide, South Australia, pp. 1–12, Personal communication; Trinkle and Gray, 2001, GPS interference mitigation; overview and experimental results. University of Adelaide, South Australia, pp. 1–14, Personal communication). This is probably the best overall chapter in the entire book with an overwhelming depth, breath, and description of very insightful details of theoretical and practical analysis, adaptive array techniques, and very useful narrow and broad principles in adaptive array beamforming in the frequency band of 100 MHz–60 GHz. Dr. Progri reveals the research and development process by demonstrating how to understand, explain, model, and simulate the four most recognized adaptive array beamforming processing techniques for interference mitigation for GRFS systems, which are as follows: (1) adaptive temporal selective attenuator (ATSA); (2) adaptive spatial selective attenuator (ASSA); (3) adaptive spatial temporal selective attenuator (ASTSA); and (4) an improved adaptive spatial temporal selective attenuator (IASTSA) (or an ASTSA with restored phase); from basic diagrams to be utilized to the principle simulation examples and make recommendations for the future final products of geolocation of RF signals. Starting with the main description and discussion of adaptive array beamforming for interference mitigation of GRFS systems in Sect. 6.2, the chapter progressively examines best adaptive array beamforming algorithm ideas, practices, and descriptions, and then continues with the best interference mitigation

I. Progri, *Geolocation of RF Signals*,
DOI 10.1007/978-1-4419-7952-0_6, © Springer Science+Business Media, LLC 2011

algorithms and practices specific to GPS and GNSS in Sect. 6.3. Next, the chapter continues with mathematical description of the ATSA mathematical model in Sect. 6.4, ASTSA mathematical model in Sect. 6.5, and the first part of this chapter is concluded with the ASTSA with restored phase mathematical model (a.k.a. IASTSA) in Sect. 6.6.

The vulnerability of the GPS signal against any kind of interference is known. Any interference signal near the band of the GPS signal can saturate the GPS receiver and at the same time can deteriorate the autocorrelation properties of the GPS signal and its PRN code, which results in loss of lock of the GPS signal. A GPS ATSA or ASTSA can be used to suppress the effects of undesired narrowband/wideband interference (and/or jamming) RF signals (or RF sources). A GPS ATSA can be modeled as a discrete finite impulse response (DFIR) filter where the filter's impulse response coefficients can be selected by minimizing the undesired signal over desired signal ratio (UDR) vs. an ASTSA, which can be modeled as a 2D array with dimensions as the number of sensors $\times$ the number of temporal delays.

Next follows a step-by-step approach of ATSA, ASSA, ASTSA, and IASTSA description and principle simulation test cases (or scenarios) for GPS (or satellite) environments in ATSA Simulations in Sect. 6.7, ATSA Implementation in Sect. 6.8, ASTSA Simulations in Sect. 6.9, and Improved ASTSA (or ASTSA with Restored Phase) Simulations in Sect. 6.10 in three important steps, which are explained in order later.

First, the UDR, which directly represents the GPS ATSA performance, depending on the *filter size, the temporal delay, sampling frequency, and desired signal frequency* [1]. A refined methodology for properly selecting the temporal delay and the sampling frequency for a given filter size and desired signal frequency is proposed, analyzed, discussed, simulated, and plotted. The simplified analysis and simulation results are validated using an example ATSA implementation on the TI C6711 digital signal processor (DSP). For a given GPS/GNSS ATSA architecture (i.e., a given desired (GPS/GLONASS/Galileo L1, L2), signal frequency and sampling frequency) optimum tap spacing can be selected, regardless of the jammer frequency [1]. We have further simplified the selection of the temporal spacing utilizing array beam pattern. Some of the lessons learned during this investigation could be particularly valuable for ATSAs or analogous systems that will operate on the L5, L1C, L2C, etc. frequencies.

Second, the analytical development of this investigation exploits the fundamental issues regarding a new adaptive methodology, the *impact of the array elements, and the number of temporal shifter delays for a given desired signal frequency* on the ASTSA performance against one broadband jammer (BBJ) and one continuous wave jammer (CWJ). The *new (or local) adaptive methodology* exploits the cross-correlation properties between a locally generated signal and the received signal. The *old (or global) adaptive technique* utilizes a pointing (or steering, or signature) vector to compute the desired set of multipliers, which act as terms of a finite impulse response digital filter. The new adaptive methodology allows restoring the phase of the signal coming out of the ASTSA. Results from several simulation experiments are presented to assess the performance of the *old (global) and new (local) adaptive techniques*. Perhaps the only later reference that discusses to certain extent these concepts is [80, 89] with one exception of an earlier reference in Fuchs [96]. We suspect that this investigation could be particularly valuable for the implementation of ASTSAs or analogous systems that will operate on the L_5 frequency as well as for ASTSAs, which are undergoing implementation for either the L_1 or the L_2 frequency.

Third, the investigation of the ASTSA considers the impact of narrowband/wideband interference sources on *GPS-like signal autocorrelation properties*. The interference effect is removed from the signal coming out of the ASTSA; moreover, this signal has better autocorrelation properties than the input signal, which is corrupted by interference. *The signal properties can be further improved if the phase of the signal coming out of the ASTSA is restored.* This chapter begins to explore the fundamental issues related to the preservation of the GPS signal structure after the interference is removed.

Chapter 6 is finally concluded with recommendations on state-of-the-art geolocation implementation as well as advanced features found in the cutting edge adaptive array algorithms to be discussed in Sects. 6.11 and 6.12.

Although the core of this research, analysis, and principle simulation examples have been published previously in [1–3], what we have added to the published data are the following:

1. I have provided the best up-to-date literature search that has helped the understanding, explanation, simulation, and interpretation of the most important concepts of the chapter material tremendously.
2. The chapter probably has the best discussion on assumptions: on where those assumptions came from; on whether there is a need to change those assumptions; and what direction should this work take if those assumptions are changed. More importantly is that this is based on the current most up-to-date literature search and survey.
3. I have refined the analysis and found a number of errors and pointed out those errors and have provided the correct closed form expressions.
4. I have expanded and provided brand new block diagrams, which give the most complete understanding of the adaptive array beamforming for interference mitigation.
5. I have added all the array factor equations and models; I have plotted array factor for four very important principle simulation examples as a very powerful tool to understand and explain the adaptive array beamforming.
6. I have added a section on future direction for research and one appendix (or Appendix B), which had provided an ease of understanding of the equations and also has facilitated the reading of the chapter's mathematical material.

It is hoped that the readers will find this chapter a far better material than the separate disconnected publications in [1–3].

6.2 Adaptive Array Beamforming for Interference Mitigation of GRFS Systems Main Description and Discussion

In this chapter, we are going to investigate the performance of a beamformer, which is a coherent combination of signals from different sensors to increase the SNR or in our case to reduce UDR [32]. Beamforming is classified into two categories: (a) narrowband and (b) wideband. *Narrowband* beamforming refers to the case in which the inverse bandwidth of the signal is large relative to the propagation time across the antenna receiver elements, which implies that all signal frequencies experience similar gain [32]. *Wideband* beamforming refers to the case in which the inverse bandwidth of the signal is smaller relative to the propagation time across the antenna receiver elements, which implied that all signal frequencies experience different gains [32, 57, 58]. An example of a dual-antenna phase-array ultra-wideband CMOS transceiver is presented in [35]. In contrast to analog beamforming, which was performed at the RF analog frequency, digital beamforming is performed at the baseband on complex-valued (in-phase and quadrature) signal components [32]. However, there is technology available or it will be made available to perform digital RF beamforming processing, which should greatly improve the performance due to the reduction of unwanted errors [33]. Or there is a possibility to implement a baseband digital beamforming in a software defined radio receiver as another alternative as discussed in [36] or taking into consideration the 3G UMTS wireless system physical layer: baseband processing hardware implementation perspective in [37]. Beamforming can be either compact system or a distributed system as in [47]. On the one hand, compact beamforming is the most traditional way of creating a beamforming that can be utilized by a single receiving unit. On the other hand, distributed beamforming can be utilized by more than one receiving unit [47]. Moreover, beamforming can be performed both at the transmitter and receiver [1–63, 88]. Receiver beamforming is referred to as a coherent combination of signals from different sensors to increase the SNR or in our case to reduce UDR [32]. Transmit beamforming refers to the technique in which an information source transmits a radio frequency signal over two or more antennas and aligns the phases of the transmissions across the antennas such that, after the propagation, the signals combine constructively at the destination [47]. In this chapter, we are going to discuss compact beamforming leaving transmit distributed beamforming form for future research and for future editions of this book. Beamforming can be also realized as a form of spatial multiplexing in which the precoder matrix performs a (single layer) beamforming function, which is also known as a *codebook-based* beamforming because it is performed as a limited set of predefined beamforming (precoder) vectors [48–52]. Beamforming can be performed at low-frequencies as low as 1 kHz [1] and as high as 60 GHz [54]; therefore, the basic principle of beamforming remains the same regardless of the frequency.

Beamforming can be performed both for active and passive arrays [1–63]; beamforming can be performed for arrays as well as subarrays [59]; and beamforming can be performed as a single beam as well as multiple-beam [1–63] in which the single-beam refers to the case when the beamformer tries to maximize the gain on a particular direction of a desired signal (or source), and the multiple beam refers to the case when the beamformer tries to maximize the gain on multiple directions of desired signals [1–63]. Beamforming can be performed having prior knowledge of the direction of the satellite [1–10] or without prior knowledge of the direction of the desired satellite when it is not available [64, 67]. Beamforming techniques can be performed generally in all applications or it can be structured around a fundamental signal feature that provides essential leverage for an effective signal processing approach [65]. Adaptive beamforming is used effectively by satellite systems to separate channels that are assigned the same narrow frequency band [68]. In CDMA systems, adaptive beamforming can increase the number of users by a factor of two to four or provide an order of magnitude improvement of omnidirectional systems [68, 72] and increase spectrum efficiency [99]. In wireless LANs, adaptive beamforming in combinations with spread-spectrum techniques achieves co-channel interference rejection and provides spatial division multiple access schemes for higher system capacity, throughput, and cope with time-varying interference [68, 70, 90; Lee and Prasanna, 1995, High throughput-rate parallel algorithms for STAP. Department of EE Systems, USC, Los Angeles, Private Communications]. When frequency invariance is required across the band, then frequency invariant beamforming, as discussed in [71], should be performed. In most applications, adaptive beamforming is the result of the far-field approximation; however, when the near-field approximation is needed [73, 74] provide a comprehensive treatment in the subject. Beamforming can be performed both for Gaussian and non-Gaussian signals [81].

6.3 Adaptive Array Beamforming for Interference Mitigation of GRFS Systems (GPS or GNSS) Main Description and Discussion

The vulnerability of the GPS signal (L1, L2, and L5), GLONASS, and Galileo to narrowband interference during code acquisition and carrier tracking is well-known. Several classical techniques have been proposed, which mitigate the effects of narrowband interference and enable the robust acquisition of the GPS signal, which are discussed in the introduction section of [1] based on [19]. Our technique uses a DFIR filter, also known as an ATSA, and is based on an eigenvalue-based minimization and maximization known as the UDR or also known as beamforming [4, 32] based on a very early MIT report work of min–max array processing from Preisig, J.C., “Adaptive matched field processing in an uncertain propagation environment” in January 1992 [84]. In Chap. 5 we have discussed extensively about the ways to deal with the recursive solution to the generalized eigen-value problem; therefore, the material discussed in Chap. 5 makes the direct implementation of this technique less complex than the others utilizing readily available DSPs to yield maximum interference suppression of an ATSA with two temporal shifters for different selections of the desired signal frequency, temporal shifter delay, and sampling frequency.

Although preliminary results of an investigation conducted by the FAA [22] appear to indicate that at low altitudes no modifications to existing systems will be needed to protect the GPS L5 service, some of the lessons learned here could be valuable for ATSA systems that will operate on the L5, L1C, and L2C GPS frequencies, GPS III, GLONASS, more recent studies conducted for the Galileo signals [23–30]. For a given GPS ATSA architecture, which consists of a given desired signal frequency and sampling frequency, an optimum temporal delay can be obtained, regardless of the interference/jamming frequency even easier using more information obtained from the array beam pattern. Although for a very simple ATSA the solution for the optimum temporal delay can be obtained in a closed form, in general this is not as difficult using more information obtained from the array beam pattern; therefore, we recommend that proper simulations be conducted to yield the optimum tap delay. Preliminary analysis and simulation results indicate that this optimum tap delay may be dependent on the ATSA architecture.

The detrimental effects of interference or hostile jamming are well known. It is reported that the navigation error can go up to 20 km when the GPS receiver is illuminated by a CW or swept CW jammer [19]. Adaptive temporal processing can be successfully utilized to mitigate the undesired impact of CW jammers [1–18]. Nevertheless, there are several advantages of utilizing a phased antenna array. Fante has demonstrated that an adaptive space–time array can successfully cancel narrowband/wideband jammers and jammer multipath [12]. Optimal performance is also achieved when mutual coupling and channel errors are present [4]. Another work conducted by Gromov et al. exploits a phased array to measure the interference direction [13, 14]. This appears to be in agreement with the experimental results obtained by Brown [17] and more recent studies conducted for the Galileo signals [23–30]. A recent investigation conducted by Progri has demonstrated a useful methodology that should be followed through analysis, simulation, and implementation to achieve the optimal performance for a given ATSA, based on UDR measure [1]. Nevertheless, UDR is not the only performance measure of an ASTSA. We are ultimately concerned about the receiver being able to track the desired pseudolite signal [1–10, 29, 30]. *We believe that the old (or global) adaptive methodology is the right approach to this issue.*

It is well known that a phased array antenna introduces a distortion to the desired Pseudolite GPS-like signal. Also the need for restoring the phase of the combined signal coming out of the phased array when processing is performed in time domain is not new. The extended replica folding acquisition search technique (XFAST) exploits the cross-correlation property of the code waveforms in such a way that the entire time uncertainty interval can be searched simultaneously [15]. XFAST performs the processing in the frequency domain, which makes it less susceptible to time delay and phase distortion introduced by jamming suppression insertion [15].

Another approach appears to improve the receiver's tracking loop to handle high dynamic stress and RF interference (RFI) conditions [16]. Under RFI conditions, this technique known as the FLL-assisted-PLL provides both the dynamic robustness due to FLL and the performance accuracy due to PLL. Nevertheless, there exists a limitation to this approach because of the saturation of the receiver due to powerful interference sources or hostile jammers.

In 2000, we proposed a technique for removing undesired interference from a GPS-like signal [4]. This technique is based on the maximization of the sensitivity in the direction of the desired signal and minimizing the sensitivity in the direction of the undesired signal. Initially, we assumed a simple structure for the pseudolite (GPS-like) signal, and the main concern is to generate an algorithm that would successfully mitigate the interference effect without mutual coupling and channel errors and for a specific structure of mutual coupling and channel errors [4].

Recently, we have proposed a more robust GPS-like (pseudolite) signal structure and we have conducted some preliminary analysis to address the most appropriate issues of the system design for indoor geolocation applications [5–10]. Dr. Progri's second book on *Indoor Geolocation Systems: Theory and Applications* will reveal even more amazing results on the novel, robust, and in need for standardization.

A preliminary investigation of the ASTSA is conducted in [2, 4]. In Progri and Michalson [4], we have proposed an uncoded desired signal, i.e., an un-modulated carrier desired signal, which is interfered with by a powerful jammer or interfering signal. This process is revisited in Michalson and Progri [2], where we have proposed and analyzed a coded carrier. The investigation of the ASTSA is heavily analytically involved, which has resulted in a very complicated close form solution [2]. The piece that is developed in this chapter is the analytical development and the simulation of the ASTSA with phase restoration. This improved ASTSA exploits the autocorrelation properties of the total received signal and the cross-correlation properties of the total received signal with a locally generated GPS-like signal. During the initial phase, we have enhanced the proposed analytical model for the ASTSA by including a phase restoration stage to account for the proposed pseudolite signal structure [10].

The proposed pseudolite signal structure can be deteriorated when strong interference sources are present. Therefore, we begin to exploit the impact of the ASTSA on the proposed signal structure; i.e., we desire not only to mitigate the undesired interference impact but also to preserve the properties of our pseudolite (GPS-like) signal. Moreover, the ASTSA can be utilized for outdoor applications, which will protect the GPS signal as part of the GPS modernization.

First, a detailed mathematical model for a GPS ATSA is provided in Sect. 6.2 where we also present the analytical ATSA performance for a given desired signal frequency, temporal shifter delay, and sampling frequency. Second, the mathematical model of the ASTSA is discussed in the ASTSA Mathematical Model in Sect. 6.5. Third, the mathematical model of an ASTSA with restored phase is presented in Sect. 6.6.

Fourth, we discuss simulations of the ATSA and present the extent to which the analytical results and the simulated results agree for an ATSA containing two temporal shifter delays in Sect. 6.7. Fifth, an implementation of the ATSA utilizing the TI C6711 DSP is presented in Sect. 6.8. Sixth, the ASTSA simulation section will depict scenarios that are useful for indoor applications in Sect. 6.9. Seventh, to assess the accuracy of the model we have considered a simple ASTSA, which has only two antenna elements and only one temporal shifter delay. For this ASTSA, we have provided a step-by-step analysis of the cross-correlation properties of the signal under investigation with the locally generated signal, which are very important for normal operation of the receiver in Sect. 6.10. An assessment of our results is provided in Sect. 6.11 along with the most up-to-date list of references. Future directions for research are provided in Sect. 6.12. Finally, in Appendix A, we present the underlying theory that enables forming an estimate of the autocorrelation matrix used in the ATSA, ASTSA, or ASTSA with restored phase; Appendix B: we present equivalent expressions with other similar publications; and in Appendices C and D, we provide refined important theorem proofs.

6.4 ATSA Mathematical Model

In the initial consider a GPS receiver with an ATSA as shown in Fig. 6.1 comes from our pristine publication [1]. Although the initial description in [1] might have been adequate for the figure 1 in [1], the added modification in Fig. 6.1 from figure 1 in [1] calls for an extensive discussion of the refined mathematical model and of the analysis. The front end (FE) section of the receiver down-converts the RF signal to the receiver's IF frequency by employing the reference signal of the local oscillator, which receives timing and clock phase information from the classic GPS receiver. Some of the functions of the classic GPS receiver are as follows: (1) signal processing/data demodulation and decoding; (2) position, navigation, and timing

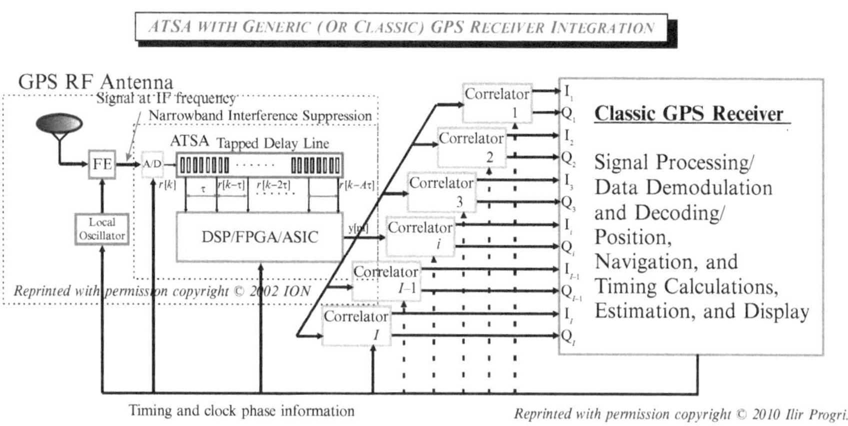

Fig. 6.1 Generic GPS receiver integrated with an ATSA. Reprint with permission © 2002 ION and 2010 Ilir Progri

calculations, estimation, and display. The IF signal enters the A/D converter, which produces the digital ATSA input signal $r[k]$. The A/D converter also receives timing and clock phase information from the classic GPS receiver. This input signal, $r[k]$, is passed through a tapped delay line, which operates at the same rate as the A/D. $(A + 1)$ samples spaced equally by τ enter the DSP where all the ATSA processing takes place. The output of the DSP is the signal, $y[m]$, which drives the early-late correlator(s). The DSP also receives timing and clock phase information from the classic GPS receiver. It is exactly this level of detail that was missing from our previous description in [1].

Consider the block diagram of an ATSA as illustrated in Fig. 6.2. The input signal is $r[k]$ and the output signal is $y[m]$ with $m = \lfloor k/M \rfloor$ and M is the number of samples required to update the output signal $y[m]$. As samples are received, the autocorrelation matrix $\hat{\mathbf{C}}[\delta]$ is updated and used to determine a set of weights, $\mathbf{m}$, based on equivalent expressions discussed in Appendix B. By adaptively and recursively optimizing these weights and applying them to the received input digital data separated by the time delay, τ, a filter is designed, which can significantly reduce the level of an undesired signal before reaching the receiver's correlator(s) also utilizing information from the receiver's timing and clock phase. Initially the receiver's timing and clock phase may not be ready available but it will greatly improve the performance when this information is available as shown in Sect. 6.10. The reminder of this section explains the processing that takes place between the input, $r[k]$, and the output, $y[m]$. Let $N \leq M$ denote the number of samples that contain useful information, which are processed in the manner explained later and where the number of samples, M, is selected as described in [8, 9]. Also, define n the index such that $n = k \bmod N$; hence, $n = \{0, 1, \ldots, N - 1\}$.

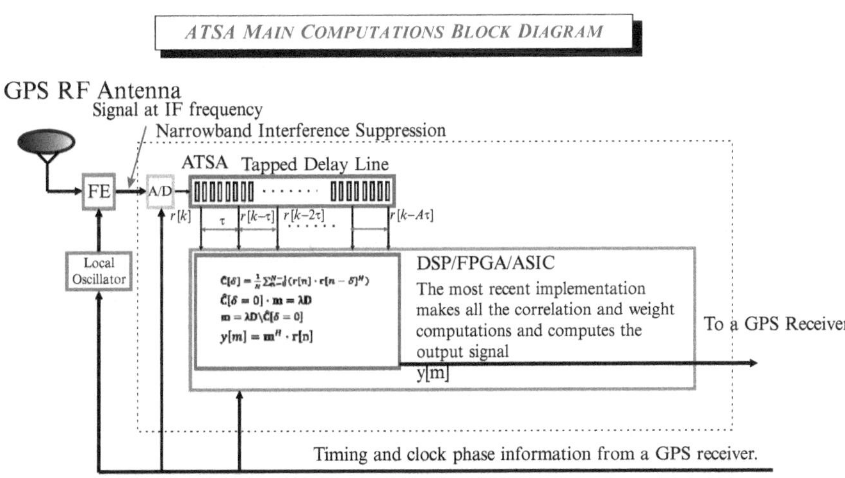

Reprinted with permission copyright © 2010 Ilir Progri.

Fig. 6.2 ATSA block diagram. Reprint with permission © 2010 Ilir Progri

Denote the total discrete received signal vector with $\mathbf{r}[n]$ coming out of the tap-delay line given by

$$\mathbf{r}[k \bmod N] = \mathbf{r}[n] \equiv r[k - 0\tau] \quad \cdots \quad r[k - A\tau]^{\mathrm{T}}. \tag{6.1}$$

The unknowns in expression (6.1) are the total number of taps, A, and the tap delay, τ, which are the tap-delay line main parameters. There is a certain interface that takes place between a tap delay line and the DSP, which is not discussed in this chapter. This is another important level of detail, which may be discussed as related to implementation details via Software Defined radio for example.

Assume that the total received signal vector $\mathbf{r}[n]$ is composed of the sum of the desired signal vector, $\mathbf{d}[n]$, with undesired signal vector, $\mathbf{u}[n]$, and uncorrelated white Gaussian noise (WGN), $\mathbf{w}[n]$, which is $N(0, \sigma_{\mathbf{w}})$. The assumption that $\mathbf{w}[n]$, which is $N(0, \sigma_{\mathbf{w}})$, is important for stationary processes. What happens when $\mathbf{w}[n]$, which is not $N(0, \sigma_{\mathbf{w}})$, is the topic of discussion for future direction of research in Sect. 6.12. The mathematical expression associated with the above assumption is

$$\mathbf{r}[n] = \mathbf{d}[n] + \mathbf{u}[n] + \mathbf{w}[n]. \tag{6.2}$$

Further, assume that the desired signal vector, $\mathbf{d}[n]$, is either a narrowband or wideband signal vector created by modulating a known carrier frequency with a pseudorandom sequence. Further material in this chapter is dedicated to greatly expand the discussion of this process. If the desired signal vector, $\mathbf{d}[n]$, is a narrowband signal, then it is assumed that the signal frequency, f_{d}, is either known or provided such as the case of FM, AM, or two-way radio signals discussed in Chap. 3. If the desired signal vector is a wideband signal, it is assumed that the carrier frequency, $f_{\mathrm{d}}^{\mathrm{c}}$, and the signal bandwidth, B_{d}, are either known or provided such as in the case of GPS L1, L2, L5, GLONASS, Galileo, Mobile communications, wireless networks, Satellite TV signals etc. also discussed in Chap. 3. P_{d} denotes the desired signal level during the sampling window N, and it is also assumed that this quantity is either known or provided. We also assume that a narrowband undesired signal vector, $\mathbf{u}[n]$, with unknown signal frequency, f_{u}, and an unknown signal level, P_{u} such as the case of a CW jammer [1–4]. The level of the WGN, denoted by P_{w}, represents the noise power spectrum density. This quantity is receiver dependent and is measured independently of the desired or undesired level but it may also be an indicator of the environment and further considerations of the environments should be taken into account for future directions of research as mentioned in Sect. 6.12. For stationary processes the noise level P_{w} is selected as the reference signal level for this simple reason: if the undesired signal level is below the noise level then there is no need to use the ATSA and conversely. For non-stationary processes pre-whitening is required to remove the coloring before this comparison analysis takes place.

The undesired signal level, P_{u}, represents interference to the desired signal, and therefore the goal of the ATSA is to minimize the detrimental effects of the undesired signal level by minimizing the level of the undesired signal that can be detected in the output $y[m]$, while maximizing the sensitivity of the system to the desired signal level, P_d based on a very early MIT report work of min–max array processing from Preisig, J.C., "Adaptive matched field processing in an uncertain propagation environment" in January 1992 [84].

Several techniques for mitigating interference have been proposed and investigated in the community [1–18]. Some of these techniques were briefly mentioned in the introduction section [1–18]. The focus of this chapter is on the known technique of applying additional temporal degrees of freedom in order to minimize the undesired signal level and at the same time maximizing the desired signal level, which is analytically written as (also known as the min–max array processing [84])

$$\mathrm{UDR} = \frac{\min_m P_y}{\max_m P_d} \quad \text{(the expression is clarified from eq. (3) in [1]).} \tag{6.3}$$

The UDR is a well-known cost function, which serves as our criterion for performance assessment subject to selecting the proper weight vector $\mathbf{m}$. The reader is reminded that in the numerator we have used the total output signal level, P_y, instead of the undesired signal level, P_u, because the total signal level is a directly measurable quantity as opposed to the undesired signal level, which is *not* a directly measurable quantity at the output of an ATSA.

The ATSA adaptively minimizes the undesired signal level and maximizes the desired signal level in four steps as illustrated in Fig. 6.2. First, we assume that the statistical properties of the system can be fully characterized by the autocorrelation matrix, $\hat{\mathbf{C}}[\delta]$, given by

$$\hat{\mathbf{C}}[\delta] = \frac{1}{N}\sum_{n=0}^{N-1}\left(\mathbf{r}[n]\cdot\mathbf{r}[n-\delta]^{\mathrm{H}}\right). \tag{6.4}$$

We have further assumed that the process is ergodic both in its mean and variance because the random noise process is WGN; therefore, for an infinite number of samples, the estimated autocorrelation matrix, $\hat{\mathbf{C}}[\delta]$, approaches the true autocorrelation matrix of the system (see Appendix A for further details on how this estimate is formed). Again for non-stationary processes further considerations of the environments should be taken into account for future directions of research in Sect. 6.12

Second, the desired signal level is maximized assuming that the desired signal frequency is known. Hence, we define the desired pointing (or steering) vector, $\mathbf{D}$, as

$$\begin{gathered}\mathbf{D} = \left[\exp(j0\omega_d\tau) \equiv 1 \quad \cdots \quad \exp(jA\omega_d\tau)\right] \\ \text{(another clarification from eq. (5) in [1]).}\end{gathered} \tag{6.5}$$

Beamforming can be performed when we have precise knowledge of the pointing (or steering or direction of arrival vector) or without requiring that knowledge as in Chap. 4 and also as discussed in [82, 97]. In Bell et al. [97], the DOA is assumed to be a discrete random variable with a known a priori probability density function (pdf) that reflects the level of uncertainty in the source DOA as opposed to our case in which we have perfect knowledge of the DOA. However, for future work, to improve the robustness of the beamformer, the methodology presented in Bell et al. [97] should be followed.

Other books use for **D** an expression that would substitute $\exp(ja\omega_d\tau)$ with $\exp(-ja\omega_d\tau)$. The reader should convince himself/herself that a conjugation in the desired pointing vector will only introduce a conjugation in the adaptive weights or multipliers therefore the final answer will not change.

The other important consideration is the discussion on the beam-steering angle θ. If we were to compare (6.5) with eq. (11.4.20) in [31] we immediately see that $2\pi d\ \sin\theta/\lambda = \omega_d\tau$, since $\lambda = c/f_d$ which yields, $\omega_d = 2\pi f_d/f_s$ and $\tau = df_s\ \sin\theta/c$, where f_s is the sampling frequency. The main question is what is d and what does it represent? d represents "imagined element spacing," which is usually $d = \lambda/2$; for the ASSA, then d represents the physical element spacing as we have discussed later in the chapter; therefore, we get

$$\tau = 0.5 f_s\ \sin\theta/f_d. \tag{6.6}$$

Equation (6.6) is particularly important as we are going to consider cases when $\tau > 1$, which can only happen when $f_s > 2f_d$ at the very least.

Third, the adaptive weights (or multipliers), $\mathbf{m} = \{m_i\}$, are computed exploiting the estimated correlation matrix, $\hat{\mathbf{C}}[\delta]$, and a desired pointing vector, **D** (see above equation) in accordance with [8, 9]

$$\hat{\mathbf{C}}[\delta = 0]\cdot\mathbf{m} = \lambda\mathbf{D} \rightarrow \mathbf{m} = \lambda\hat{\mathbf{C}}[\delta = 0]/\mathbf{D} \quad \text{(another clarification from eq.(6) in [1])}, \tag{6.7}$$

where each m_i serves as a impulse response coefficient of a DFIR filter.

Fourth, the ATSA output signal, $y[m]$, with or without timing and clock phase information as shown in Fig. 6.2, is determined from

$$y[m] = \mathbf{m}^{\mathrm{H}}\cdot\mathbf{r}[n] \quad \text{(another clarification from eq.(7) in [1])}. \tag{6.8}$$

A simplified expression for the UDR is derived in [4] and is rederived here for the sake of simplicity (also known as the min–max array processing [84])

$$\mathrm{UDR} = \frac{\min_m P_y}{\max_m \sum_a P_d} = \frac{1}{AP_d\mathbf{D}^{\mathrm{H}}\cdot\tilde{\mathbf{m}}}, \tag{6.9}$$

where $\tilde{\mathbf{m}} = \mathbf{m}/\lambda$ are the normalized weights (or multipliers). Further in the reminder of the chapter we use **m** to denote the normalized weights. (See appendix B for a complete derivation of (6.9).)

Some discussion on the desired power $\sum_{a} P_{d}$ based on the eq. (11.5.4) in [31] is that $\sum_{a} P_{d} = AP_{d}$, where P_{d} is the desired signal level of one antenna element and one tap (i.e., nonadaptive array system) i.e., UDR is inverse proportional with the number of taps, A.

Equation (6.9) is equivalent to that of the journal article [1] (eq. 8 [1]), which drives the need to rederive all the remaining equations in Progri et al. [1] and check those equations for accuracy and report other equivalent expressions in Appendix B. We do not suspect any major differences on the results; however, given this opportunity we have to provide our readers with the most accurate and up to date information.

Initially, we will restrict our analysis to a simple ATSA with only two temporal delays and derive later in the chapter the more general case that was not derived in the journal article [1]. Therefore, we have new information added to the chapter, which was not published in the journal article, and we hope that this information will be useful to our readers.

6.4.1 Principle Illustration Example 1

Equations (9) through (22) in [1] provide all the material needed to explain the introduction in principle illustration example 1.

The expression for the inner product between the multiplier vector and the desired pointing vector based on the equivalent equation as discussed in Appendix B is

$$\mathbf{D}^{\mathrm{H}} \cdot \mathbf{m} = \alpha \sum_{i=1}^{3} \sum_{l=1}^{3} \left(c_{il}^{-1} \exp[-j(i-l)\omega_{\mathrm{d}}\tau]\right) \quad \text{(beam response or beam pattern).} \tag{6.10}$$

Expression (6.10) is different from eq. (23) in [1] can be simplified further to

$$\mathbf{D}^{\mathrm{H}} \cdot \mathbf{m} = \alpha \left(\sum_{i=1}^{3} c_{ii}^{-1} + \beta \right) \quad \text{(beam response or beam pattern),} \tag{6.11}$$

where (6.11) is different from eq. (24) in [1]

$$\beta = 2(\mathrm{Re}[c_{12}^{-1}] + \mathrm{Re}[c_{23}^{-1}])\chi + \mathrm{Re}[c_{13}^{-1}]\delta, \tag{6.12}$$

$$\chi = \cos(\omega_{\mathrm{d}}\tau), \tag{6.13}$$

$$\delta = \cos(2\omega_{\mathrm{d}}\tau). \tag{6.14}$$

There was an error fixed in (6.12), which is different from eq. (25) in [1].

To simplify our analysis we set $\omega_{\mathrm{d}} = \omega_{\mathrm{u}}$, $A_{\mathrm{d}} = A_{\mathrm{u}} = 1$, $\theta_{\mathrm{d}} = \theta_{\mathrm{u}} = 0$, and $\sigma_{\mathrm{w}}^2 = 1$. The expression for the UDR is then reduced to (leaving all the intermediate steps to the reader)

$$\mathrm{UDR} = \frac{13}{21 - 8\delta - 4\varepsilon}, \tag{6.15}$$

where

$$\varepsilon = \cos(4\omega_{\mathrm{d}}\tau) \quad \text{(is the same as eq. (29) in [1]).} \tag{6.16}$$

We now observe the biggest correction of UDR from eq. (28) of [1]. From this point onward I believe that the recalculation of all the results and simulations of [1] is necessary because we are performing calculations for a different UDR.

The partial derivative of the UDR with respect to τ can be used to obtain a relationship that produces the minimum and maximum values of the UDR. It can be easily verified that

$$\frac{\partial \mathrm{UDR}}{\partial \tau} = 0 \tag{6.17}$$

is equivalent to (leaving all the details to the reader)

$$\sin(2\omega_{\mathrm{d}}\tau) = 0 \; (\text{or} \cos(2\omega_{\mathrm{d}}\tau) = \pm 1) \quad \text{or} \quad \cos(2\omega_{\mathrm{d}}\tau) = -0.5. \tag{6.18}$$

Luckily we have the same answer as in eq. (31) in [1] which also yields

$$2\omega_{\mathrm{d}}\tau = l\pi \quad \text{or} \quad 2\omega_{\mathrm{d}}\tau = 2l\pi \pm 2\pi/3. \tag{6.19}$$

To find out which one of these solutions produces the minimum and which one the maximum we check the following

$$\left.\frac{\partial^2 \mathrm{UDR}}{\partial \tau^2}\right|_{2\omega_{\mathrm{d}}\tau = l\pi} = -\frac{32\omega_{\mathrm{d}}^2(\delta + 2\varepsilon)}{(21 - 8\delta - 4\varepsilon)^2} < 0 \quad \text{and} \quad \left.\frac{\partial^2 \mathrm{UDR}}{\partial \tau^2}\right|_{2\omega_{\mathrm{d}}\tau = 2l\pi \pm 2\pi/3} > 0, \tag{6.20}$$

which indicates that $2\omega_{\mathrm{d}}\tau = l\pi$ produces the maximum UDR and that $2\omega_{\mathrm{d}}\tau = 2l\pi \pm 2\pi/3$ produces the minimum UDR.

Further, given the sampling frequency, f_{s}, and the frequency of the desired signal before sampling, f_{d}, the following expression for ω_{d} can be determined

$$\omega_{\mathrm{d}} = 2\pi \frac{f_{\mathrm{d}}}{f_{\mathrm{s}}}. \tag{6.21}$$

Therefore, considering (6.16) through (6.21) the minimization of the UDR is based upon satisfying the conditions of

$$6f_{\mathrm{d}}\tau = (3l \pm 1)f_{\mathrm{s}} \quad \text{and} \quad l = \{0, 1, 2, \ldots\}. \tag{6.22}$$

and the maximization of the UDR is based upon the fulfillment of this criterion

$$4f_{\mathrm{d}}\tau = lf_{\mathrm{s}} \quad \text{and} \quad l = \{0, 1, 2, \ldots\}. \tag{6.23}$$

To keep the amount of data processing reasonable for this example we, without loss of generality, will use audio frequencies to demonstrate how the math works and provide an example that can be implemented using the TI C6711 DSP. The UDR as a function of the temporal shifter delay, τ, for $f_{\mathrm{d}} = 1$ kHz and $f_{\mathrm{s}} = \{6,60,600\}$ kHz is illustrated in Fig. 6.3. The optimum tap delay is proportional to the sampling frequency, which is consistent with the results shown in Fig. 6.3.

It can be easily observed from the analysis (see (6.23)) and from Fig. 6.3 that the smallest optimum shifter delay that minimizes the UDR corresponds to

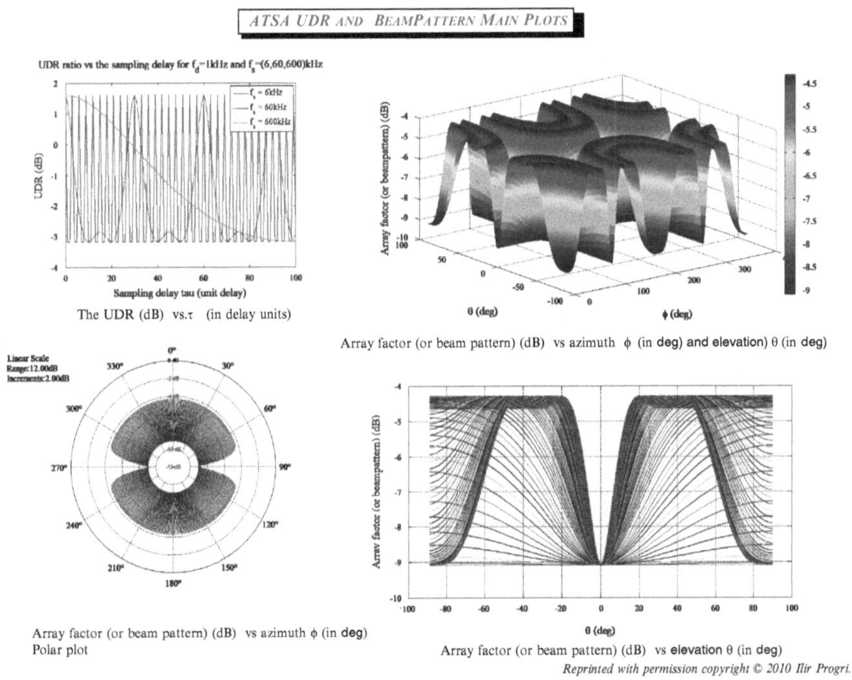

Fig. 6.3 UDR and beam pattern (or array factor) main plots for principle simulation example 1. Reprint with permission © 2010 Ilir Progri

$\tau_o = \{1,10,100\}$. For this particular example, the sampling frequency ought to be at least six times higher than the desired signal frequency to enable a selection of optimum temporal delay of 1 or greater. For a different example, the relation of the sampling frequency and desired signal frequency for selecting the optimum τ might be different.

Figure 6.3 also illustrates the array beam pattern (or beam response, or beam factor, etc.) which is a standard tool for analyzing the performance of an array as a function of azimuth ϕ (in $0° \leq \phi \leq 360°$) and elevation (in $-90° \leq \theta \leq 90°$). These plots are shown here for the first time and the main intent is to add the value of the material published in [1] and at the same time to facilitate the reading and understanding of the material on adaptive array beamforming for interference mitigation for GRFS systems.

Beamforming (or the beam response should be independent of the frequency, time delay, sampling frequency, etc.; it should only depend on the array size (or number of taps), which is expressed in the steering vector (or pointing vector) as given by (6.5), and the frontend bandwidth, which is expressed in the weight vector (or multipliers) as given by (6.7). As such beam response (or beam pattern) is defined as

$$\mathbf{B}(\phi,\theta) = \mathbf{D}^{\mathrm{H}}(\phi,\theta)\cdot\mathbf{m} = \mathbf{m}^{\mathrm{H}}\cdot\mathbf{D}(\phi,\theta) \quad \text{(beam response or beam pattern).} \tag{6.24}$$

In general, the results of this work would have been understood a lot easier had we plotted the bean pattern or the beam response as indicated in Fig. 6.3. As shown in Fig. 6.3, we have nulls in the azimuth ϕ (in $\phi = 90°$ and $\phi = 270°$) and elevation (in $\theta = -90°$, $\theta = 0°$, and $\theta = 90°$). These results are consistent with our findings.

Apparently, the 3D and 2D representation of the array factor plots as shown in Figs. 6.3 and 6.4 is a way that people from the *IEEE Transactions on Antennas and Propagation* [85] and people from *PIERS Online Journal* [98] really like it. In addition to that we have added a polar plot.

Figure 6.4 illustrates the same for uniform weights (or multipliers **m**) as given by equation below

$$\mathbf{m} = \begin{bmatrix} m_1 & m_2 & m_3 \end{bmatrix}^{\mathrm{T}} = 1/\sqrt{3}\begin{bmatrix} 1 & 1 & 1 \end{bmatrix}^{\mathrm{T}}. \tag{6.25}$$

This is the most common form of the beam response, which is shown in many books such as [31] only for azimuth ϕ (in $\phi = 0°$). But we have shown the beam response for all values of azimuth and elevation, and we are showing polar plot for azimuth and 3D plot for both azimuth and elevation. There is a very simple explanation why the results of Fig. 6.4 are very different from those of Fig. 6.3 because the weights used to produce Fig. 6.3 are different from those used to produce Fig. 6.4. More plots are more discussions on array beam response, which are presented later in the chapter.

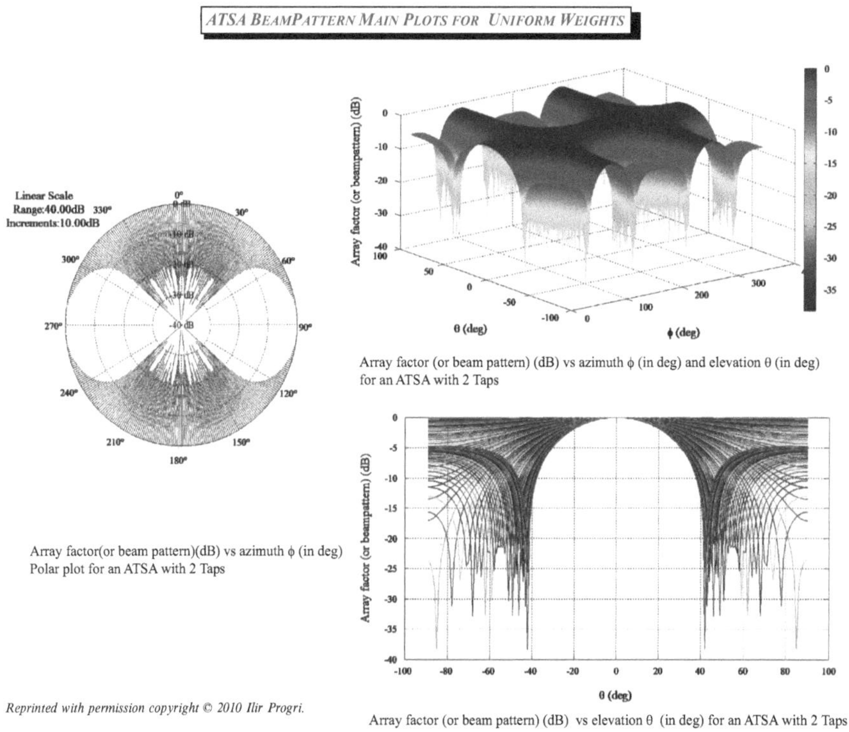

Fig. 6.4 ATSA beam pattern (or array factor) main plots for an ATSA with 2 taps and uniform weights (or multipliers). Reprint with permission © 2010 Ilir Progri

6.5 ASTSA Mathematical Model

STAP (or in the form of ASTSA) involves adaptively (or dynamically) adjusting the two-dimensional space–time filter response in an attempt at maximizing output SINR, and consequently, receiver signal reception under heavy interference and jamming [56, 59]. In the space–frequency adaptive processing (SFAP) [86] the processing is done in the frequency domain, whereas in the STAP processing the weights and the adaptive signal processing is done in the time domain.

The main objective of this section is to develop the basic theory of ATSA, ASSA, and ASTSA as it relates to GRFS systems for interference mitigation (or suppression) following groundbreaking adaptive array development by Howells, Applebaum, and Widrow, and since Brenna and Reed who introduced STAP to the airborne radar community in a 1973 [56, 59] recent advancement of high-speed, high performance, DSPs make STAP-based radar and GRFS systems possible on manned and unmanned airborne platforms and spaceborne satellites and also ground systems [1–10, 56, 59]. STAP can be efficiently and effectively performed in any of the 39 engagement scenarios described in Chap. 2. The given references

[62, 63] are success stories in building active multichannel array radars with the capability of recording the outputs of more than four quadrants.

This section is based on the works that were initially published in [2, 3–5] in which the main emphasis is given in improving the mathematical model, offering detailed explanations of the main principles, providing a novel block-diagram as illustrated in Fig. 6.5, and including the model for the beam response and beam pattern. Taking the above into consideration, a generic ASTSA block diagram with classic GPS receiver is presented in Fig. 6.5 [4].

Denote the desired, received signal at the operational frequency as seen by the kth antenna array element, $\forall k \in \{1, \ldots, A\}$, by Progri and Michalson [5]

$$s_k^{iI}(t) = A_i c(n, \tilde{t}_{\mathrm{d}}) \cos(\tilde{\omega}_{\mathrm{o}}^{\mathrm{d}} \tilde{t}_{\mathrm{d}} + \alpha_k^{\mathrm{d}} + \phi_{\mathrm{d}}), \tag{6.26}$$

$$s_k^{iQ}(t) = A_i d_l(\tilde{t}_{\mathrm{d}}) p(\tilde{t}_{\mathrm{d}}) c(l, \tilde{t}_{\mathrm{d}}) \sin(\tilde{\omega}_{\mathrm{o}}^{\mathrm{d}} \tilde{t}_{\mathrm{d}} + \alpha_k^{\mathrm{d}} + \phi_{\mathrm{d}}). \tag{6.27}$$

The undefined terms in equations (1) and (2) are defined in the following order: (1) $A_i = \sqrt{P_i}$, P_i is the power received at the antenna element from the ith

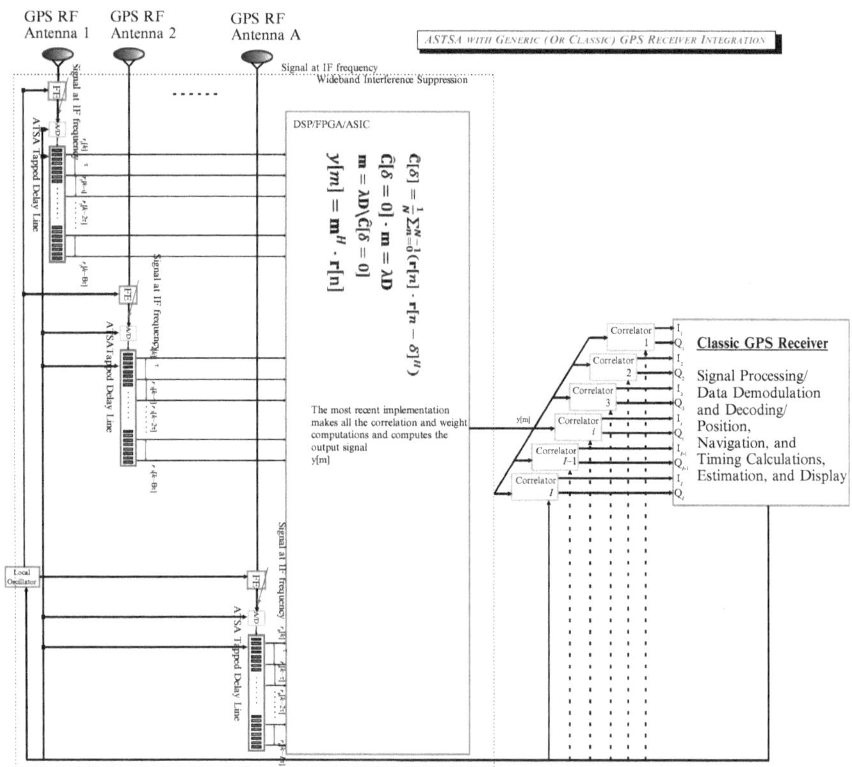

Fig. 6.5 The ASTSA with classic (or generic) GPS receiver integration block diagram. Reprint with permission © 2010 Ilir Progri

transmitter (W/Hz); (2) t is the time duration; (3) $\tilde{t}_\mathrm{d} = t - \tau_i^\mathrm{d}$, $\tau_i^\mathrm{d} = R_i^\mathrm{d}(t)/c$ is the time difference (delay) between the ith transmitter and the receiver, $R_i^\mathrm{d}(t)$ is the geometric range (m), and c is the speed of light (m/s); (4) $c(l, \tilde{t}_\mathrm{d})$ is the pseudo random (or PN) code; (5) $d_\mathrm{l}(t)$ is the data stream coded at R_d data rate; l and n correspond to a given ith transmitter for a unique pair of PN codes; (6) $p(t)$ is an appropriate code [5], which improves the signal cross-correlation properties, resolves the bit timing clock, and reduces the spectral densities. At this stage, our analyses are simplified and do not consider the impact of this sequence; (7) $\tilde{\omega}_\mathrm{o}^\mathrm{d} = \omega_\mathrm{o}^\mathrm{d}(1 \pm \dot{R}_i^\mathrm{d}(t)/c)$, $\omega_\mathrm{o}^\mathrm{d} = 2\pi f_\mathrm{o}^\mathrm{d}$ with f_o^d the operational carrier frequency and $\dot{R}_i^\mathrm{d}(t)$ is the geometric range rate (m/s) between the ith transmitter and the array reference element; (8) α_k^d is the phase shift of the signal as seen by the kth antenna array's element with respect to the array's reference element; (9) ϕ_d is of course the initial phase of the desired signal.

Similarly, the undesired signal at the operational frequency as seen by the kth antenna array's element, $\forall k \in \{1, \cdots, A\}$, can be written as [2]

$$u_k(t) = U(\tilde{t}_\mathrm{u}, W)\exp[j(\tilde{\omega}_\mathrm{o}^\mathrm{u}\tilde{t}_\mathrm{u} + \alpha_k^\mathrm{u} + \phi_\mathrm{u})]. \tag{6.28}$$

The reader is reminded that in our notation convention, the subscript/superscript d defines the desired signal components of the signal; the subscript/superscript u denotes the undesired signal components of the signal; and subscript o is assigned to the operational frequency of the signal. Therefore, the only undefined terms in equation (3) are the modulated/unmodulated amplitude of the undesired signal $U(\tilde{t}_\mathrm{u}, W)$ and the bandwidth W.

Both the desired and undesired signal components are corrupted by noise due to transmission through a nonuniform environment, the effect of which is approximated as a white noise, $\sigma_0 v_k(t)$, where $\sigma_0 = \sqrt{N_0}$ is the noise standard deviation, N_0 is the thermal noise PSD (W/Hz).

Therefore, the total noisy signal at the operational frequency as seen by the kth antenna array element, $\forall k \in \{1, \cdots, A\}$, is denoted by

$$r_k(t) = s_k^i(t) + u_k^j(t) + \sigma_0 v_k(t). \tag{6.29}$$

The detrimental impact of the undesired signal over the desired signal can be investigated by employing the autocorrelation and the cross-correlation properties of the desired reference signal, $d(t)$, and the received signal vector, $\mathbf{r}(t)$, which are determined from

$$d^{iI}(t) = A_i c(n, t_\mathrm{d})\cos(\omega_\mathrm{o}^\mathrm{d} t_\mathrm{d} + \phi_\mathrm{d}), \tag{6.30}$$

$$d^{iQ}(t) = A_i d_l(t_\mathrm{d})p(t_\mathrm{d})c(l, t_\mathrm{d})\sin(\omega_\mathrm{o}^\mathrm{d} t_\mathrm{d} + \phi_\mathrm{d}), \tag{6.31}$$

$$\mathbf{r}(t) = [\; r_1(t - 0\tau) \quad \cdots \quad r_1(t - B\tau) \quad \cdots \quad r_A(t - 0\tau) \quad \cdots \quad r_A(t - B\tau) \;]^\mathrm{T}. \tag{6.32}$$

Denote with $y(t)$ the output of the ASTSA, which can be computed from

$$y(t) = \mathbf{m}^{\mathrm{H}} \cdot \mathbf{r}(t). \tag{6.33}$$

Denote with $\mathbf{C}(\delta)$ the autocorrelation matrix, which is defined as [2]

$$C(\delta) = E[\mathbf{r}(t) \cdot \mathbf{r}^{\mathrm{H}}(t - \delta)] \tag{6.34}$$

which manifests the properties given in [2]. Contrast (6.34) with (6.4). The autocorrelation given by (6.4) is an approximation of the autocorrelation matrix given by (6.34). For more detail, the reader should refer to Appendix A.

Denote with $\mathbf{d}(\delta)$ the cross-correlation vector, which is determined from

$$\mathbf{d}(\delta) = E[\mathbf{r}(t) \cdot x^*(t - \delta)] = \mathbf{C}(\delta) \cdot \mathbf{m}(\delta). \tag{6.35}$$

Before the desired signal is jammed or interfered with, we know the direction of the desired signal that the receiver is tracking. Therefore, we would desire that in the presence of interference the output of the ASTSA could form a beam in the direction of the desired signal. This can be accomplished if the adaptive computation methodology is established as follows

$$\mathbf{C}(\delta = 0) \cdot \mathbf{m}(\delta = 0) = \mathbf{p}_i, \tag{6.36}$$

where $\mathbf{p}_i$ is the desired pointing vector in the direction of the ith source. Further down this section we are going to assume that $\delta = 0$; i.e., we are going to consider the optimum values.

The appropriate performance criterion is given as the ratio of the total signal power over the reference signal power as [2] (also known as the min–max array processing [84])

$$\mathrm{UDR} = \frac{\min_m P_{\mathrm{y}}}{\max_m P_{\mathrm{d}}} = \frac{E[|\mathbf{m}^{\mathrm{H}} \cdot \mathbf{r}(t)|^2]}{E(|\tilde{s}_i(t)|^2)}, \tag{6.37}$$

where

$$\tilde{s}_i(t) = \mathbf{m}^{\mathrm{H}} \cdot \mathbf{s}_i(t) \tag{6.38}$$

is the beam formed if the only signal were to be the desired signal and

$$\mathbf{s}_i(t) = [\ s_1^i(t - 0\tau) \quad \cdots \quad s_1^i(t - B\tau) \quad \cdots \quad s_A^i(t - 0\tau) \quad \cdots \quad s_A^i(t - B\tau)\ ]^{\mathrm{T}} \tag{6.39}$$

denotes the desired vector.

Again, compare and contrast UDR given by (6.37) with UDR given by (6.3). The UDR given by (6.37) is the mean value of the (6.37) given by (6.3) (also known as the min–max array processing [84]).

It can be shown that the numerator of (6.37) is of the form

$$\min_{m} P_{\mathrm{y}} = E[|\mathbf{m}^{\mathrm{H}} \cdot \mathbf{r}(t)|^{2}] = \mathbf{m}^{\mathrm{H}} \cdot \mathbf{C} \cdot \mathbf{m}. \tag{6.40}$$

The denominator of the expression (6.37) can be written as

$$\max_{m} P_{\mathrm{d}} = E(|\tilde{s}_i(t)|^2) = \mathbf{m}^{\mathrm{H}} \cdot \mathbf{s}_i(t) = P_i(\mathbf{m}^{\mathrm{H}} \cdot \mathbf{p}_i)(\mathbf{m}^{\mathrm{H}} \cdot \mathbf{p}_i)^{\mathrm{H}}. \tag{6.41}$$

Combining expressions (6.37), (6.40), and (6.41), yields (also known as the min–max array processing [84]) (see also Appendix B)

$$\mathrm{UDR} = \frac{\min_m P_{\mathrm{y}}}{\max_m P_{\mathrm{d}}} = \frac{E[|\mathbf{m}^{\mathrm{H}} \cdot \mathbf{r}(t)|^2]}{E(|\tilde{s}_i(t)|^2)} = \frac{1}{P_{\mathrm{d}}(\mathbf{m}^{\mathrm{H}} \cdot \mathbf{p}_i)^{\mathrm{H}}} = \frac{1}{P_{\mathrm{d}} \mathbf{p}_i^{\mathrm{H}} \cdot \mathbf{m}}. \tag{6.42}$$

We are already familiar with (6.42) (see [2]); nevertheless, the structure of the pointing vector would be different.

Previously, we have proposed an algorithm for computing the desired set of multipliers [2], which reads

1. Compute a simplified expression for the steady-state correlation matrix, $\mathbf{C}(\delta = 0)$, given by (6.34)
2. Find an appropriate expression for calculating the desired pointing vector, $\mathbf{p}_i$
3. Find a simplified expression for the desired set of multipliers based on (6.36)
4. Estimate the UDR ratio according to (6.42)

We follow the procedure by computing the diagonal and the off-diagonal elements of autocorrelation matrix, $\mathbf{C}(\delta = 0)$, for $ii = f(k, m)$

$$c_{ii,ii}(0) = E[r_k(t - m\tau) r_k^*(t - m\tau)]. \tag{6.43}$$

Based on the definition of the received signal (see (6.29)), an expression for computing the diagonal elements of the autocorrelation matrix can be written as

$$c_{ii,ii}(0) = |s_k^i(t - m\tau) + u_k^j(t - m\tau)|^2 + \sigma_0^2. \tag{6.44}$$

Expression (6.44) can be further written as

$$c_{ii,ii}(0) = |s_k^i(t - m\tau)|^2 + |u_k^j(t - m\tau)|^2 + 2\mathrm{Re}\{s_k^i(t - m\tau)\} u_k^{j*}(t - m\tau) + \sigma_0^2. \tag{6.45}$$

Next, we provide an expression for the off-diagonal elements of the autocorrelation matrix, for $ii = f(k,m)$ and $jj = f(b,e)$

$$\begin{aligned} c_{ii,jj}(0) &= E[r_k(t-m\tau)r_k^*(t-e\tau)] \\ &= [s_k^i(t-m\tau)+u_k^j(t-m\tau)][s_k^{i*}(t-m\tau)+u_k^{j*}(t-m\tau)] \\ &= s_k^i(t-m\tau)s_k^{i*}(t-m\tau)+s_k^i(t-m\tau)u_k^{j*}(t-m\tau) \\ &\quad +u_k^j(t-m\tau)s_k^{i*}(t-m\tau)+u_k^j(t-m\tau)u_k^{j*}(t-m\tau). \end{aligned} \tag{6.46}$$

Next, we proceed to determine an appropriate expression for the desired pointing vector. The desired pointing vector can be determined from the desired vector as below:

$$\mathbf{s}_i(t) = A_i\mathbf{p}_i(t) = A_i[\mathbf{q}_i(t)e^{j\varepsilon} + \mathbf{g}_i(t)e^{-j\varepsilon}], \tag{6.47}$$

where

$$\varepsilon = \tilde{\omega}_{\rm o}^{\rm d}\tilde{t}_{\rm d} + \phi_{\rm d}, \tag{6.48}$$

$$\mathbf{q}_i(t) = [\; q_1^i(\tilde{t}_{\rm d}) \quad \cdots \quad q_1^i(\tilde{t}_{\rm d}^B) \quad \cdots \quad q_A^i(\tilde{t}_{\rm d}) \quad \cdots \quad q_A^i(\tilde{t}_{\rm d}^B)\;]^{\rm T}, \tag{6.49}$$

$$\tilde{t}_{\rm d}^B = t - B\tau, \tag{6.50}$$

$$q_k^i(\tilde{t}_{\rm d} - l\tau) = \frac{c(n,\tilde{t}_{\rm d}-l\tau) + \tilde{c}(l,\tilde{t}_{\rm d}-l\tau)}{2}e^{j\gamma_k^{{\rm d}l}}, \tag{6.51}$$

$$\gamma_k^{{\rm d}l} = \alpha_k^{\rm d} - \tilde{\omega}_{\rm o}^{\rm d} l\tau, \tag{6.52}$$

$$\tilde{c}(l,\tilde{t}_{\rm d}-l\tau) = d_l(\tilde{t}_{\rm d}-l\tau)p(\tilde{t}_{\rm d}-l\tau)c(l,\tilde{t}_{\rm d}-l\tau) \tag{6.53}$$

and

$$\mathbf{g}_i(t) = [\; g_1^i(\tilde{t}_{\rm d}) \quad \cdots \quad g_1^i(\tilde{t}_{\rm d}^B) \quad \cdots \quad g_A^i(\tilde{t}_{\rm d}) \quad \cdots \quad g_A^i(\tilde{t}_{\rm d}^B)\;]^{\rm T}. \tag{6.54}$$

$$g_k^i(\tilde{t}_{\rm d} - l\tau) = \frac{c(n,\tilde{t}_{\rm d}-l\tau) - \tilde{c}(l,\tilde{t}_{\rm d}-l\tau)}{2}e^{-j\gamma_k^{{\rm d}l}}. \tag{6.55}$$

In light of the pointing vector's new expression, we can obtain the following expression for the desired set of multipliers

$$\begin{aligned} \mathbf{m}(t) &= \mathbf{C}^{-1}\cdot\mathbf{p}_i(t) = \mathbf{C}^{-1}\cdot[\mathbf{q}_i(t)e^{j\varepsilon} + \mathbf{g}_i(t)e^{-j\varepsilon}] \\ &= \mathbf{C}^{-1}\cdot\mathbf{q}_i(t)e^{j\varepsilon} + \mathbf{C}^{-1}\cdot\mathbf{g}_i(t)e^{-j\varepsilon}. \end{aligned} \tag{6.56}$$

The above expression can be further written as

$$\mathbf{m}(t) = \mathbf{n}_i(t)e^{j\varepsilon} + \mathbf{w}_i(t)e^{-j\varepsilon}, \tag{6.57}$$

where

$$\mathbf{n}_i(t) = \mathbf{C}^{-1} \cdot \mathbf{q}_i(t), \tag{6.58}$$

$$\mathbf{w}_i(t) = \mathbf{C}^{-1} \cdot \mathbf{g}_i(t). \tag{6.59}$$

Hence, to obtain the UDR ratio, we perform the following inner product

$$\begin{aligned}\mathbf{p}_i^{\mathrm{H}}(t) \cdot \mathbf{m}(t) &= [\mathbf{q}_i^{\mathrm{H}}(t)e^{-j\varepsilon} + \mathbf{g}_i^{\mathrm{H}}(t)e^{j\varepsilon}] \cdot [\mathbf{n}_i(t)e^{j\varepsilon} + \mathbf{w}_i(t)e^{-j\varepsilon}] \\ &= \mathbf{q}_i^{\mathrm{H}}(t) \cdot \mathbf{n}_i(t) + \mathbf{q}_i^{\mathrm{H}}(t) \cdot \mathbf{w}_i(t)e^{-2j\varepsilon} + \mathbf{g}_i^{\mathrm{H}}(t) \cdot \mathbf{n}_i(t)e^{2j\varepsilon} + \mathbf{g}_i^{\mathrm{H}}(t) \cdot \mathbf{w}_i(t).\end{aligned} \tag{6.60}$$

It is easy to verify that (see Appendix B for details)

$$\mathbf{g}_i^{\mathrm{H}}(t) \cdot \mathbf{n}_i(t) = \mathbf{w}_i^{\mathrm{H}}(t) \cdot \mathbf{q}_i(t) = [\mathbf{q}_i^{\mathrm{H}}(t) \cdot \mathbf{w}_i(t)]^{\mathrm{H}} \tag{6.61}$$

which enables (6.61) to be written as

$$\mathbf{p}_i^{\mathrm{H}}(t) \cdot \mathbf{m}(t) = \mathbf{q}_i^{\mathrm{H}}(t) \cdot \mathbf{n}_i(t) + 2|\mathbf{g}_i^{\mathrm{H}}(t) \cdot \mathbf{n}_i(t)|\cos[2(\varepsilon + \beta)] + \mathbf{g}_i^{\mathrm{H}}(t) \cdot \mathbf{w}_i(t), \tag{6.62}$$

where

$$\beta = \arg[\mathbf{g}_i^{\mathrm{H}}(t) \cdot \mathbf{n}_i(t)]. \tag{6.63}$$

The open expression for a general case of the autocorrelation matrix reads

$$\mathbf{C} = \begin{bmatrix} c_{11} & c_{12} & \cdots & c_{1N} \\ c_{12}^* & c_{22} & & c_{2N} \\ \vdots & & \ddots & \vdots \\ c_{1N}^* & c_{2N}^* & \cdots & c_{NN} \end{bmatrix} \tag{6.64}$$

where, $N = AB$ denotes the size (or better the rank) of the autocorrelation matrix, $\mathbf{C}$.

The diagonal elements of the correlation matrix can be determined from

$$c_{ii,ii}(0) = |x_k(t_{m\tau})|^2 + \sigma_0^2, \quad \forall ii \in \{1,2,\cdots,N\}, \forall k \in \{1,2,\cdots,A\}, \forall m \in \{1,2,\cdots,B\}, \tag{6.65}$$

where, $x_k(t_{m\tau})$ is the noiseless signal; hence, its analytical expression reads

$$x_k(t_{m\tau}) = s_k^i(t - m\tau) + u_k^j(t - m\tau). \tag{6.66}$$

Based on (6.64) the expression for the off-diagonal term of the autocorrelation matrix can be written as

$$c_{ii,jj}(0) = x_k(t_{m\tau})x_b^*(t_{e\tau}). \tag{6.67}$$

First, according to Theorem 1 (see Appendix C) the determinant of the autocorrelation matrix $\mathbf{C}$ can be computed from

$$|\mathbf{C}| = \sigma_0^{2N-2}\left(\sum_{k=1}^{N}|x_k(t_{m\tau})|^2 + \sigma_0^2\right). \tag{6.68}$$

Next, based upon the results of Theorem 2 (see Appendix D) the inverse of the autocorrelation matrix $\mathbf{C}$ is computer from the following expression

$$\mathbf{C}^{-1} = \frac{-1}{\sigma_0^2|\mathbf{C}|}\begin{bmatrix} -c_{11}^{-1} & \sigma_0^{2N-2}c_{12} & \cdots & \sigma_0^{2N-2}c_{1N} \\ \sigma_0^{2N-2}c_{12}^* & -c_{22}^{-1} & & \sigma_0^{2N-2}c_{2N} \\ & \vdots & \ddots & \vdots \\ \sigma_0^{2N-2}c_{1N}^* & \sigma_0^{2N-2}c_{2N}^* & \cdots & -c_{NN}^{-1} \end{bmatrix}, \tag{6.69}$$

where

$$c_{ii,ii}^{-1} = |\mathbf{C}| - \sigma_0^{2N-2}|x_k(t_{m\tau})|^2. \tag{6.70}$$

The expression of the desired vector of multipliers can be written as

$$n_{ii}^i(t) = \frac{1}{\sigma_0^2|\mathbf{C}|}[-\sigma_0^{2N-2}c_{1,ii}^*q_1^i(\tilde{t}_\mathrm{d}) - \cdots + c_{ii,ii}^{-1}q_k^i(\tilde{t}_\mathrm{d}^m) - \cdots - \sigma_0^{2N-2}c_{ii,N}q_A^i(\tilde{t}_\mathrm{d}^B)], \tag{6.71}$$

$$w_{ii}^i(t) = \frac{1}{\sigma_0^2|\mathbf{C}|}[-\sigma_0^{2N-2}c_{1,ii}^*g_1^i(\tilde{t}_\mathrm{d}) - \cdots + c_{ii,ii}^{-1}g_k^i(\tilde{t}_\mathrm{d}^m) - \cdots - \sigma_0^{2N-2}c_{ii,N}g_A^i(\tilde{t}_\mathrm{d}^B)]. \tag{6.72}$$

There were two errors in eqs. (46) and (47) in [2], which are correctly represented in (6.71) and (6.72) and drives the need to rewrite and check all the remaining equations for accuracy.

Next, we look at the product, $\mathbf{q}_i^\mathrm{H}(t)\cdot\mathbf{n}_i(t)$, which can be written as

$$\mathbf{q}_i^\mathrm{H}(t)\cdot\mathbf{n}_i(t) = \sum_{ii=1}^{N}[q_k^{i*}(\tilde{t}_\mathrm{d}^m)n_{ii}^i(t)], \tag{6.73}$$

where

$$q_k^{i*}(\tilde{t}_{\mathrm{d}}^m)n_{ii}^i(t) = \frac{q_k^{i*}(\tilde{t}_{\mathrm{d}}^m)}{\sigma_0^2|\mathbf{C}|}\begin{bmatrix} -\sigma_0^{2N-2}c_{1,ii}^* q_1^i(\tilde{t}_{\mathrm{d}}) - \cdots + c_{ii,ii}^{-1} q_k^i(\tilde{t}_{\mathrm{d}}^m) \\ - \cdots - \sigma_0^{2N-2} c_{ii,N} q_A^i(\tilde{t}_{\mathrm{d}}^B) \end{bmatrix}. \tag{6.74}$$

Equations (6.73) and (6.74) are in fact equivalent expressions of eqs. (48) and (49) in [2]. Moreover (6.73) and (6.74) also utilize the revised expressions of (6.71) and (6.72).

Similarly, the product $\mathbf{g}_i^{\mathrm{H}}(t) \cdot \mathbf{w}_i(t)$ can be expressed as

$$\mathbf{g}_i^{\mathrm{H}}(t) \cdot \mathbf{w}_i(t) = \sum_{ii=1}^{N} [g_k^{i*}(\tilde{t}_{\mathrm{d}}^m)w_{ii}^i(t)], \tag{6.75}$$

where

$$g_k^{i*}(\tilde{t}_{\mathrm{d}}^m)w_{ii}^i(t) = \frac{g_k^{i*}(\tilde{t}_{\mathrm{d}}^m)}{\sigma_0^2|\mathbf{C}|}\begin{bmatrix} -\sigma_0^{2N-2}c_{1,ii}^* g_1^i(\tilde{t}_{\mathrm{d}}) - \cdots + c_{ii,ii}^{-1} g_k^i(\tilde{t}_{\mathrm{d}}^m) \\ - \cdots - \sigma_0^{2N-2} c_{ii,N} g_A^i(\tilde{t}_{\mathrm{d}}^B) \end{bmatrix}. \tag{6.76}$$

And finally, the product $\mathbf{g}_i^{\mathrm{H}}(t) \cdot \mathbf{n}_i(t)$ reads

$$\mathbf{g}_i^{\mathrm{H}}(t) \cdot \mathbf{n}_i(t) = \sum_{ii=1}^{N} [g_k^{i*}(\tilde{t}_{\mathrm{d}}^m)n_{ii}^i(t)], \tag{6.77}$$

where

$$g_k^{i*}(\tilde{t}_{\mathrm{d}}^m)n_{ii}^i(t) = \frac{g_k^{i*}(\tilde{t}_{\mathrm{d}}^m)}{\sigma_0^2|\mathbf{C}|}\begin{bmatrix} -\sigma_0^{2N-2}c_{1,ii}^* q_1^i(\tilde{t}_{\mathrm{d}}) - \cdots + c_{ii,ii}^{-1} q_k^i(\tilde{t}_{\mathrm{d}}^m) \\ - \cdots - \sigma_0^{2N-2} c_{ii,N} q_A^i(\tilde{t}_{\mathrm{d}}^B) \end{bmatrix}. \tag{6.78}$$

The output beam given by (6.33) can be computed from

$$y(t) = \mathbf{m}^{\mathrm{H}} \cdot \mathbf{r}(t) = \mathbf{n}_i^{\mathrm{H}}(t)e^{-j\varepsilon} + \mathbf{w}_i^{\mathrm{H}}(t)e^{j\varepsilon}. \tag{6.79}$$

There was a conjugate error in eq. (54) in [2] which is fixed in (6.79) in this chapter. This might introduce some variations on the results. Therefore, we will have to rerun most or all the results that we will be able to and observe any variations.

Where

$$\mathbf{n}_i^{\mathrm{H}}(t) \cdot \mathbf{r}(t) = \sum_{ii=1}^{N} [n_{ii}^{i*}(t)r_k(\tilde{t}_{\mathrm{d}}^m)] \tag{6.80}$$

and

$$n_{ii}^{i*}(t)r_k(\tilde{t}_{\mathrm{d}}^m) = \frac{r_k(\tilde{t}_{\mathrm{d}}^m)}{\sigma_0^2|\mathbf{C}|}\begin{bmatrix} -\sigma_0^{2N-2}c_{1,ii} q_1^{i*}(\tilde{t}_{\mathrm{d}}) - \cdots + c_{ii,ii}^{-1} q_k^{i*}(\tilde{t}_{\mathrm{d}}^m) \\ - \cdots - \sigma_0^{2N-2} c_{ii,N}^* q_A^{i*}(\tilde{t}_{\mathrm{d}}^B) \end{bmatrix} \tag{6.81}$$

and

$$\mathbf{w}_i^{\mathrm{H}}(t) \cdot \mathbf{r}(t) = \sum_{ii=1}^{N} [w_{ii}^{i*}(t) r_k(\tilde{t}_{\mathrm{d}}^m)] \tag{6.82}$$

and

$$w_{ii}^{i*}(t) r_k(\tilde{t}_{\mathrm{d}}^m) = \frac{r_k(\tilde{t}_{\mathrm{d}}^m)}{\sigma_0^2 |\mathbf{C}|} \begin{bmatrix} -\sigma_0^{2N-2} c_{1,ii} g_1^{i*}(\tilde{t}_{\mathrm{d}}) - \cdots + c_{ii,ii}^{-1} g_k^{i*}(\tilde{t}_{\mathrm{d}}^m) \\ - \cdots - \sigma_0^{2N-2} c_{ii,N}^{*} g_A^{i*}(\tilde{t}_{\mathrm{d}}^B) \end{bmatrix}. \tag{6.83}$$

We have converted all the equations up to here. We shall see how the reruns will appear and more importantly see if there are any variations from our initial runs. More importantly this provides a unique opportunity how to provide extended analyses, presentation, derivations, simulations, etc. of very important concepts that I was only able to partially study, analyze, present, publish some 10 years ago in [1–5].

In the following section, we discuss an ASTSA with Restored Phase as an example of an adaptive spatial and temporal interference selective attenuator for Geolocation of RF Signals systems for GNSS applications [3]. While the initial primary reference is [3], additional material is added to better connect the material with the rest of the chapter and also with the rest of the book.

6.6 Improved ASTSA (or ASTSA with Restored Phase) Mathematical Model

Integration of adaptive beamforming with GPS and other navigation and sensors is a very important technology for fulfilling a number of requirements in very benign (heavy multipath, loss of GPS signals, heavily jammed) heterogonous environments in which *accurate positioning and timing information* is essential [83]. It is precisely this information that an ASTSA with restored phase is able to provide better than an ASTSA without restored phase. This section and the simulation results in Sect. 6.10 are intended to explain this much needed capability.

A generic ASTSA with restored phase is presented in Fig. 6.6. In this section, we will inherit the same signal model that is proposed in [2].

Denote with $y(t)$ the output of the ASTSA, which can be computed (6.33) from [2]

$$y(t) = \mathbf{m}^{\mathrm{H}} \cdot \mathbf{r}(t). \tag{6.84}$$

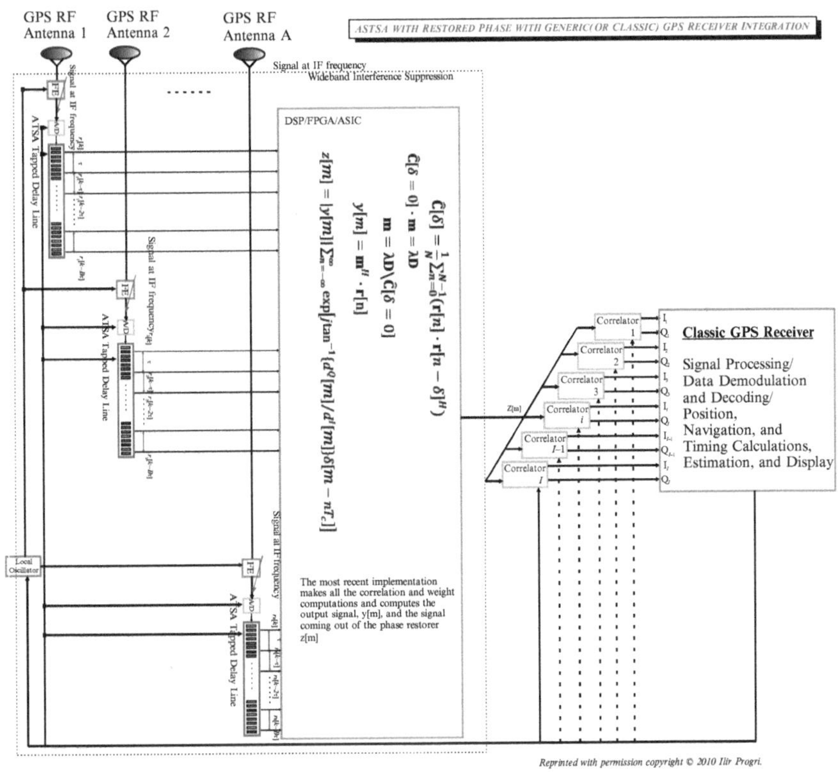

Fig. 6.6 The block diagram of a generic ASTSA with phase restorer. Reprint with permission © 2010 Ilir Progri

Denote with $d(t)$ the desired reference signal, the mathematical expression of which is given by

$$d^I(t) = c(n,t)\cos(\omega_B t + \phi_\mathrm{d}) \tag{6.85}$$

$$d^Q(t) = c(l,t)\cos(\omega_B t + \phi_\mathrm{d}). \tag{6.86}$$

During the phase restoration process, the phase of the output signal is been restored to the phase of a locally generated signal. The signal coming out of the phase restorer, $z(t)$, can be expressed as

$$z(t) = |y(t)| \sum_{n=-\infty}^{\infty} \exp[j\tan^{-1}\{d^Q(t)/d^I(t)\}\delta(t - nT_\mathrm{c})]. \tag{6.87}$$

This signal is then cross-correlated with the locally generated signal given by (6.86) and ratio of the cross-correlation peaks forms the test statistic at the observation point 6 (see Fig. 6.7).

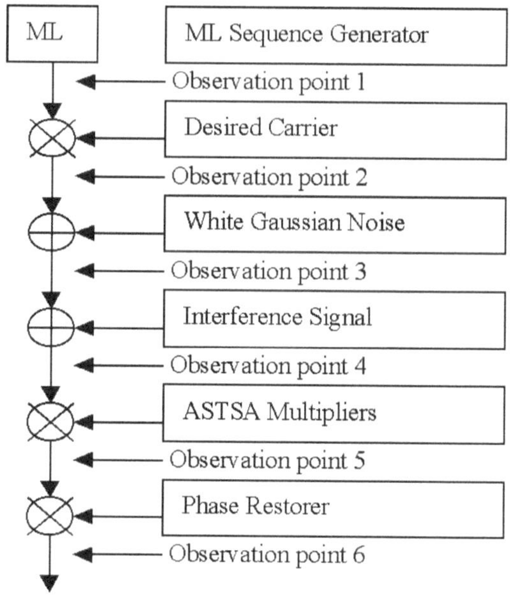

Fig. 6.7 The simulation block diagram of the 2E-1T ASTSA with phase restorer. Reprint with permission © 2010 Ilir Progri

6.7 ATSA Simulations

The ATSA presented in the previous section was initially simulated using MATLAB. We intend to integrate the capability of the ATSA with the receiver and navigation modules of this simulator in the near future. The simulation results presented here are simply used to validate the analytical results and to serve as a baseline for evaluating our ATSA implementation.

In the simulation, the total input signal is modeled as the combination of a wideband desired signal, one narrowband undesired signal, and WGN, in accordance with

$$r[k] = \frac{2A_{\mathrm{d}}}{B_{\mathrm{d}}k}\sin\left(\frac{B_{\mathrm{d}}k}{2}\right)e^{2\pi j(f_{\mathrm{d}}^{\mathrm{c}}/f_{\mathrm{s}})k+\phi_{\mathrm{d}}j} + A_{\mathrm{u}}e^{2\pi j(f_{\mathrm{u}}/f_{\mathrm{s}})k+\phi_{\mathrm{u}}j} + A_{\mathrm{w}}w[k], \tag{6.88}$$

where A_{d}, A_{u}, and A_{w} are the magnitudes of the desired signal, the undesired signal, and the noise respectively. The variables $f_{\mathrm{d}}^{\mathrm{c}}$ and f_{u} are the center frequencies of the desired signal and the undesired signal, respectively; ϕ_{d} and ϕ_{u} are the initial phase shifts of the desired and undesired signals; f_{s} is the sampling frequency; B_{d} represents bandwidth of the desired signal; and $w[k]$ is white noise with zero mean and unit variance.

There was a tiny typo in eq. (37) of [1] which we have corrected in (6.89). Moreover, we have rewritten eq. (37) of [1] in a much nice and more legible form as in (6.89).

The ATSA filter is simulated as a discrete time system, which receives one sample of the input signal and outputs one sample of the output signal at a time. First, the input signal sample is inserted into a buffer of length N. Second, the autocorrelation matrix $\mathbf{C}[n]$ is estimated through an update procedure given by

$$\mathbf{C}[n] = \mathbf{C}[n-1] + \frac{1}{N}(\mathbf{r}_\mathrm{S}[n] \cdot \mathbf{r}_\mathrm{S}^\mathrm{T}[n] - \mathbf{r}_\mathrm{E}[n] \cdot \mathbf{r}_\mathrm{E}^\mathrm{T}[n]). \tag{6.89}$$

In expression (6.90) the starting signal vector, $\mathbf{r}_\mathrm{S}[n]$, and the ending input signal vector, $\mathbf{r}_\mathrm{E}[n]$, are given by

$$\begin{aligned} \mathbf{r}_\mathrm{S}[n] &= [\, r[k] \quad r[k-\tau] \quad r[k-2\tau] \,]^\mathrm{T}, \\ \mathbf{r}_\mathrm{E}[n] &= [\, r[\tilde{k}+2\tau] \quad r[\tilde{k}+\tau] \quad r[\tilde{k}] \,]^\mathrm{T} \end{aligned} \tag{6.90}$$

with $\tilde{k} = k - N$, where N is the number of samples in the average and τ is the tap delay parameter. We considered an ATSA with 2 taps (or shifter delays). Third, we utilized a pointing (steering) vector for a real desired signal given by

$$\begin{gathered} \mathbf{D} = [\, e^{0\pi j (f_\mathrm{d}^\mathrm{c}/f_\mathrm{s})\tau} \equiv 1 \quad e^{2\pi j (f_\mathrm{d}^\mathrm{c}/f_\mathrm{s})\tau} \quad e^{4\pi j (f_\mathrm{d}^\mathrm{c}/f_\mathrm{s})\tau} \,] \\ \text{(another clarification from eq.40 in [1]).} \end{gathered} \tag{6.91}$$

This vector is computed prior to the iteration process. We estimated the autocorrelation matrix in accordance with (6.90). Fourth, the set of multipliers is computed by solving (6.7). The scale factor, λ, normalizes the multipliers to ensure that the output signal has the same magnitude as the input signal is determined from

$$\lambda = \mathrm{LPF}\left(\frac{1}{||\mathbf{m}[k]||}\right), \tag{6.92}$$

where LPF indicates a low-pass filter and $||\mathbf{m}||$ indicates the Euclidean norm of $\mathbf{m}$. A simple first-order infinite impulse response (IIR) filter with a selectable time constant serves as the low-pass filter for (6.93).

There was an error in referencing eq. (37) in [1] in the paragraph right above (6.93), which is been correctly referenced in (6.90).

Finally, the output signal is computed as the dot product of the weight vector and the input signal vector

$$y[m] = \mathbf{m}^\mathrm{H} \cdot \mathbf{r}_\mathrm{S}[n]. \tag{6.93}$$

The MATLAB program also computes the UDR based on expression (6.9).

Plotting UDR as a function of the adjustable parameter τ and a range of the undesired signal frequencies allows optimizing the performance of the ATSA. For systems that employ only one real channel, it is possible to mitigate the effect of the undesired signal level by replacing the "exp" with "cos" in expressions (6.89) and (6.92). The behavior of complex spreading and real spreading were treated previously to show a slight difference leading to an extra term in the real codes case [92]. Although the approach presented here is different from the approach suggested in Montalbano et al. [92], the rest of the simulation and implementation are essentially the same. The reader is reminded that the ATSA performance utilizing complex signals is entirely different from the ATSA performance utilizing real signals because complex signals will entirely change the signal structure. For example, consider the product $e^{-j\omega\tau}e^{j\omega\tau} = 1$, which is independent of τ. Now consider the product $\cos(-\omega\tau)\cos(\omega\tau) = \cos^2(\omega\tau)$, which is a function of τ. This simple example illustrates why an ATSA model that operates on complex signals will have an entirely different performance than an ATSA model that operates on real signals.

6.7.1 Principle Illustration Example 2

It would be very tedious to analytically change the undesired signal frequency and repeat principle illustration example 1, found in the ATSA Mathematical Model section. Principle illustration example 2 is designed to assess the performance of the ATSA when the undesired signal frequency is changing provided that the mathematical model of the ATSA is kept the same as that of principle illustration example 1. Therefore, this example utilizes some of the settings of example 1 in the previous section. The desired signal center frequency, bandwidth, and level (or power) are set at $f_d^c = 1$ kHz, $B_d = 3$ kHz, and $P_d = -30$ dB, respectively. The undesired signal level is set at 40 dB. In this example, three different sampling frequencies are used ($f_s = \{6,60,600\}$ kHz). Using these parameters, for an undesired signal frequency in the range between 0 and 3 kHz, we summarize the simulation results in Fig. 6.8. In Fig. 6.8, the UDR values in dB are given in the vertical bar. Note that the input UDR is 70 dB and that the optimum output UDR is 25 dB, which corresponds to the optimum $\tau = \{1,10,100\}$ and this is in complete agreement with the analysis developed earlier.

As shown in Fig. 6.8, the tap delay changes from 1 to 10, the undesired signal frequency changes from 0 to 3 kHz, and the sampling frequency remains at 6 kHz. The UDR is computed for every value of the tap delay and undesired signal frequency and is shown in the contour plot in Fig. 6.8. The largest achievable value of the UDR is 66 dB, which are the peaks in the figure, and the smallest value is close to 25 dB, which is the background of the figure. If we where to take a cut of Fig. 6.8 at the undesired signal frequency of 1 kHz, then we would essentially produce the results of Fig. 6.3 with the exception that the minimum and maximum values of UDR are different than those of Fig. 6.3, but they occur at the same values of τ as those of Fig. 6.3.

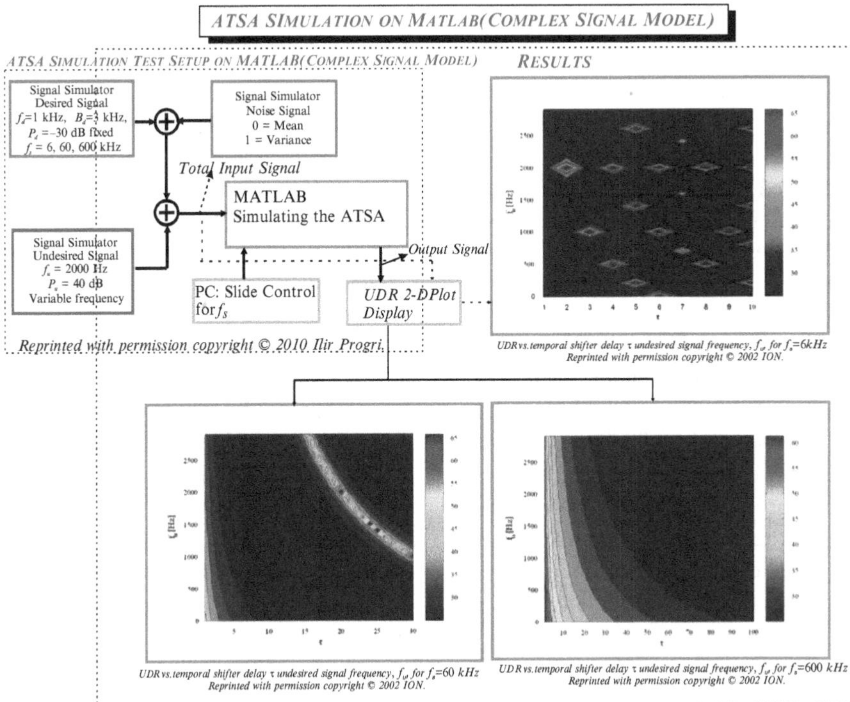

Fig. 6.8 Principle illustration example 2: ATSA simulation on MATLAB (complex signal model). Reprint with permission © 2002 ION and 2010 Ilir Progri

In Fig. 6.8, the tap delay changes from 1 to 30, the undesired signal frequency changes from 0 to 3 kHz, while the sampling frequency remains at 60 kHz. The UDR is computed for every pair of τ and f_u and is shown in the contour plot in Fig. 6.8.

Although the UDR changes from 25 to 66 dB just as in Fig. 6.8, the same minimum and maximum values of UDR in Fig. 6.8 occur for different values of τ. However, for $\tau = 10$, the minimum value of UDR is achieved despite the undesired signal frequency.

In Fig. 6.8, the tap delay changes from 1 to 100, the undesired signal frequency changes from 0 to 3 kHz, while the sampling frequency remains at 600 kHz. The UDR is computed for every pair of τ and f_u and is shown in the contour plot in Fig. 6.8. Although the UDR changes from 25 to 62 dB just as in Fig. 6.8, the same minimum and maximum values of UDR in Fig. 6.8 occur for different values of τ. However, for $\tau = 100$, the minimum value of UDR is achieved despite the undesired signal frequency.

Also, the optimum values of τ corresponding to Fig. 6.8 are $\{1, 10, 100\}$. Note that for Fig. 6.8, the sampling frequency changes from 6, to 60, to 600 kHz; i.e., by a factor of 10 and the optimum τ changes also by a factor of 10.

6.7.2 Principle Illustration Example 3

So far we have demonstrated a consistency between the results obtained from theory with those obtained in simulation. To demonstrate the same consistency from the results obtained in implementation utilizing the mathematical model discussed in the previous section and previous principle simulation examples, we need to conduct an experiment that would repeat principle illustration example 2 using a readily available DSP at the time of illustration. However, the readily available TI C6711 DSP at the time of illustration contained a single input and a single output channel.

Since an ATSA which models a complex input signal and a complex output signal requires dual input and output channels, we were restricted to chose one of the two alternatives: (1) changing the ATSA model to real input and output signals or (2) not to show any implementation results. We decided to follow the first alternative; therefore, example 3 demonstrates the performance of an ATSA that uses real input and output signals. In this situation, the undesired and desired signal levels are set to 0 and −20 dB, respectively, and the desired and sampling frequencies were set at 2.667 kHz and $f_s = \{8{,}80{,}800\}$ kHz. The simulation results of this example are depicted in Fig. 6.9.

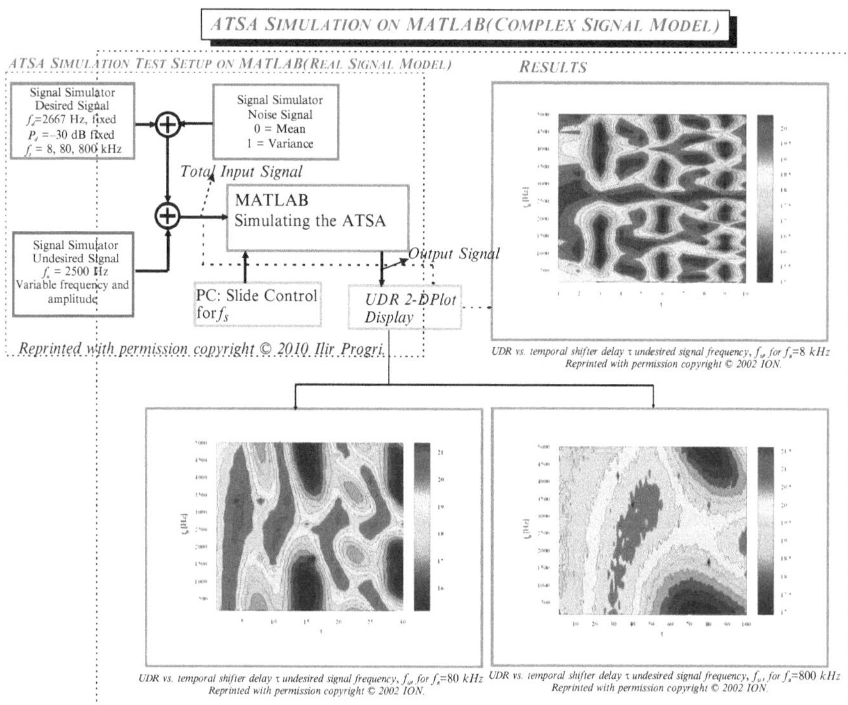

Fig. 6.9 Principle illustration example 3: ATSA simulation on MATLAB (real signal model). Reprint with permission © 2002 ION and 2010 Ilir Progri

As shown in Fig. 6.9, the tap delay changes from 1 to 10, the undesired signal frequency changes from 0 to 5 kHz, and the sampling frequency remains at 8 kHz. The UDR is computed for every value of the tap delay and undesired signal frequency and is shown in the contour plot in Fig. 6.9. The maximum value of UDR is 20.5 dB and the smallest value is 15 dB. If we where to take a cut of Fig. 6.9 at the undesired signal frequency of 2.667 kHz, then we would essentially observe the smallest value of UDR corresponds to $\tau = 3$.

Again in Fig. 6.9, the tap delay changes from 1 to 30, the undesired signal frequency changes from 0 to 5 kHz, while the sampling frequency remains at 80 kHz. The UDR is computed for every pair of τ and f_u and is shown in the contour plot in Fig. 6.9. Although the UDR changes from 15 to 20.5 dB just like in Fig. 6.9, the same minimum and maximum values of UDR in Fig. 6.9 occur for different values of τ. In Fig. 6.9 for $\tau = 15$, the minimum value of UDR is achieved despite the undesired signal frequency.

Again in Fig. 6.9, the tap delay changes from 1 to 100, the undesired signal frequency changes from 0 to 5 kHz, while the sampling frequency remains at 800 kHz. The UDR is computed for every pair of τ and f_u and is shown in the contour plot in Fig. 6.9. Although the UDR changes from 17 to 21.5 dB just like in Fig. 6.9, the same minimum and maximum values of UDR in Fig. 6.9 occur for different values of τ. In Fig. 6.9 for $\tau = 75$, the minimum value of UDR is achieved despite the undesired signal frequency.

Note that for this example, the input UDR is $20 \log(3) = 21.9$ dB and the output optimum UDR is $\{15, 16, 17\}$ dB. We did not observe more dramatic changes here because the noise level and the undesired signal level are both set to a power level of 0 dB. This shows that the ATSA has difficulty driving an undesired signal below the noise level. Shortly we will see an implementation in which a much more dramatic difference in the ATSA performance is observed.

Also, the optimum values of τ corresponding to Fig. 6.9 are $\{3, 15, 75\}$. Note that for these figures the sampling frequency changes from 8, to 80, to 800 kHz; i.e., by a factor of 10; however, the optimum τ changes by a factor of 5 as opposed to 10 for the complex signal model.

6.8 ATSA Implementation

Although there can be found a number of digital beamforming implementations aiming to GNSS, very few technical detail about the implementation is available (Fernández-Prades, Closas, and Arribas, 2009, Implementation of digital beamforming in GNSS receivers, personal communications). This section of this chapter was designed specifically to reveal some of those technical implementation challenges.

The ATSA system described in this chapter was implemented at audio frequencies using a TI TMS320C6711GFN DSP, which supports floating-point instructions as well as fixed point instructions. The software was written in C and compiled using TI's Code Composer Studio.

To ensure that the computational burden of the algorithm fit within the performance limitations of the TI processor, a fixed second order ATSA system was implemented. On the basis of our simulations, such a system should be capable of attenuating one narrowband interference signal and preserving one narrowband desired signal. All calculations were done in single floating-point precision. The resulting ATSA filter was capable of operating in real-time at 8,000 samples/s, which allows the algorithm to operate in real-time, since this is also the maximum sampling frequency of the A/D converter on the DSP board.

Figure 6.10 illustrates the test setup for the ATSA system and Table 1 in [1] lists the equipment used in this setup. Two signal generators are connected in parallel through a BNC Tee connector, producing the sum of the desired and undesired sinusoidal signals. The desired signal frequency of 2.6667 kHz was selected and hard-coded into the ATSA's software for this experiment. References [34, 36–46, 53, 55] provide good examples for future interface implementations that will utilize

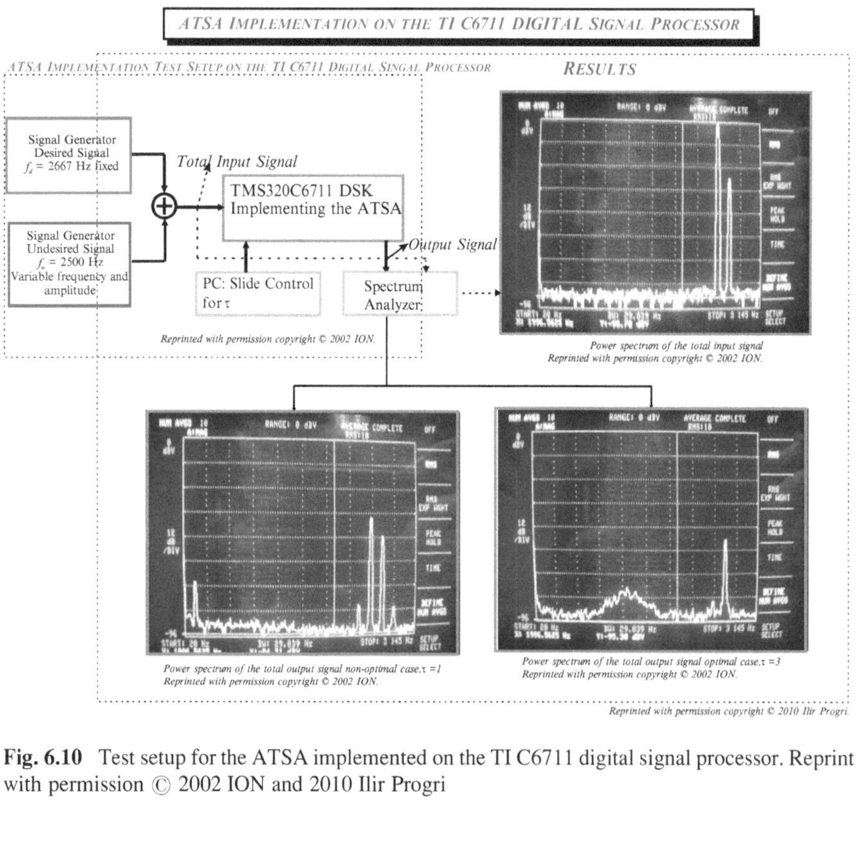

Fig. 6.10 Test setup for the ATSA implemented on the TI C6711 digital signal processor. Reprint with permission © 2002 ION and 2010 Ilir Progri

field programmable gate arrays (FPGA), application specific standard processors (ASSP), general purpose processors (GPP), or DSPs.

The signal generator is designed to allow the undesired signal to vary both in frequency and amplitude; nevertheless, for every run the signal amplitude and frequency are kept fixed. The combined signal is plugged into the input jack of the DSP board. The output from the DSP board is the processed signal, which is fed into the spectrum analyzer for observation. The DSP board's interface to the PC allows monitoring and debugging the DSP application. Additionally, the PC interface can provide input to the DSP. A "slider" control was implemented on the PC to adjust the ATSA filter parameter τ. The τ slider ranges from 0 to 5. A setting of 0 indicates nonadaptive (pass-through) operation in which no filtering occurs and the output signal is the same as the input signal. All other settings indicate the sample delay in multiples of the sampling period. By default τ is set at 3; this corresponds to the optimal temporal delay for this particular combination of the desired frequency and the sampling frequency.

The data collected using the spectrum analyzer (Fig. 6.10) was compared with the MATLAB simulation results of principle illustration example 3 and Fig. 6.10. Selecting $\tau = 0$ (pass-through) allows observing the spectrum of the combined desired and undesired signals on the spectrum analyzer. By adjusting the amplitude and frequency controls on the signal generators, we selected a signal pattern that was previously simulated in MATLAB (see principle illustration example 3).

6.8.1 *Principle Illustration Example 4*

The purpose of this example is to validate the simulation results of principle illustration example 3. The sensitivity of the undesired signal is set at 0 dB and that of the desired signal at −30 dB. This is intended to demonstrate the true capability of the ATSA filter in removing large undesired signals. First, the spectrum of the input signal is shown in Fig. 6.10. As shown in Fig. 6.10, the noise level is around −90 dB, the desired signal level is at −30 dB, and the undesired signal level at 0 dB. This is the nonadaptive case.

Second, the power spectrum of the output signal for the nonoptimal adaptive example (i.e., for $\tau = 1$) is shown in Fig. 6.10. In Fig. 6.10, the noise level is again at −90 dB; however, the desired signal level is at −48 dB and the undesired signal level at −40 dB.

And third, the power spectrum of the output signal for the optimal adaptive filter operation (i.e., for $\tau = 3$) is shown in Fig. 6.10. In this figure, the noise level is again at −90 dB; however, the desired signal level is at −55 dB and the undesired signal level at −90 dB. Thus by comparing the results in Fig. 6.10, we have achieved a 55 dB decrease in the UDR.

6.9 ASTSA Simulations

In this section, we have considered three principle simulation scenarios. In the first principle simulation scenario we analyze the performance of the ATSA, in the second the performance of the ASSA, and in the third the performance of the ASTSA.

6.9.1 *Principle Simulation Example 5.1*

The adaptive-spatial-temporal filtering method can be employed either using the decision-feedback equalization or without [91]. The method that uses the decision-feedback equalization relies on channel estimation (or without the DOA information) vs. the method that does not use the decision-feedback equalization performs the adaptive-spatial-temporal filtering only based on the information on the direction of arrival of the desired signal. The latter is the one that we are considering in the principle simulation example.

The fifth principle simulation scenario consists of simulating an ATSA with up to 6 taps. The performance methodology is given in terms of the UDR ratio (dB) vs. simulation time (μs) against one BBJ with 20 MHz bandwidth and 94 dB power, employing the *new* (*local*) *and old* (*global*) algorithms (see Fig. 6.11 (left) and (right)). For the *old* (*or global*) adaptive methodology the reader may refer to [4].

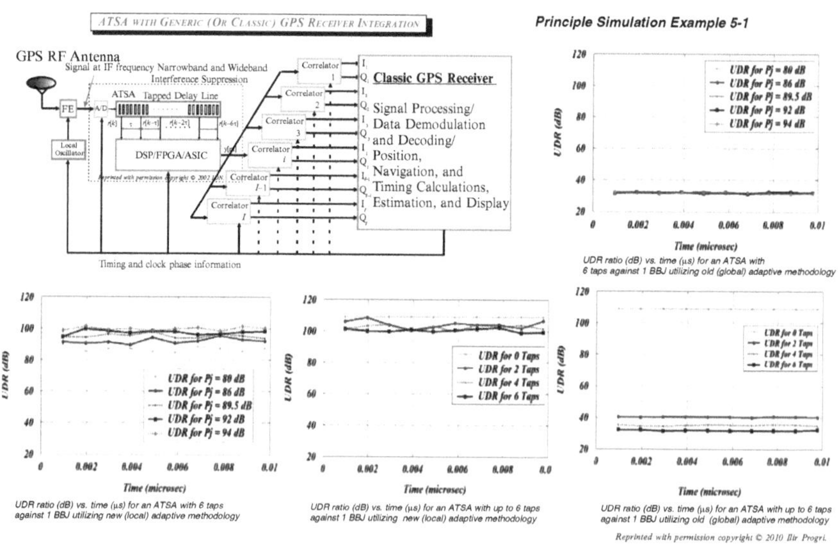

Fig. 6.11 Principle simulation example 5.1. Reprint with permission © 2002 ION and 2010 Ilir Progri

For this principle simulation example, the input UDR is almost 109 dB, because the jammer power is 94 dB and the desired signal power is −15 dB; hence, total input UDR is 109 dB. For 6 taps, when the *new* (*or local*) adaptive methodology is used then the UDR ratio becomes almost 100 dB as opposed to 32 dB when the *old* (*or global*) methodology is applied. It appears that the ATSA, which exploits the *old adaptive* methodology, outperforms by 68 dB the ATSA, which exploits the *new* (*of local*) adaptive methodology when trying to mitigate wide band interference.

For this experiment, the input UDR changes from 95 to 109 dB; however, the number of taps remains 6. When employing the *new* (*or local*) adaptive methodology, the UDR changes from 88 to 100 dB when the jammer power changes from 80 to 94 dB. This implies the ATSA, which employs the *new* (*or local*) adaptive methodology is not capable or optimized to mitigate wide band interference.

The ATSA, which exploits the *old* (*or global*) adaptive methodology, appears to provide a 32-dB UDR ratio when the jammer power changes from 80 to 94 dB. This indicates that the ATSA, which exploits the old adaptive methodology, has reached the optimal performance. As a final remark we note that the *old* (*or global*) ATSA outperforms the *new* (*local*) ATSA by 55–68 dB against one BBJ with 20 MHz bandwidth and power changing from 80 to 94 dB.

6.9.2 *Principle Simulation Example 5.2*

The principle simulation example 5.2 is designed to explore the performance of the *old* (*global*) *and new* (*local*) ATSA against CWJ when number of taps changes from 0 to 6 (see Fig. 6.12 (left) and (right)). The input UDR is 109 dB.

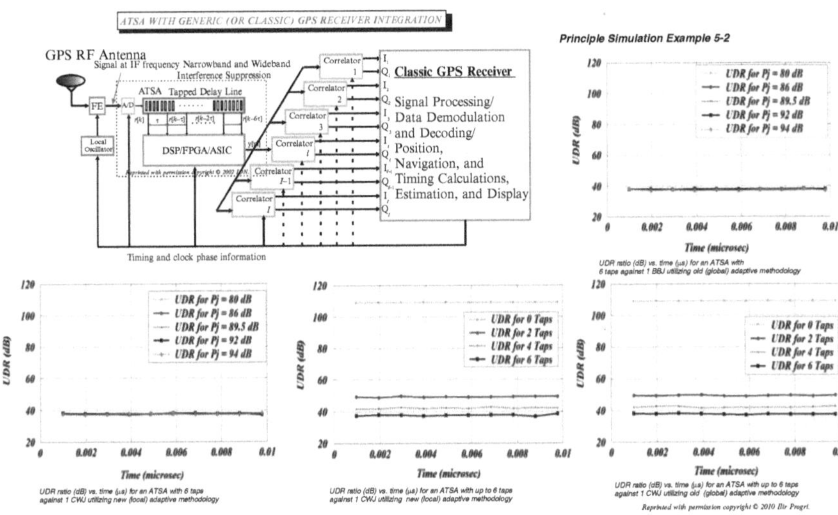

Fig. 6.12 Principle simulation example 5.2. Reprint with permission © 2002 ION and 2010 Ilir Progri

It appears that UDR ratio corresponding to both the *old* (*global*) *and the new* (*local*) ATSA reaches 38 dB when the number of taps reaches 6.

The principle simulation example 5.2 is also designed to check the optimality of the ATSA; therefore, we changed the jammer power from 80 to 94 dB (see Fig. 6.12 (left) and (right)). For this experiment, the input UDR changes from 95 to 109 dB; however, the UDR corresponding to both the *new* (*local*) *and the old* (*global*) ATSA remains the same at 38 dB, which implies that both the old and the new ATSA have reached the optimal performance.

Based on the results of these four experiments, we conclude that the *new* (*or local*) adaptive methodology is optimized against the CWJ and not for BBJ, as opposed to the *old* (*or global*), which is an optimal technique for both.

6.9.3 Principle Simulation Example 5.3

Principle simulation example 5.3 consists of plotting the main beam response for an ATSA with 6 taps for uniform weights, which is the most common method to understand and explain array beamforming. Figure 6.13 depicts the array beam response for an ATSA with 6 taps and uniform weights in 1 3D plot and 2 2D plots (one of which is a polar plot). The 3D beam response plot is important because it gives an idea of how the beam response is in 3D in both azimuth and elevation. The 2D polar plot in azimuth/elevation gives an idea of how the main beam is in azimuth and elevation.

Apparently the 3D and 2D representation of the array factor plots as shown in Fig. 6.13 is a way that people from the *IEEE Transactions on Antennas and Propagation* [85] and people from *PIERS Online Journal* [98] really like it. In addition to that we have added a polar plot.

The number of nulls is equal to the number of taps. In principle an ATSA can create nulls just as deep as an ASSA or ASTSA with equal number of degrees of freedom.

As we are going to see later for example 6.3 for an ASSA and example 7.3 for an ASTSA, the number of number of nulls is equal to the number of degrees of freedom -1.

6.9.4 Principle Simulation Example 6.1

When the desired signal and interfering signal occupy the same temporal frequency band the temporal, the conventional temporal filtering approach will be ineffective in separating the desired signal from the interference signal [66]. Since the desired signal and the jamming signals originate from different locations, then the spatial separation can be exploited to separate the desired signal from the interference signal by means of an ASSA. This technique is effective as long as the satellite and jammer are sufficiently separated in angle and do not both fall within the mainbeam

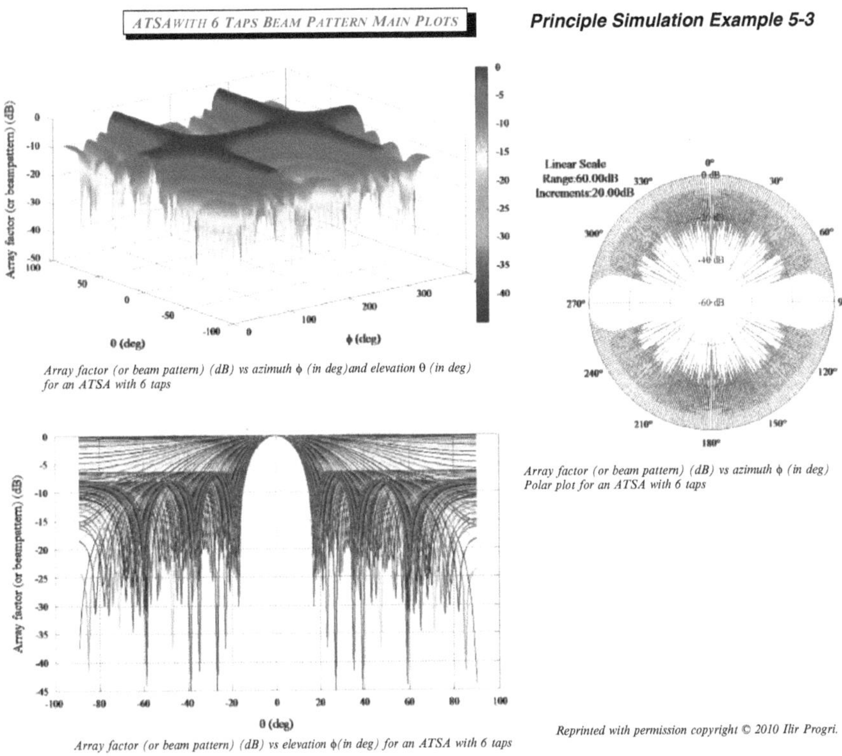

Fig. 6.13 Principle simulation example 5.3: ATSA with 6 taps beam pattern main plots. Reprint with permission © 2010 Ilir Progri

of the receive antenna [77]. In the limit, when the satellite signal and jammer are aligned, the spatially adaptive processor cannot cancel the jammer; i.e., an ASSA will not be able to suppress the jamming power, an ASTSA can be used instead as illustrated in examples 7.1 and 7.2. However, when spatio-temporal processing cannot be employed to precancel the interference coming from the jammer, significant performance can still be achieved by spatial filtering only [92].

The principle simulation scenario (or example) 6.1 consists of simulating an ASSA with up to 4 sensors. The performance methodology is given in terms of the UDR ratio (dB) vs. simulation time (μs) against one BBJ employing the new and old adaptive methodology (see Fig. 6.14 (left) and (right)).

The input UDR is 109 dB. We observe that for both the *new* (*local*) *and old* (*global*) ASSA, the UDR ratio decreases with the increase of the number of sensors (or antenna elements). For an array with 4 sensors, the UDR ratio reaches 42 dB for both the new (local) and the old (global) ASSA. This implies that both the new (local) and the old (global) ASSA provide up to 65-dB improvement in the UDR ratio, which is 10 dB higher than the UDR ratio of the old ATSA against the same jammer. Caution must be shown about the result obtained with the old (global) ATSA, because it appears to be the only scenario that outperforms the ASSA!

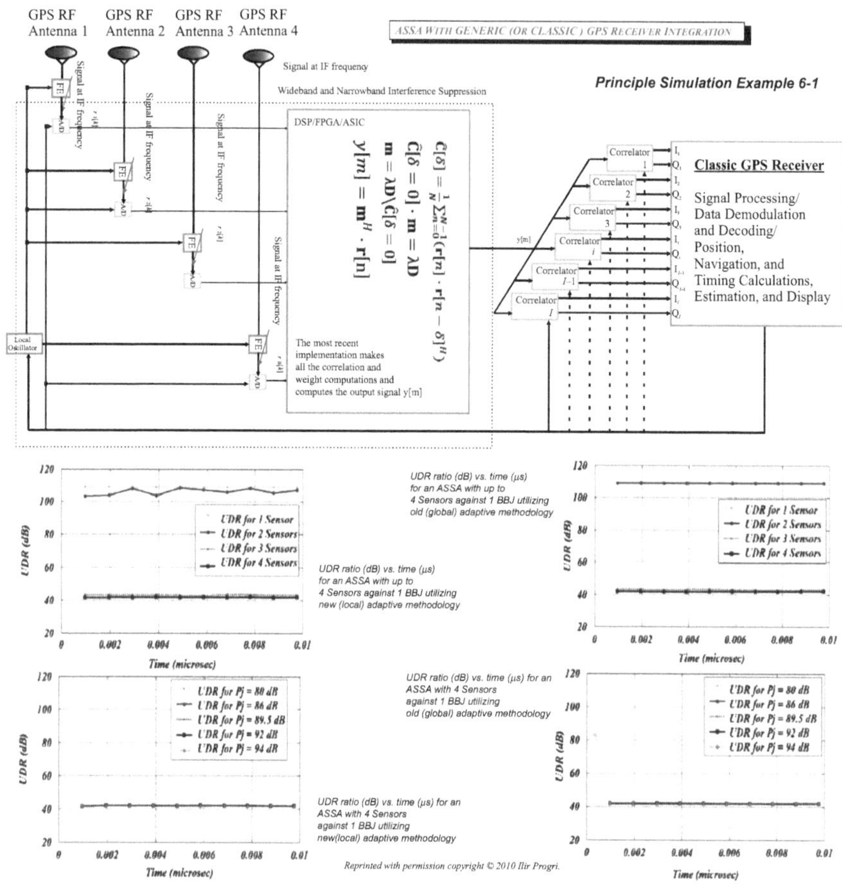

Fig. 6.14 Principle simulation example 6.1. Reprint with permission © 2010 Ilir Progri

Next we varied the jammer power to assess the optimality of the adaptive methodology. Figure 6.14 (left) and (right) present the performance of the same ASSA against one BBJ exploiting the new and old adaptive methodology for an array with 4 sensors.

We observe that even though the input UDR changes from 95 to 109 dB, the output UDR has reached 42 dB, which implies that both the new and the old ASSA perform equally well and have achieved the optimal performance.

6.9.5 Principle Simulation Example 6.2

Next, we repeat the principle simulation example 6.1 for an ASSA against one 1 CWJ. Thus, for the third experiment we vary the number of sensors from 1 to 4 and keep the jammer power equal to 94 dB. The result of this work is shown in Fig. 6.15 (left) and (right). With this experiment we intend to show the sensitivity of the UDR ratio vs. the increase of the number of sensors from 1 to 4. Even for this experiment the input UDR is 110 dB and the output UDR changes as a function of the number of sensors.

For both the *old* (*global*) *and the new* (*local*) ASSA, we observe a dramatic change of the UDR ratio when the number of sensors changes from 2 to 3 by about 95 dB. When the number of sensors changes from 3 to 4 there is about 1 dB improvement of the UDR ratio. This implies that both the *new* (*local*) *and the old* (*global*) ASSA appears to perform equally well against one CWJ.

The forth experiment of scenario 2 depicts the optimality of the ASSA against one CWJ. For this experiment, we changed the jammer power from 80 to 94 dB and kept the number of sensors equal to 4 (see Fig. 6.15 (left) and (right)).

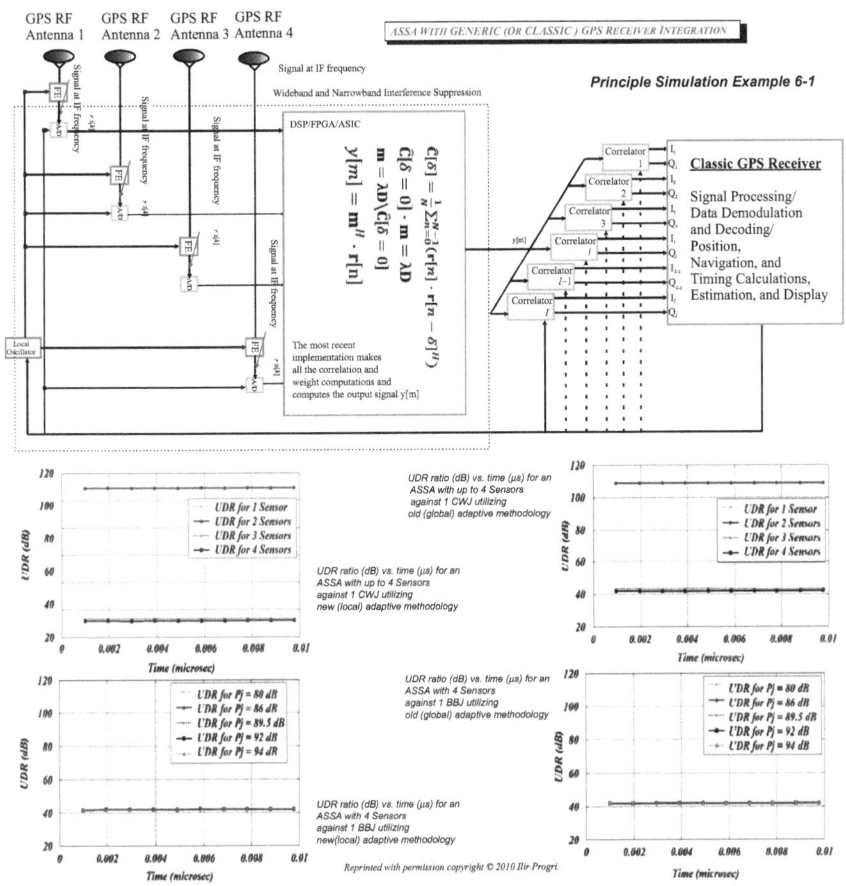

Fig. 6.15 Principle simulation example 6.2. Reprint with permission © 2010 Ilir Progri

Although the input UDR changes from 95 to 109 dB, the output UDR remains the constant at 12 dB, which infers that both the *old (or global) and the new (local)* ASSA have reached their optimal performance.

6.9.6 Principle Simulation Example 6.3: ASSA with 4 Sensors Beampattern Main Plots

Principle simulation example 6.3 illustrates an ASSA with 4 sensors beam response for uniform weights as depicted in Fig. 6.16. The number of nulls equals $4 - 1 = 3$.

Passive ranging for air-to-air, air-to-sea, and sea-to-sea applications is illustrated in [75] and also in Chap. 2. The relative azimuth (ϕ) and elevation (θ) angles are defined from the tracking platform. These bearings applications are kept in mind while performing the beam-pattern of the principle simulation example 6.3.

Apparently the 3D and 2D representation of the array factor plots as shown in Fig. 6.16 is a way that people from the *IEEE Transactions on Antennas and Propagation* really like it [85]. In addition to that we have added a polar plot.

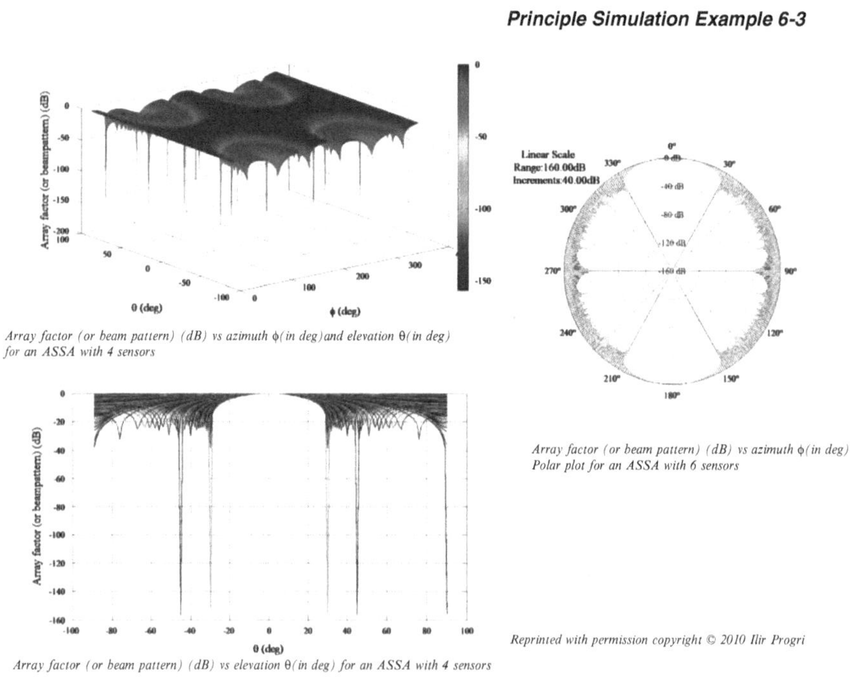

Fig. 6.16 Principle simulation example 6.3: ASSA with 4 sensors beam pattern main plots. Reprint with permission © 2010 Ilir Progri

6.9.7 Principle Simulation Example 7.1

STAP is a signal processing technique that was originally developed for detecting slow-moving targets, using airborne radars [77]. Similarly in passive adaptive array signal processing, ASTSA is designed to suppress interference and find the direction of arrival of *slow moving jammers* using ASTSA arrays. There are a great deal of similarities and differences of the signal processing techniques that take place in both the radar STAP world and also in the ASTSA world. Although Lapierre et al. [77] provides a great description of the general principle of STAP in radar, we are going to provide here the general principle of the ASTSA and let the reader recognize the differences.

The general principle of the ASTSA is as follows. It is assumed that there are J jammers randomly distributed in the terrain generally within a few degrees of elevation. The pulses from these jammers are received at each of the A elements (or sensors) of an antenna array. Separate receiver chains (or taps) are attached to each of the array elements. The received signals are sampled at a series of B successive ranges (or distances) also referred to as range gates [77]. STAP processing is applied to an $A \times B$ matrix of samples collected at each such range typically called a snapshot. The ensemble of snapshots of all successive ranges is called a data cube and contains all the information about the jammers detection within a coherent processing interval (CPI) [77].

Although STAP has been known at least since 1980s, the field has regained a major regain of interest in the 1990s mainly as the result of significant increase in computational power [77]. Much of the research in the 1990s focused on two major topics of interests: (1) the design of computationally efficient adaptive methods to reduce the computational load of the STAP processor; (2) the design of methods to mitigate barrage jammers (which emits very wide bandwidth jamming signals) [77]. This is exactly the view that we have employed for designing the principle simulation examples 7.1 and 7.2. Also the correlation of errors as the function of the frequency will affect the sidelobes performance of the antenna in both space and time [93]. In this edition of the book, we have not considered these effects, perhaps in the future editions; nevertheless, the main principles discussed here will still apply to those cases.

The principle simulation example 7.1 consists of simulating an ASTSA with up to 4 sensors and 6 taps. The performance criterion is given in terms of the UDR ratio (dB) vs. simulation time (μs) against one BBJ exploiting the *new* (*or local*) *and old* (*global*) adaptive technique (see Fig. 6.17 (left) and (right)).

The input UDR is 109 dB when the number of sensors changes from 1 to 4 and the number of taps changes from 0 to 6. While the number of sensors is incremented by one, the number of taps is incremented by 2. *As a general remark, the old* (*global*) *ASTSA outperforms the new* (*or local*) *ASTSA when the number taps and the sensors increase by the same number.*

For an ASTSA with 6 taps and 4 sensors, the UDR ratio reaches 33/25 dB when employing the *new* (*local*)/*old* (*global*) adaptive methodology against one BBJ.

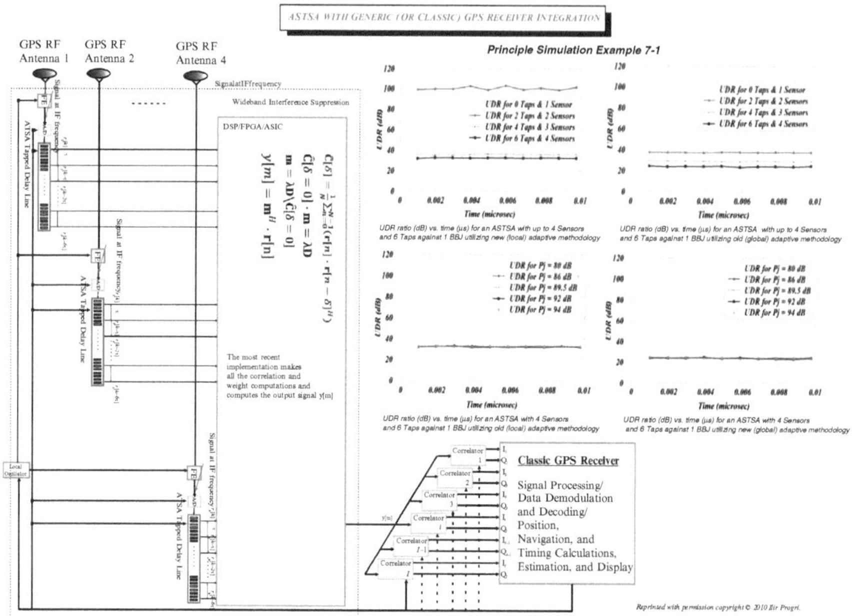

Fig. 6.17 Principle simulation example 7.1. Reprint with permission © 2010 Ilir Progri

This implies that the *old* (*global*) ASTSA outperforms the *new* (*local*) ASTSA by 8 dB in terms of the UDR ratio.

The second experiment is designed to depict whether the *new* (*local*) *or old* (*global*) ASTSA have reached their optimal performance. Figure 6.17 (left) and (right) presents the performance of the same ASTSA against one BBJ utilizing the *new* (*local*) *or old* (*global*) adaptive methodology when the jammer power changes from 80 to 94 dB. It can be easily observed that both the *new* (*local*) *and the old* (*global*) ASTSA have reached their optimal performance at 33/25 dB when the input UDR changes from 95 to 109 dB.

The third principle simulation example is designed to explore the performance of the ASTSA against 1 CWJ when the number of taps changes from 0 to 6 by an increment of 2 and the number of sensors changes from 1 to 4 by 1.

The input UDR is again 109 dB. We note almost identical performance between the *new* (*local*) *or old* (*global*) ASTSA. When the number of taps reaches 6 and the number of sensors reaches 4, the UDR ration becomes almost 2.5 dB, which implies that we get about 107.5 dB improvements in terms of the UDR ratio.

6.9.8 Principle Simulation Example 7.2

The principle simulation example 7.2 of the chapter depicts the optimality of the ASTSA methodology against one CWJ as depicted in Fig. 6.18. For this principle

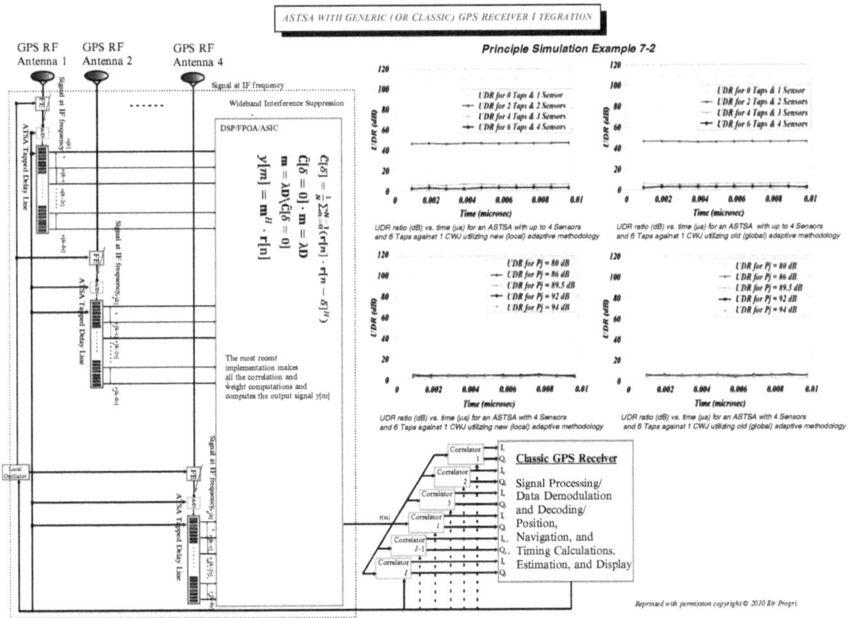

Fig. 6.18 Principle simulation example 7.2. Reprint with permission © 2010 Ilir Progri

simulation example, the jammer power changes from 80 to 94 dB. It can be easily observed in Fig. 6.18 that both the *new* (*local*) *and old* (*global*) ASTSA with 6 taps and 4 sensors have reached their optimal performance.

6.9.9 Principle Simulation Example 7.3: ASTSA with 6 Taps and 4 Sensors Beampattern Main Plots

The last and final antenna array beam response plots are those obtained from principle simulation example 7.3 for an ASTSA with 6 taps, 4 sensors, and uniform weights as shown in Fig. 6.19. For this ASTSA, the way the taps are modeled is to form a 2D array in which on the x-axis are the taps and on the y-axis are the sensors.

For this particular ASTA, the number of degrees of freedom is equal to the max (4,6) which is equal to 6. In this case, we have better redundancy; i.e., we get better suppression of the jammer power or interference due to averaging or the min $(4.6) = 4$. Apparently the 3D and 2D representation of the array factor plots as shown in Fig. 6.19 is a way that people from the *IEEE Transactions on Antennas and Propagation* [85] and people from *PIERS Online Journal* [98] really like it. In addition to that we have added a polar plot.

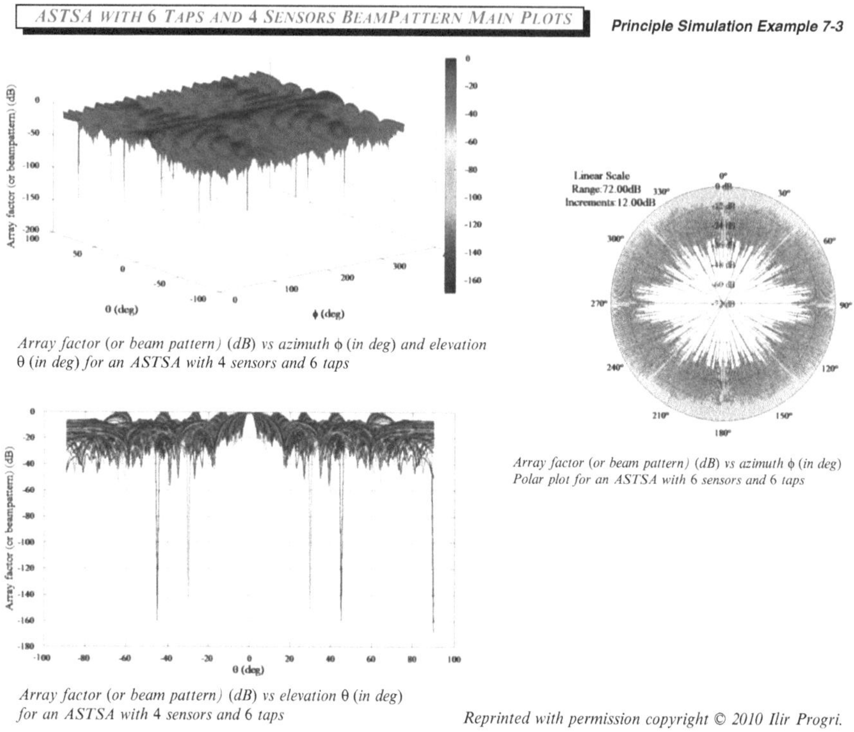

Fig. 6.19 Principle simulation example 7.3: ASTSA with 6 taps and 4 sensors beam pattern main plots. Reprint with permission © 2010 Ilir Progri

6.10 Improved ASTSA (or ASTSA with Restored Phase) Simulations

In one particular application, an adaptive spatial and temporal filter can act as a novel adaptive antenna array for DS/CDM code-acquisition [76]. With a spatially designed structure the spatial filter can act as a beamformer suppressing interference while the adaptive temporal can act as a code-delay estimator [76]. Another important application is the adaptive antenna arrays that enable systems to meet the aggressive accuracy and integrity requirements for piloted and autonomous aircraft landing while simultaneously providing significant attenuation of radio frequency interference [79] unlike previous publications that required expensive calibrations and were able to provide DOA angle accuracy estimation to within 1–2° (Trinkle and Gray, 2001, Adaptive antenna arrays for GPS interference localisation. University of Adelaide, South Australia, pp. 1–12, Personal communication; Trinkle and Gray, 2001, GPS interference mitigation; overview and experimental results. University of Adelaide, South Australia, pp. 1–14, Personal communication).

We have designed three experiments to assess the theoretical performance of the improved ASTSA with restored phase unit. The Mathworks™ Software MATLAB® student version 5.3 was used for simulation purposes.

6.10.1 Principle Simulation Example 8: ASTSA with Two Antennae and One Tap (2E-1T) and One CW and Ideal Phase Restorer

This is the simplest ASTSA that we are considering; nevertheless, it is complicated enough to provide useful information about the desired signal structure properties when narrowband interference is present and when narrowband interference is removed. We have assumed that we restore the phase to the ideal noiseless signal.

The block diagram of the simulation software is illustrated in Fig. 6.7. Initially, we generate a maximum length (ML) sequence of length 1,023 [18]. This sequence provides a 60-dB ratio between the maximum autocorrelation peak and its out-of-phase autocorrelation peak in absolute value. Therefore, 60 dB is the upper bound for the ratio between the cross-correlation peaks. This is the first data point of interest.

The GPS signal frequency at L_1, 1575.42 MHz, is used as the desired carrier frequency. Only the in-phase channel of this carrier is modulated employing the ML sequence of length 1,023 bits. The autocorrelation function of this signal is the second data point of interest. We suspect that the ratio of the cross-correlation peaks would be the same as that of the observation point 1.

WGN corrupts the composite signal. The noise power is assumed to be 0 dB and the power of the desired carrier is assumed to be -15 dB. The autocorrelation function of this signal is the third data point of interest. The ratio of the cross-correlation peaks will be significantly degraded at this point.

The signal at point 3 is corrupted further with a narrowband interference signal. The carrier frequency of the interference signal is the same as that of the desired signal and the power of the interference signal is set at 40 dB. The autocorrelation function serves as the forth data point of interest.

The ASTSA's multipliers are applied to the input signal to remove the interference effect. The autocorrelation function of this signal is the fifth data point of interest. The phase of this signal is restored and the autocorrelation function the signal with restored phase is the last data point of interest.

In Fig. 6.20 we present the autocorrelation function of the ML sequence during 1 ms time interval. The maximum autocorrelation peak is 1 and the minimum is $-1/1{,}023$, which produces a ratio of about 60.2 dB.

The autocorrelation function of the noiseless and interference-free input signal (see at point 2 in Fig. 6.7) is shown in Fig. 6.20. Although the autocorrelation peak is reduced, the "secondary" autocorrelation peak is reduced at the same rate; hence, the ratio of the cross-correlation peaks is 60.2 dB at the observation point 2.

The autocorrelation function of the noisy and interference-free input signal (see at point 3 in Fig. 6.7) is displayed in Fig. 6.20. The receiver would attempt to track

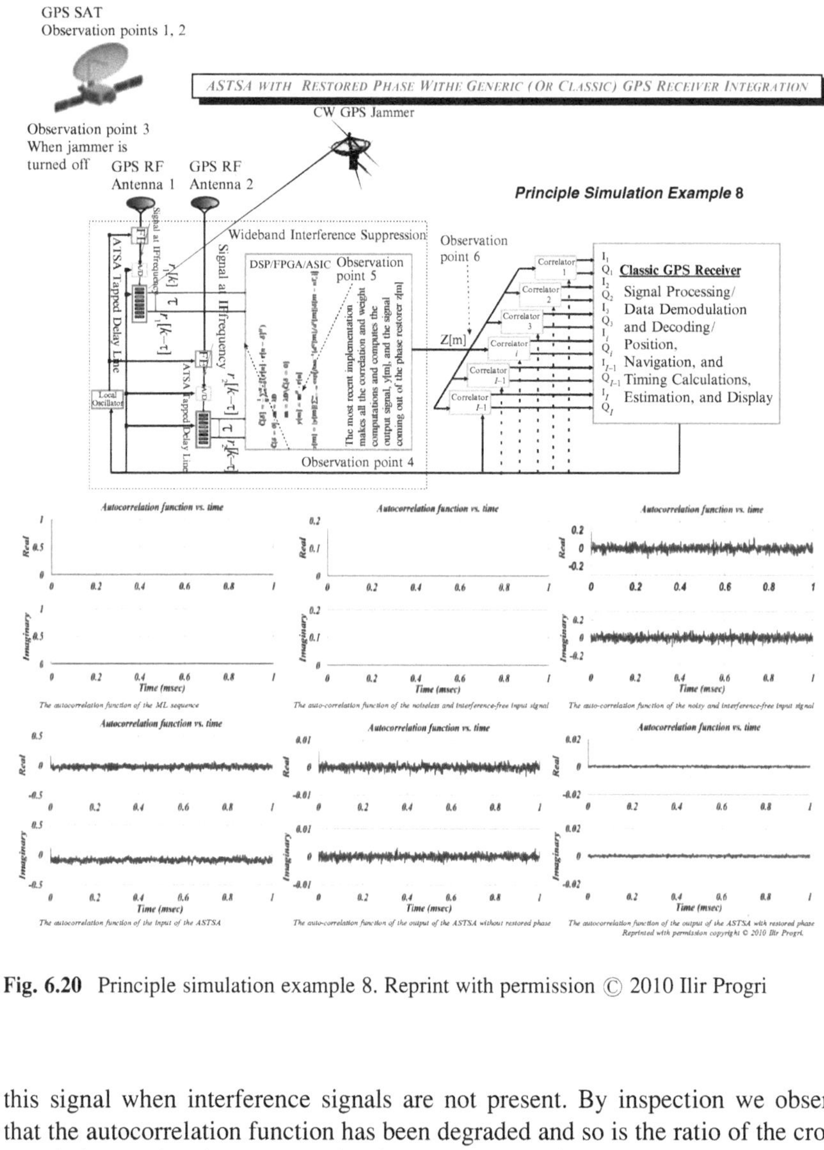

Fig. 6.20 Principle simulation example 8. Reprint with permission © 2010 Ilir Progri

this signal when interference signals are not present. By inspection we observe that the autocorrelation function has been degraded and so is the ratio of the cross-correlation peaks. On average the signal-to-noise ratio is about 13.75 dB ($1\sigma = 2.84$ dB), which is good enough for the receiver to acquire the desired signal. The autocorrelation function of the input of the ASTSA (see point 4 in Fig. 6.7) is shown in Fig. 6.20. The cross-correlation properties of this signal with the ML sequence are deteriorated even further, which results in a much lower signal-to-noise ratio (its average is 9.86 dB ($1\sigma = 4.09$ dB), see column 4 of Table 6.1).

Although the receiver may have enough dynamic range (over 55 dB) to avoid saturation from the jamming signal, the signal-to-noise ratio is being degraded and hence the acquisition of the desired signal is uncertain.

Table. 6.1 Performance results, SNR (dB) of the 2E-1T ASTSA against one CWJ source. Reprint with permission © 2010 Ilir Progri

	Observation points					
	1	2	3	4	5	6
Samples	60.20	60.20	11.18	14.3	16.72	34.13
	60.20	60.20	17.77	10.2	16.7	34.23
	60.20	60.20	15	7.38	5.387	34.92
	60.20	60.20	16.87	12.9	19.81	34.73
	60.20	60.20	13.99	7.7	15.14	34.61
	60.20	60.20	13.46	9.97	20.12	34.84
	60.20	60.20	16.25	1.03	19.77	33.99
	60.20	60.20	10.25	13.2	16.42	33.98
	60.20	60.20	9.305	7.84	16.58	34.55
	60.20	60.20	13.38	13.9	21.14	34.51
μ	60.20	60.20	13.75	9.86	16.78	34.45
σ	0.00	0.00	2.84	4.09	4.48	0.34

The autocorrelation function of the output of the ASTSA (see point 5 in Fig. 6.7) is shown in Fig. 6.20. The cross-correlation properties of this signal with the ML sequence are improved, which results in a better signal-to-noise ratio (its average is 16.78 dB, $1\sigma = 4.48$ dB) than the input signal properties (see column 5 of Table 6.1). Although the signal-to-noise ratio is improved, it has a much higher standard deviation compared with case 3 when interference is not present. This may lead to unreliable tracking of the GPS signal.

The autocorrelation function of the output of the ASTSA with restored phase (see point 6 in Fig. 6.7) is shown in Fig. 6.20. The cross-correlation properties of this signal with the ML sequence are improved, which results in a much better signal-to-noise ratio (its average is 34.45 dB with $1\sigma = 0.35$ dB) than the input signal properties (see column 6 of Table 6.1). Note that restoring the phase to the ideal signal phase appears to improve the signal-to-noise ratio by a factor of 2.5 in dB. Although this would be very desirable, it is impractical, because of the lack of knowledge of the phase of the ideal signal.

6.10.2 Principle Simulation Example 9: The 2E-1T ASTSA and One WB Interference Source and Ideal Phase Restorer

The desired and noise parameters are the same as those utilized for the first principle simulation scenario. However, the interference source has a 20-MHz bandwidth with a center frequency of 1575.42 MHz. From this point, the procedure followed was the same as the procedure of principle simulation scenario 1. We have assumed that we restore the phase to the ideal noiseless signal. The snapshot cross-correlation function for points 1–3 are not presented for this principle simulation scenario, because they are the same as those presented in principle simulation scenario 1 (see Fig. 6.21).

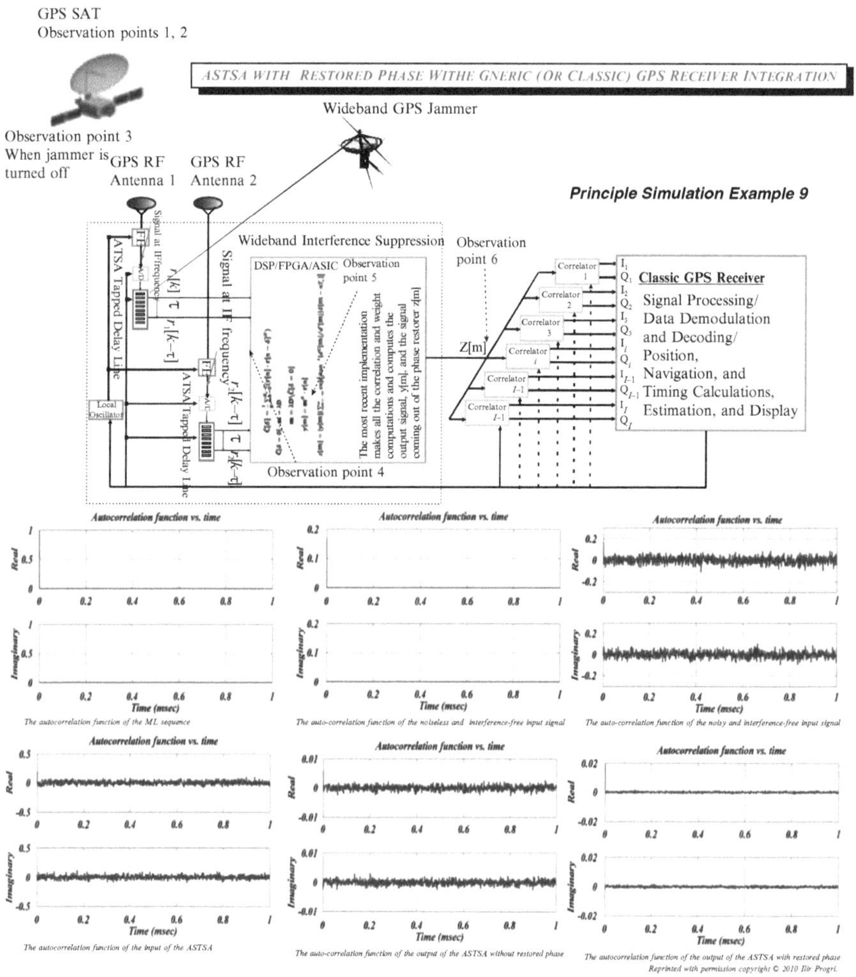

Fig. 6.21 Principle simulation example 9. Reprint with permission © 2010 Ilir Progri

The autocorrelation function of the input of the ASTSA (see point 4 in Fig. 6.7) is shown in Fig. 6.21. The cross-correlation properties of this signal with the ML sequence are deteriorated even further, which results in a much lower signal-to-noise ratio (its average is 13.55 dB with $1\sigma = 1.66$ dB, see column 4 of Table 6.2). On the basis of the signal-to-noise ratio values, the acquisition of the desired signal is uncertain.

The autocorrelation function of the output of the ASTSA (see point 5 in Fig. 6.7) is shown in Fig. 6.21. The cross-correlation properties of this signal with the ML sequence are improved, which results in a better signal-to-noise ratio (its average is 17.55 dB with $1\sigma = 4.97$ dB) than the input signal properties (see column 5 of Table 6.2).

Table. 6.2 Performance results, SNR (dB) of the 2E-1T ASTSA against one WEJ source. Reprint with permission © 2010 Ilir Progri

	Observation points					
	1	2	3	4	5	6
Samples	60.20	60.20	15.51	12.02	19.58	34.52
	60.20	60.20	15.63	13.78	17.65	34.10
	60.20	60.20	11.35	16.35	4.17	34.07
	60.20	60.20	14.04	13.81	20.98	34.07
	60.20	60.20	15.00	10.87	16.34	34.24
	60.20	60.20	15.45	13.21	20.99	34.13
	60.20	60.20	15.73	13.54	20.41	34.57
	60.20	60.20	11.84	14.35	17.96	34.25
	60.20	60.20	13.53	11.96	17.47	33.98
	60.20	60.20	15.04	15.57	19.96	34.04
μ	60.20	60.20	14.31	13.55	17.55	34.20
σ	0.00	0.00	1.60	1.66	4.97	0.20

A higher standard deviation of the signal-to-noise ratio suggests an unreliable acquisition of the desired signal.

The autocorrelation function of the output of the ASTSA with restored phase (see point 6 in Fig. 6.21) is shown in Fig. 6.21. The cross-correlation properties of this signal with the ML sequence are improved, which results in a much better signal-to-noise ratio (its average of 34.20 dB with $1\sigma = 0.20$ dB) than the input signal properties (see column 6 of Table 6.2).

Note that even for this principle simulation scenario restoring the phase appears to improve the signal-to-noise ratio by a factor of 2.5 in dB. Even though restoring the phase of the desired signal to the ideal signal is very desirable, it is impractical. We are going to seek an alternate route to restore the ASTSA output signal phase to the phase of a locally generated, desired signal.

6.10.3 *Principle Simulation Example 10: The 2E-1T ASTSA and One WB Interference Source and a Realistic Phase Restorer*

The desired and noise parameters are the same as those utilized for the second principle simulation scenario. The procedure of principle simulation scenario 2 is followed for this principle simulation scenario as well. We have assumed that we restore the phase to the locally generated noiseless signal, just as we do in a standard GPS receiver. There are two uncertainties associated with this technique: (1) there is a random absolute phase difference between the locally generated signal and the ideal desired signal because they are driven with unsynchronized clocks; (2) this absolute phase difference has a jitter because of the unequal clocks drift rates.

Note that we have ignored third and higher order clock statistics because we can get by with clocks that have good short-term stability. For simulation purposes the

Table. 6.3 Performance results, SNR (dB) of the 2E-1T ASTSA against one wideband interference source and a realistic phase restorer. Reprint with permission © 2010 Ilir Progri

	Observation points					
	1	2	3	4	5	6
Samples	60.20	60.20	15.55	14.09	17.60	33.04
	60.20	60.20	13.70	14.40	17.69	33.86
	60.20	60.20	12.65	12.78	6.85	17.84
	60.20	60.20	14.58	14.01	20.27	33.99
	60.20	60.20	11.90	11.41	17.66	32.55
	60.20	60.20	13.17	15.80	21.14	34.10
	60.20	60.20	16.49	16.45	20.15	34.23
	60.20	60.20	13.88	16.32	16.08	33.11
	60.20	60.20	13.46	15.80	19.12	34.39
	60.20	60.20	12.05	15.49	20.05	34.06
μ	60.20	60.20	13.74	14.65	17.66	32.12
σ	0.00	0.00	1.47	1.64	4.12	5.05

bias component was selected as a random quantity uniformly distributed in the $(0, 2\pi)$ interval and the jitter component was assumed a random component with a 2-mm standard deviation. The snapshot runs are not shown, because they do not provide any additional information.

The values corresponding to the signal-to-noise ratio for observation points 3 through 5 are almost the same as those of the previous run. Nevertheless, the signal-to-noise ratio of the output of the ASTSA with restored phase (its average is 32.12 dB with standard deviation of 5.05 dB) is degraded compared with the signal-to-noise ratio of point 6 of Table 6.3. This appears to suggest that restoring the phase to a locally generated signal can provide an improvement of the signal-to-noise ratio by a factor of 2.

6.11 Summary and Conclusions

6.11.1 ATSA Summary and Conclusions

The ATSA can be used successfully to mitigate the undesired effect of a finite number of narrow band interference sources by providing a substantial improvement of the UDR. For a second order ATSA a 45 or 55 dB improvement of the UDR is achieved when the input UDR is 70 dB and the noise floor is 40 or more dB below the input UDR as shown in principle illustration examples 2 and 4.

For a second order, dual channel, ATSA when the narrowband desired/undesired signals are modeled as complex values, a well-defined relationship exists among the desired signal frequency, sampling frequency, and tap delay, which produces the minimum and maximum values of the UDR. This result appears to be insensitive of both the desired/undesired signal levels and the undesired signal frequency. Although in our analysis we considered a narrowband desired signal, we obtained identical findings utilizing MATLAB simulation where we employed a wideband desired signal; therefore, to the extent that our analysis and simulation agree, this result is also insensitive of the desired signal model. This leads us to speculate that the same performance can be obtained for a PRN code modulated on L1, L2, or L5 GPS frequencies.

Because of the complexity (tedious work) and the lack of attractiveness, we did not provide an analysis tool for the real case. Nevertheless, we simulated and implemented the real case because of constraints imposed by TI C6711 DSP that only has a single input and a single output channel. Moreover, the observed ATSA performance according to the MATLAB simulation was validated by our experimental findings. They both suggest that a minimum tap delay of 3 appears to be the optimal value for the example presented in the principle illustration example 3 and implementation when the sampling frequency is set at 8 kHz (see Figs. 6.9 and 6.10).

Therefore, to achieve the full capability of this attenuator and eliminate an ad hoc design procedure, care must be taken in digitally implementing the ATSA, which ultimately results in total performance improvement and cost reduction. This chapter is an attempt to identify some of the critical parameters influencing the performance of the ATSA and provides a detailed methodology for investigating the ATSA.

The most important message communicated from this work can be formulated as follows:

> For a given number of temporal shifters (delays) and for a given desired signal structure (i.e., wideband with known bandwidth and center frequency) a proper shifter delay can be selected in accordance with the sampling frequency to utilize the full capability of the ATSA in mitigating a finite number of undesired, narrowband sources, which is equal to the number of temporal shifters (taps).
>
> A general rule for finding for selecting τ given the sampling rate, f_s, the desired signal frequency, f_d, and the number of tap delays, A; is given by (6.6) nevertheless, based on their experience, they suggest that proper simulations should be conducted to assess each and every application independently and carefully [1]. (This result is new and was not included in our previous publication [1]!)

For example, assuming that input signal frequency at the IF band is 40 MHz then an A/D and a tapped delay line of at least 80 MHz and a DSP driven by 1.2 GHz clock (assuming 30 instructions per sample) would be required for implementing an ATSA in a GPS receiver.

The lessons learned here remain to be pursued and verified in more sophisticated ATSAs, which operate in L1, L2, L5, L3, L4, and L1C frequency. If proven successful this methodology can become a valuable tool for designing and implementing an ATSA, which will operate in the L5 and L1C frequency in the near future.

6.11.2 ASTSA Summary and Conclusions

We have conducted a preliminary investigation of an ASTSA, which either/both temporal or/and spatial degrees of freedom to mitigate wide/narrow band interference.

We have compared an ASTSA, which employs two adaptive techniques: the first forms a signature vector exploiting the information from a local generated signal, and the second employs the well-known pointing (or a global) vector [4].

We observe that the *old* (*global*) adaptive methodology outperforms the *new* (*local*) adaptive methodology for any of the ATSA, ASSA, or ASTSA with more than 1 sensor and more than one temporal shifter against one BBJ. Utilizing the *old* (*global*) adaptive methodology yields 33, 42, and 25 dB output UDR corresponding to an ATSA with 6 taps, an ASSA with 4 sensors, and ASTSA with 6 taps and 4 sensors. When exploiting the *new* (*local*) adaptive methodology, the numeric performance values are nonoptimized, 42, and 33 dB corresponding to an ATSA with 6 taps, an ASSA with 4 sensors, and ASTSA with 6 taps and 4 sensors.

Both adaptive criteria appear to perform equally well against one CWJ by providing the output UDR equal to 37, 12, and 2.5 dB corresponding to an ATSA with 6 taps, an ASSA with 4 sensors, and ASTSA with 6 taps and 4 sensors, respectively.

However, the UDR criterion is not the only performance measure of an ASTSA. We are ultimately concerned about the receiver ability to track the desired pseudolite (GPS-like) signal. We believe that the *old* (*or global*) adaptive methodology intends to resolve this issue. The future investigation will aim toward this goal.

6.11.3 Improved ASTSA Summary and Conclusions

We have started to investigate our proposed pseudolite (GPS-like) signal properties when subject to narrowband/wideband interference signals and employed an ASTSA, which mitigates these interference signals.

We have proposed additional improvements of the ASTSA to achieve better SNR. These improvements consist of the following:

1. Employing a locally generated signal to generate the cross-correlation vector, which is then used to yield the desired set of the ASTSA multipliers.
2. Restoring the ASTSA output signal phase to the phase of the locally generated signal.

This technique appears to double the SNR against any wideband/narrowband interference. We have not addressed the receiver reaction in the presence of phase restoration vs. in the absence of the phase restoration. Future work will address this concern and assess the ASTSA against multiple wideband/narrowband and pulse interference sources.

6.12 Future Direction for Research

There are tremendous interests in the area of the adaptive array beamforming for interference mitigation, multipath mitigation, etc. for GRFS systems. The very simple fact that there are a huge number of recent studies in the forms of master thesis, Ph.D. dissertation, journal articles, white papers, etc. indicate that this area is well on its way of further ground breaking research and development. Based on the above we propose the following list of research and development in the area of adaptive array beamforming for interference mitigation for GRFS systems:

1. We need to review all the adaptive array beamforming principles (i.e., the theory and practice) for moving jammers especially for fast moving jammers. Now we need to include in the jammer profile both signal amplitude and phase variations as the result of the jammer movement.
2. Although in Chap. 2 we laid out 39 configurations for effective ranges going from 100 m to 100,000 km, we need to analyze these heterogeneous environments one by one and propose typical GRFS systems that will be able to address the growing demand of the jamming and interference suppression requirements (example [78]). We may find out that in some of these environments and novel configurations, stationarity may disappear and therefore the conventional covariance estimation methods may not be used. These have given rise to the knowledge-aided STAP, which attempts to remove as much of the heterogeneity from the snap-shots prior to using the conventional estimation methods; this falls in the general domain of the knowledge-aided signal processing, which is done by using prior knowledge, typically stored in databases [77, 95, 99, 101].
3. The study of arbitrary 3D antenna arrays such as the surface of conformal antenna array (CAA) and concentric ring arrays (CRA) that follows the surface of the carrying platform, such as the fuselage of an airplane, or the side of a balloon, or the surface of an aircraft carrier, etc. ([77, 94], example [78], Optimizing the array geometry such as in [87, 94]). Improve the range of beamforming to process weak signals.
4. Also of great interests is an array of arrays because one array may not be sufficient to provide all the information required and to address all the growing requirements; thus, an array of arrays with well-defined interfaces and integration of all arrays might be able to address all growing requirements for all the moving arbitrary and increasing jammers and jammer waveforms.
5. Perhaps more complex configurations of space–time–frequency–code-adaptive processing should be exploited, which have not been proposed before.

Acknowledgments I thank you for reading the book and I hope this is a good and useful resource to you, your company, and your colleagues.

Appendix A

Assuming that the only impairment in the system is WGN $\mathbf{w}[n]$, with statistics $N(0, \mathbf{R_w})$, (zero mean covariance matrix $\mathbf{R_w}$) and we can modify as follows

$$\mathbf{w}[n] = \mathbf{r}[n] - (\mathbf{d}[n] + \mathbf{u}[n]). \tag{6.94}$$

The probability density function for the vector $\mathbf{w}_i \equiv \mathbf{w}[n+1]$ looks like [31]

$$f(\mathbf{w}_i) \equiv \frac{1}{\pi^A |\mathbf{R}_\mathbf{w}|} e^{-\mathbf{w}_i^\mathrm{H} \cdot \mathbf{R}_\mathbf{w}^{-1} \cdot \mathbf{w}_i}. \tag{6.95}$$

Assuming that we have N available independent measurements, the joint distribution function for $\mathbf{w}_i$, $\forall i \in \{1, 2, \cdots, N\}$, looks as

$$f(\mathbf{W}) = \prod_{i=1}^{N} f(\mathbf{w}_i) \equiv \prod_{i=1}^{N} \frac{1}{\pi^A |\mathbf{R}_\mathbf{w}|} e^{-\mathbf{w}_i^\mathrm{H} \cdot \mathbf{R}_\mathbf{w}^{-1} \cdot \mathbf{w}_i} = \frac{1}{\pi^{NA} |\mathbf{R}_\mathbf{w}|^N} e^{-\sum_{i=1}^{N} \mathbf{w}_i^\mathrm{H} \cdot \mathbf{R}_\mathbf{w}^{-1} \cdot \mathbf{w}_i}, \tag{6.96}$$

where

$$\mathbf{W} = \begin{bmatrix} \mathbf{w}_1 & \cdots & \mathbf{w}_N \end{bmatrix}^\mathrm{T}. \tag{6.97}$$

We seek to estimate the matrix $\mathbf{R}_\mathbf{w}$ using the maximum likelihood ML estimation; therefore, consider

$$L = \log(f(\mathbf{W})) = N \log(|\mathbf{R}_\mathbf{w}|^{-1}) - \sum_{i=1}^{N} \mathbf{w}_i^H \cdot \mathbf{R}_\mathbf{w}^{-1} \cdot \mathbf{w}_i + C, \tag{6.98}$$

where, the constant C is simply

$$C = -NA \log \pi. \tag{6.99}$$

The best estimation of $\mathbf{R}_\mathbf{w}$ can be obtained when

$$\frac{\partial L}{\partial \mathbf{R}_\mathbf{w}^*} = 0 \quad \text{or} \quad \frac{\partial L}{\partial \mathbf{R}_\mathbf{w}^{-1*}} = 0. \tag{6.100}$$

Applying (6.99) into (6.101) produces

$$\hat{\mathbf{R}}_\mathbf{w} = \frac{1}{N} \sum_{i=1}^{N} (\mathbf{w}_i \cdot \mathbf{w}_i^\mathrm{H}). \tag{6.101}$$

Taking the expected value on both sides of (6.102) yields

$$E\{\hat{\mathbf{R}}_\mathbf{w}\} = \frac{1}{N} \sum_{i=1}^{N} E(\mathbf{w}_i \cdot \mathbf{w}_i^\mathrm{H}) = \mathbf{R}_\mathbf{w}. \tag{6.102}$$

Next we derive an expression for the auto-covariance matrix of the estimator $\hat{\mathbf{R}}_\mathbf{w}$; hence, consider the calculation

$$
\begin{aligned}
\mathrm{Cov}\{\hat{\mathbf{R}}_{\mathbf{w}}\} &= E\{\hat{\mathbf{R}}_{\mathbf{w}}^{2}\} - E^{2}\{\hat{\mathbf{R}}_{\mathbf{w}}\} = \frac{1}{N^{2}} E\left\{\left(\sum_{i=1}^{N} \mathbf{w}_i \cdot \mathbf{w}_i^{\mathrm{H}}\right)\left(\sum_{j=1}^{N} \mathbf{w}_j \cdot \mathbf{w}_j^{\mathrm{H}}\right)\right\} - \mathbf{R}_{\mathbf{w}}^{2} \\
&= \frac{1}{N^{2}} E\left\{\sum_{i=1}^{N}\sum_{j=1}^{N} (\mathbf{w}_i \cdot \mathbf{w}_i^{\mathrm{H}})(\mathbf{w}_j \cdot \mathbf{w}_j^{\mathrm{H}})\right\} - \mathbf{R}_{\mathbf{w}}^{2} \\
&= \frac{1}{N^{2}} \sum_{i=1}^{N}\sum_{j=1}^{N} E\{(\mathbf{w}_i \cdot \mathbf{w}_i^{\mathrm{H}})(\mathbf{w}_j \cdot \mathbf{w}_j^{\mathrm{H}})\} - \mathbf{R}_{\mathbf{w}}^{2},
\end{aligned}
\tag{6.103}
$$

where

$$
\begin{aligned}
\sum_{i=1}^{N}\sum_{j=1}^{N} E\{(\mathbf{w}_i \cdot \mathbf{w}_i^{\mathrm{H}})(\mathbf{w}_j \cdot \mathbf{w}_j^{\mathrm{H}})\} &= \sum_{i=1}^{N} E\{(\mathbf{w}_i \cdot \mathbf{w}_i^{\mathrm{H}})^2\} + \sum_{i=2}^{N}\sum_{j=1, j\neq i}^{N} E\{(\mathbf{w}_i \cdot \mathbf{w}_i^{\mathrm{H}})(\mathbf{w}_j \cdot \mathbf{w}_j^{\mathrm{H}})\} \\
&= \sum_{i=1}^{N} E\{(\mathbf{w}_i \cdot \mathbf{w}_i^{\mathrm{H}})^2\} + 2\sum_{i=2}^{N} E\{(\mathbf{w}_i \cdot \mathbf{w}_i^{\mathrm{H}})(\mathbf{w}_1 \cdot \mathbf{w}_1^{\mathrm{H}})\} \\
&\quad + \cdots + 2E\{(\mathbf{w}_N \cdot \mathbf{w}_N^{\mathrm{H}})(\mathbf{w}_{N-1} \cdot \mathbf{w}_{N-1}^{\mathrm{H}})\}.
\end{aligned}
\tag{6.104}
$$

There was an error in eq. (53) of [1] which we have corrected in (6.105). Although that error does not change the answer, it is best to have it in the correct form in (6.105).

Using the identity for the joint normal complex RVs [31]

$$
E\{(\mathbf{w}_i \cdot \mathbf{w}_i^{\mathrm{H}})(\mathbf{w}_j \cdot \mathbf{w}_j^{\mathrm{H}})\} = E\{\mathbf{w}_i \cdot \mathbf{w}_i^{\mathrm{H}}\}E\{\mathbf{w}_j \cdot \mathbf{w}_j^{\mathrm{H}}\} + E^{2}\{\mathbf{w}_i \cdot \mathbf{w}_j^{\mathrm{H}}\} \tag{6.105}
$$

and knowing that $E\{\mathbf{w}_i \cdot \mathbf{w}_j^{\mathrm{H}}\} = 0$, for $i \neq j$, thus

$$
\mathrm{Cov}\{\hat{\mathbf{R}}_{\mathbf{w}}\} = \frac{2\mathbf{R}_{\mathbf{w}}^{2}}{N^{2}}[N + (N-1) + \cdots + 1] - \mathbf{R}_{\mathbf{w}}^{2} \rightarrow \mathrm{Cov}\{\hat{\mathbf{R}}_{\mathbf{w}}\} = \frac{\mathbf{R}_{\mathbf{w}}^{2}}{N}. \tag{6.106}
$$

It can be shown that the auto-covariance matrix of this estimator reached the Cramer–Rao Bound; thus, $\hat{\mathbf{R}}_{\mathbf{w}}$ provides an efficient estimate of $\mathbf{R}_{\mathbf{w}}$; i.e., when N going to infinity the $\mathrm{Cov}\{\hat{\mathbf{R}}_{\mathbf{w}}\}$ goes to zero; and therefore, the estimator $\hat{\mathbf{R}}_{\mathbf{w}}$ reaches the true estimate of $\mathbf{R}_{\mathbf{w}}$.

Assuming that the desired and undesired signals are purely deterministic then

$$
\hat{\mathbf{C}} = \alpha \hat{\mathbf{R}}_{\mathbf{w}} \quad \text{with } \alpha = \frac{\hat{\mathbf{C}}}{\hat{\mathbf{R}}_{\mathbf{w}}}. \tag{6.107}
$$

In this case the coefficient α can be either a real or complex scalar.

We conclude this appendix by reminding the reader that for most practical applications of adaptive array beamforming for interference mitigation for GRFS systems, a robust and accurate estimate of the covariance matrix is very important. Although Chap. 5 is a superb source of the robust techniques based on recursive Cholesky and recursive generalized eigenvalue solution, another robust sample covariance estimate technique is discussed in Wipf and Nagarajan [69].

Appendix B: Equivalent Expressions with Other Similar Publications

The main purpose of this appendix is to derive all equivalent expressions that we have derived in this chapter with similar ones found in similar publications. We would like to provide our reader with an opportunity to see (or visualize) the material in many different ways.

The first proof!

A simplified expression for the UDR is derived in [4] and is rederived here for the sake of simplicity (also known as the min–max array processing [84])

$$\begin{aligned}\mathrm{UDR} &= \frac{\min_{\mathbf{m}} P_{\mathrm{y}}}{\max_{\mathbf{m}} \sum_{\mathrm{a}} P_{\mathrm{d}}} = \frac{\mathbf{m}^{\mathrm{H}} \cdot \hat{\mathbf{C}}[\delta = 0] \cdot \mathbf{m}}{\sum_{\mathrm{a}} P_{\mathrm{d}} |\mathbf{m}^{H} \cdot \mathbf{D}|^2} = \frac{\mathbf{m}^{\mathrm{H}} \cdot \lambda \mathbf{D}}{A P_{\mathrm{d}} |\mathbf{m}^{\mathrm{H}} \cdot \mathbf{D}|^2} \\ &= \frac{\lambda}{A P_{\mathrm{d}} \mathbf{D}^{\mathrm{H}} \cdot \mathbf{m}} = \frac{1}{A P_{\mathrm{d}} \mathbf{D}^{\mathrm{H}} \cdot \mathbf{m}/\lambda} = \frac{1}{A P_{\mathrm{d}} \mathbf{D}^{\mathrm{H}} \cdot \tilde{\mathbf{m}}},\end{aligned} \tag{6.108}$$

where $\tilde{\mathbf{m}} = \mathbf{m}/\lambda$ are the normalized weights (or multipliers). Further in the reminder of the chapter we use $\mathbf{m}$ to denote the normalized weights.

Similarly using the property that

$$\hat{\mathbf{C}}[\delta = 0] = \hat{\mathbf{C}}^{\mathrm{H}}[\delta = 0] \quad \text{(because the matrix is Hermitian symmetric)} \tag{6.109}$$

then the UDR expression becomes

$$\begin{aligned}\mathrm{UDR} &= \frac{\min_{\mathbf{m}} P_{\mathrm{y}}}{\max_{\mathbf{m}} \sum_{a} P_{\mathrm{d}}} = \frac{\mathbf{m}^{\mathrm{H}} \cdot \hat{\mathbf{C}}^{\mathrm{H}}[\delta = 0] \cdot \mathbf{m}}{\sum_{a} P_{\mathrm{d}} |\mathbf{m}^{\mathrm{H}} \cdot \mathbf{D}|^2} = \frac{\lambda \mathbf{D}^{\mathrm{H}} \cdot \mathbf{m}}{A P_{\mathrm{d}} |\mathbf{m}^{\mathrm{H}} \cdot \mathbf{D}|^2} \\ &= \frac{\lambda}{A P_{\mathrm{d}} \mathbf{m}^{\mathrm{H}} \cdot \mathbf{D}} = \frac{1}{A P_{\mathrm{d}} \mathbf{m}/\lambda^{\mathrm{H}} \cdot \mathbf{D}} = \frac{1}{A P_{\mathrm{d}} \tilde{\mathbf{m}}^{\mathrm{H}} \cdot \mathbf{D}}\end{aligned} \tag{6.110}$$

which is the equivalent expression in [1].

The Second proof!

Combining expressions (6.37), (6.40), and (6.41), (based on the result of (6.109)) yields (also known as the min–max array processing [84])

$$\mathrm{UDR} = \frac{\min_{\mathbf{m}} P_{\mathrm{y}}}{\max_{\mathbf{m}} P_{\mathrm{d}}} = \frac{E[|\mathbf{m}^{\mathrm{H}} \cdot \mathbf{r}(t)|^2]}{E(|\tilde{s}_i(t)|^2)} = \frac{1}{P_{\mathrm{d}}(\mathbf{m}^{\mathrm{H}} \cdot \mathbf{p}_i)^{\mathrm{H}}} = \frac{1}{P_{\mathrm{d}}\mathbf{p}_i{}^{\mathrm{H}} \cdot \mathbf{m}}. \tag{6.111}$$

An equivalent expression of (6.112) can be obtained as follows

$$\mathrm{UDR} = \frac{\min_{\mathbf{m}} P_{\mathrm{y}}}{\max_{\mathbf{m}} P_{\mathrm{d}}} = \frac{E[|\mathbf{m}^{\mathrm{H}} \cdot \mathbf{r}(t)|^2]}{E(|\tilde{s}_i(t)|^2)} = \frac{1}{P_{\mathrm{d}}\mathbf{m}^{\mathrm{H}} \cdot \mathbf{p}_i} \tag{6.112}$$

which is the same as one found in [2].

The third proof!

It is easy to verify (6.61)

$$\begin{aligned} \mathbf{g}_i^{\mathrm{H}}(t) \cdot \mathbf{n}_i(t) &= \mathbf{w}_i^{\mathrm{H}}(t) \cdot \mathbf{C}^{\mathbf{H}} \cdot \mathbf{n}_i(t) = \mathbf{w}_i^{\mathrm{H}}(t) \cdot \mathbf{C} \cdot \mathbf{n}_i(t) = \mathbf{w}_i^{\mathrm{H}}(t) \cdot \mathbf{q}_i(t) \\ &= [\mathbf{q}_i^{\mathrm{H}}(t) \cdot \mathbf{w}_i(t)]^{\mathrm{H}} \end{aligned} \tag{6.113}$$

which complete the proof.

Appendix C: Important Theorem Proofs

Theorem 1

For the matrix given by (6.64) prove that its determinant is given by (6.68).

Proof of Theorem 1

The total mathematical induction is used to prove this Theorem. When $N = 2$ the expression for the determinant of the autocorrelation matrix is given by

$$|\mathbf{C}| = \sigma_0^{4-2}\left(\sum_{k=1}^{2}|x_k(t_{m\tau})|^2 + \sigma_0^2\right) = \sigma_0^2\left(\sum_{k=1}^{2}|x_k(t_{m\tau})|^2 + \sigma_0^2\right), \tag{6.114}$$

which is the same as expression (43).

Next, assume that the determinant of the autocorrelation matrix (6.64) is indeed given by (6.68). We will prove that when the size (or more precisely the rank) of $\mathbf{C}$ is $N + 1$ then the expression for the determinant of $\mathbf{C}$ is

$$|\mathbf{C}| = \sigma_0^{2N}\left(\sum_{k=1}^{N+1}|x_k(t_{m\tau})|^2 + \sigma_0^2\right). \tag{6.115}$$

Employing Kramer's Rule we can compute the determinant of $\mathbf{C}$ as follows

$$|\mathbf{C}| = \sum_{i=1}^{N} (-1)^{i+N+1} c_{i,N+1}^{*} M_{i,N+1}^{*} + c_{N+1,N+1} M_{N+1,N+1}, \tag{6.116}$$

where

$$M_{i,N+1}^{*} = \begin{vmatrix} c_{1,1} & c_{1,2} & \cdots & c_{1,N} \\ c_{1,2}^{*} & c_{2,2} & & c_{2,N} \\ \vdots & & \ddots & \vdots \\ c_{1,i-1}^{*} & c_{2,i-1}^{*} & & c_{i-1,N} \\ c_{1,i+1}^{*} & c_{2,i+1}^{*} & \ddots & c_{i+1,N} \\ \vdots & & & \vdots \\ c_{1,N+1}^{*} & c_{2,N+1}^{*} & & c_{N,N+1}^{*} \end{vmatrix} \tag{6.117}$$

and

$$M_{N+1,N+1} = \begin{vmatrix} c_{11} & c_{12} & \cdots & c_{1N} \\ c_{12}^{*} & c_{22} & & c_{2N} \\ \vdots & & \ddots & \vdots \\ c_{1N}^{*} & c_{2N}^{*} & \cdots & c_{NN} \end{vmatrix}. \tag{6.118}$$

There were two errors in eqs. (61) and (62) of [2], which we have corrected in (6.117) and (6.118). Although this does not change the end results, it is nice to have the correct equations in the book to help the readers to follow along.

We compute the following product

$$c_{N+1,N+1} M_{N+1,N+1} = |\mathbf{C}| + \sigma_0^{2N-2} |x_{N+1}(t_{m\tau})|^2 \sum_{k=1}^{N} |x_k(t_{m\tau})|^2. \tag{6.119}$$

It can be shown that

$$\sum_{i=1}^{N} (-1)^{i+N+1} c_{i,N+1}^{*} M_{i,N+1}^{*} = -\sigma_0^{2N-2} |x_{N+1}(t_{m\tau})|^2 \sum_{k=1}^{N} |x_k(t_{m\tau})|^2 \tag{6.120}$$

which completes the proof.

Appendix D: Important Theorem Proofs

Theorem 2

For the matrix given (6.64) prove that its inverse is given by (6.69).

Proof of Theorem 2

Assume that its inverse is given by (6.69), it can be shown that product of (6.69) with the original matrix is indeed an identity matrix that completes the proof.

References

1. Progri, I.F., Michalson, W.R., Bogdanov, G., and Ramanna, V.C., "An investigation of a GPS adaptive temporal selective attenuator," *NAVIGATION J. Inst. Nav.*, vol. 49, nr. 3, pp. 137–147, fall 2002.
2. Michalson, W.R., and I.F. Progri, "An investigation of the adaptive spatial and temporal selective attenuator," in *Proc. 14th Inter. Tech. Mtg. Sat. Div. ION (ION-GPS 2001)*, Salt Lake City, UT, pp. 1985–1996, Sep. 2001.
3. Progri, I.F, and Michalson, W.R., "An improved adaptive spatial temporal selective attenuator," in *Proc. 14th Inter. Tech. Mtg. Sat. Div. ION (ION-GPS 2001)*, Salt Lake City, UT, pp. 932–938, Sep. 2001.
4. Progri, I.F., and Michalson, W.R., "Adaptive spatial and temporal selective attenuator in the presence of mutual coupling and channel errors," in *Proc. 13th Inter. Tech. Mtg. Sat. Div. ION (ION GPS 2000)*, Salt Lake City, UT, pp. 462–470, Sep. 2000.
5. Progri, I.F., and Michalson, W.R., "An investigation of the pseudolite's signal structure for indoor applications," in *Proc. 57th Ann. Mtg. ION*, Albuquerque, NM, pp. 453–462, June 2001.
6. Progri, I.F., and Michalson, W.R., "The impact of the proposed pseudolite's signal structure on the receiver's phase and code error," in *Proc. 57th Ann. Mtg. ION*, Albuquerque, NM, pp. 414–422, June 2001.
7. Progri, I.F., and Michalson, W.R., "An alternative approach to multipath and near-far problem for indoor geolocation systems," in *Proc. 14th Inter. Tech. Mtg. Sat. Div. ION (ION GPS 2001)*, Salt Lake City, UT, pp. 1434–1467, Sep. 2001.
8. Progri, I.F., Michalson, W.R., and Bromberg, M.C., "A comparison between the recursive Cholesky and MGSO algorithms," in *Proc. 2002 Nat. Tech. Mtg. ION*, San Diego, CA, pp. 655–665, Jan. 2002.
9. Progri, I.F., Michalson, W.R., and Bromberg, M.C., "A solution to the recursive generalized eigenvalue problem," in *Proc. 58 Ann. Mtg. CIGTF 21st Guid. Test Sym.*, Albuquerque, NM, pp. 154–162, June 2002.
10. Progri, I.F., "An assessment of indoor geolocation systems," *Ph.D. dissertation*, Worcester Polytechnic Institute, Worcester, MA, May 2003.
11. Johnston, K.D., and Miller, M.M., "CW jamming effects on a COTS C/A code GPS receiver," in *Proc. 12th Inter. Tech. Mtg. Sat. Div. ION (ION GPS 1999)*, Nashville, TN, pp. 1169–1178, Sep. 1999.
12. Fante, R.L., and Vaccaro, J.J., "Enhanced anti-jam capability for GPS receivers," in *Proc. 11th Inter. Tech. Mtg. Sat. Div. ION (ION GPS 1998)*, Nashville, TN, pp. 251–254, Sep. 1998.
13. Gromov, K., Akos, D., Pullen, S., Enge, P., Parkinson, B., and Pervan, B., "Interference direction finding for aviation applications of GPS," in *Proc. 12th Inter. Tech. Mtg. Sat. Div. ION (ION GPS 1999)*, Nashville, TN, pp. 115–124, Sep. 1999.
14. Gromov, K., Akos, D., Pullen, S., Enge, P., and Parkinson, B., "GIDL: generalized interference detection and localization system," in *Proc. 13th Inter. Tech. Mtg. Sat. Div. ION (ION GPS 2000)*, Salt Lake City, UT, pp. 447–457, Sep. 2000.
15. Yang, C., Vasquez, J., and Chaffee, J., "Frequency-domain Doppler search and jam-suppression for fast direct P(Y)-code acquisition," in *Proc. 12th Inter. Tech. Mtg. Sat. Div. ION (ION GPS 1999)*, Nashville, TN, pp. 1157–1168, Sep. 1999.

16. Ward, P.W., "Performance comparisons between FLL, PLL and a novel FLL-assisted-PLL carrier tracking loop under RF interference conditions," in *Proc. 11th Inter. Tech. Mtg. Sat. Div. ION (ION GPS 1998)*, Nashville, TN, pp. 783–795, Sep. 1998.
17. Brown, A., and Gerein, N., "Test results from a digital P(Y) code beamsteering receiver for multipath minimization," in *Proc. 57th Ann. Mtg. ION*, Albuquerque, NM, pp. 872–878, June 2001.
18. Proakis, J.G., and Salehi, M., *Communication Systems Engineering*. New Jersey: Prentice-Hall, 1995.
19. Iltis, R.A. and Hanson, G.A., "C/A code tracking and acquisition with interference rejection using the extended Kalman filter," in *Proc. 1999 Nat. Tech. Mtg. 19th Bien. Guid. Sym.*, San Diego, CA, pp. 881–889, Jan. 1999.
20. Zhodzishky, M., Cherniavsky, D., Kirsanov, A., Vorobiev, M., Prasolov, V., Zhdanov, A., and Ashjaee, J., "In-band interference suppression for GPS/GLONASS," in *Proc. 11th Inter. Tech. Mtg. Sat. Div. ION (ION GPS-1998)*, Nashville, TN, pp. 769–773, Sep. 1998.
21. Capozza, P.T., Holland, B.J., Hopkinson, T.M., Li, C., Moulin, D., Pacheco, P., and Ritkin, R.,"Measured effects of a narrowband interference suppressor on GPS receivers," in *Proc. 1999 Ann. Mtg. ION*, Cambridge, MA, June 1999.
22. Hegarty, CH. and Van Dierendonck, A.J., "Civil GPS/WAAS signal design and interference environment at 1176.45 MHZ: Results of RTCA SC159 WG1 activities," in *Proc. 12th Inter. Tech. Mtg. Sat. Div. ION (ION GPS-1999)*, Nashville, TN, Sep., 1999.
23. Powe, M., and Owen, J.I.R., "The European GNSS L5/E5 interference environment and the performance of pulsed interference mitigation techniques," *Proc. 17th Inter. Tech. Mtg. Sat. Div. ION (ION GNSS 2004)*, Long Beach, CA, pp. 122–131, Sep. 2004.
24. Godefroy, B., Poncelet, M., and Chenebault, J., "Analysis of potential CW interference effects caused by UWB devices on GNSS receivers," in *Proc. 17th Inter. Tech. Mtg. Sat. Div. ION (ION GNSS 2004)*, Long Beach, CA, pp. 187–196, Sep. 2004.
25. Kelly, J.T., "SPS v. PPS: why military GPS applications require "military" GPS," *Proc. 17th Inter. Tech. Mtg. Sat. Div. ION (ION GNSS 2004)*, Long Beach, CA, pp. 533–543, Sep. 2004.
26. Giraud, J., Busquet, C., Bauer, F., and Flament, D., "Pulsed interference and GALILEO sensor stations (GSS)," in *Proc. 18th Inter. Tech. Mtg. Sat. Div. ION (ION GNSS 2005)*, Long Beach, CA, pp. 914–925, Sep. 2005.
27. Skidmore, T.A., Cohenour, C., van Graas, F., DiBenedetto, M., Peterson, B., "JPALS tactical system performance evaluation using a controlled reception pattern antenna," in *Proc. 19th Inter. Tech. Mtg. Sat. Div. ION (ION GNSS 2006)*, Fort Worth, TX, pp. 738–753 Sep. 2006.
28. Esbri-Rodriguez, O., Philipakkis, M., Konovaltsev, A., Antreich, F., Martel, C., and Moore, D., "Antenna-based multipath and interference mitigation for aeronautical applications: present and future," in *Proc. 19th Inter. Tech. Mtg. Sat. Div. ION (ION GNSS 2006)*, Fort Worth, TX, pp. 754–762, Sep. 2006.
29. Raimondi, M., Julien, O., Macabiau, C., Bastide, F., "Mitigating pulsed interference using frequency domain adaptive filtering," in *Proc. 19th Inter. Tech. Mtg. Sat. Div. ION (ION GNSS 2006)*, Fort Worth, TX, pp. 2251–2260, Sep. 2006.
30. Martineau, A., Macabiau, C., Julien, O., Nikiforov, I., and Roturier, B., "GPS/Galileo RAIM performance in presence of multiple pseudorange failures due to interference," in *Proc. 19th Inter. Tech. Mtg. Sat. Div. ION (ION GNSS 2006)*, Fort Worth, TX, pp. 3045–3056, Sep. 2007.
31. Manolakis, D.G., Ingle, V.K., and Kogon, S.M., *Statistical and Adaptive Signal Processing: Spectral Estimation, Signal Modeling, Adaptive Filtering and Array Processing*. McGraw-Hill Higher Education, Boston: MA, 2000.
32. Blostein, S.D., and Leib, H., "Multiple antenna systems: their role and impact in future wireless access," *IEEE Com. Mag.*, vol. 41, nr. 7, pp. 94–101, Jul. 2003.
33. Muhammad, K., Staszewski, R.B., and Leipold, D., "Digital RF processing: toward low-cost reconfigurable radios," *IEEE Com. Mag.*, vol. 43, nr. 8, pp. 105–113, Aug. 2005.

34. Pucker, L., "Which architectural model should I adopt in my wireless communications systems: well, once again, it depends...," *IEEE Com. Mag.*, vol. 43, nr. 12, (*IEEE Radio Com.* vol. 2, nr. 4) pp. S4–S9, Dec. 2005.
35. Sever, I., Lo, S., Ma, S.-P., Jang, P., Zou, A., Arnott, C., Ghatak, K., Schwartz, A., Huynh, L., and Nguyen, T., "A dual-antenna phase-array ultra-wideband CMOS transceiver," *IEEE Com. Mag.*, vol. 44, nr. 8, pp. 102–110, Aug. 2006.
36. Begheri, R., Mirzaei, A., Heidari, M.E., Chehrazi, S., Lee, M., Mekhemar, M., Tang, W.K., and Abidi, A.A., "Software-defined radio receiver: dream to reality," *IEEE Com. Mag.*, vol. 44, nr. 8, pp. 111–118, Aug. 2006.
37. Hawwar, Y., Farag, E., Vanakayala, S., Pauls, R., Yang, X., Subramanian, S., Sadhanala, P., Yang, L., Wang, B., Li, Z., Chen, H., Lu, Z., Clark, D., Fosket, T., Mallela, P., Shelton, M., Laurens, D., Salaun, T., Gougeon, L., Aubourg, N., Morvan, H., Henaff, N.L., Prat, G., Charles, F., Creach, C., Calvez, Y., and Butel, P., "3G UMTS wireless system physical layer: baseband processing hardware implementation perspective," *IEEE Com. Mag.*, vol. 44, nr. 9, pp. 52–58, Sep. 2006.
38. Dohler, M., Mclaughlin, S., Laurenson, D., and Aghvami, A.H., "Implementable wireless access for B3G networks – part II: MIMO receiver architectures," *IEEE Com. Mag.*, vol. 45, nr. 3, pp. 93–97, Mar. 2007.
39. Dohler, M., Mclaughlin, S., Sweeney, P., and Hanzo, L., "Implementable wireless access for B3G networks – part III: complexity reducing transceiver structures," *IEEE Com. Mag.*, vol. 45, nr. 3, pp. 98–104, Mar. 2007.
40. Klumperink, E.A.M., Shrestha, R., Mensink, E., Arkesteijn, V.J., and Nauta, B., "Polyphase multipath radio circuits for dynamic spectrum access," *IEEE Com. Mag.*, vol. 45, nr. 5, pp. 104–112, May 2007.
41. Minden, G.J., Evans, J.B., Searl, L.S., DeParlo, D., Rajbanshi, R., Guffey, J., Chen, Q., Newman, T.R., Petty, V.R., Weilding, F., Peck, M., Cordill, B., Dalta, D., Barker, B., and Agah, A., "An agile radio for wireless innovation," *IEEE Com. Mag.*, vol. 45, nr. 5, pp. 113–121, May 2007.
42. Soecher, E., and Chang, M.-C.F., "Can RF help CMOS processors?" *IEEE Com. Mag.*, vol. 45, nr. 8, pp. 104–111, Aug. 2007.
43. Womac, M., "Designing a single-chip tuner for DVD-H," *IEEE Com. Mag.*, vol. 45, nr. 8, pp. 122–127, Aug. 2007.
44. Smulders, P., Yang, H., and Akkermans, I., "On the design of low-Cost 60-GHz radios for multigigabit-per-second transmission over short distances," *IEEE Com. Mag.*, vol. 45, nr. 12, pp. 44–51, Dec. 2007.
45. Dickens, M.L., Dunn, B.P., and Laneman, J.N., "Design and implementation of a portable software radio," *IEEE Com. Mag.*, vol. 46, nr. 8, pp. 58–66, Aug. 2008.
46. Hilbert, J.L., "RF-MEMS for wireless communications," *IEEE Com. Mag.*, vol. 46, nr. 8, pp. 68–74, Aug. 2008.
47. Mudumbai, R., Brown III, D.R., Madhow, U., and Poor, V.H., "Distributed transmit beamforming: challenges and recent progress," *IEEE Com. Mag.*, vol. 47, nr. 2, pp. 102–110, Feb. 2009.
48. Astely, D., Dahlman, E., Furuskar, A., Jading, Y., Lindstrom, M., and Parkvall, S., "LTE: the evolution of mobile broadband," *IEEE Com. Mag.*, vol. 47, nr. 4, pp. 44–51, Apr. 2009.
49. Bourdeau, G., Panicker, J., Guo, N., Chang, R., Wang, N., and Vrzic, S., "Interference coordination and cancellation for 4G networks," *IEEE Com. Mag.*, vol. 47, nr. 4, pp. 74–81, Apr. 2009.
50. Jungnickel, V., Schellmann, M., Thiele, L., Wirth, T., Haustein, T., Koch, O., Zirwas, W., and Schulz, E., "Interference-aware scheduling in the multiuser MIMO-OFDM downlink," *IEEE Com. Mag.*, vol. 47, nr. 6, pp. 56–66, Jun. 2009.
51. Li, Q., Lin, X.E., Zhang, J., and Roh, W., "Advancement of MIMO technology in WiMAX: from IEEE 802.16d/e/j to 802/16m," *IEEE Com. Mag.*, vol. 47, nr. 6, pp. 100-107, Jun. 2009.

52. Berger, S., Kuhn, M., Wittneben, A., Unger, T., and Klein, A., "Recent advances in amplify-and-forward two-hop relaying," *IEEE Com. Mag.*, vol. 47, nr. 7, pp. 50–56, Jul. 2009.
53. Gonzales, C.R.A., Dietrich, C.B., and Reed, J.H., "Understanding the software communications architecture," *IEEE Com. Mag.*, vol. 47, nr. 9, pp. 50–57, Sep. 2009.
54. Park, M., Gopalakrishnan, P., and Roberts, R., "Interference mitigation techniques in 60 GHz wireless networks," *IEEE Com. Mag.*, vol. 47, nr. 12, pp. 34–40, Dec. 2009.
55. Kim, J., Hyeon, S., and Choi, S., "Implementation of an SDR system using graphics processing unit," *IEEE Com. Mag.*, vol. 48, nr. 3, pp. 156–162, Mar. 2010.
56. Melvin, W.L., "A STAP overview," *IEEE A & E Sys. Mag.*, vol. 19, nr. 1, pp. 19–35, Jan. 2004.
57. Cantrell, B., Rao, J., Tavik, G., Dorsey, M., and Krishevsky, V., "Wideband array antenna concept," *IEEE A & E Sys. Mag.*, vol. 21, nr. 1, pp. 9–12, Jan. 2006.
58. Kalinin, A.V., "Anechoic chamber wideband antenna measurements," *IEEE A & E Sys. Mag.*, vol. 21, nr. 1, pp. 21–24, Jan. 2006.
59. Nickel, U., "Overview of generalized monopulse estimation," *IEEE A & E Sys. Mag.*, vol. 21, nr. 6, part. 2, pp. 27–56, Jun. 2006.
60. Huang, Y., "WiMAX dynamic beamforming antenna" *IEEE A&E Sys. Mag.*, vol. 23, nr. 8, pp. 26–31, Aug. 2008.
61. Lentilhac, S., "UAV flight plan optimized for sensor requirements" *IEEE A&E Sys. Mag.*, vol. 25, nr. 2, pp. 11–14, Feb. 2010.
62. Milin, J.-L., Moore, S., Burger, W., Triboulloy, P.-Y., Royden, M., and Gerster, J., "AMSAR-a European success story in AESA radar" *IEEE A&E Sys. Mag.*, vol. 25, nr. 2, pp. 21–28, Feb. 2010.
63. Moore, S., "UK airborne AESA radar research" *IEEE A&E Sys. Mag.*, vol. 25, nr. 2, pp. 29–35, Feb. 2010.
64. Zheng, Y., "Adaptive antenna array processing for GPS receivers," *Master of Engineering Thesis*, University of Adelaide, South Australia, July 2008.
65. Forsythe, K.W., "Utilizing waveform features for adaptive beamforming and direction finding with narrowband signals," *Lincoln Lab. J.*, vol. 10, nr. 2, pp. 99–126, 1997.
66. Jeyanthi, K.M.A., and Kabilan, A.P., "A simple adaptive beamforming algorithm with interference suppression" *Int. J. Eng. Technol.*, vol. 1, no. 1, pp. 67–70, Apr. 2009.
67. Gu, Y.J., Shi, Z.G., and Chen, K.S., "Robust adaptive beamforming for steering vector uncertainties based on equivalent DOAs method," *Prog. Electromagn. Res.* PIER vol. 79, pp. 277–290, 2008.
68. Lee, J.-Y., Liu, H.-C., and Samueli, H., "A digital adaptive beamforming QAM demodulator IC for high bit-rate wireless communications," *IEEE J. Solid-State Circuits*, vol. 33, nr. 3, pp. 367–377, Mar. 1998.
69. Wipf, D., and Nagarajan, S., "Beamforming using the relevance vector machine," in *Proc. 24th Inter. Conf. Mach. Learn.*, Corvalis, OR, pp. 1023–1030, 2007.
70. Jordan, M., Schmitt, L., Ascheid, G., and Meyr, H., "Prediction of downlink SNR for opportunistic beamforming," in *Proc. IEEE WCNC*, unkown, pp. 2211–2215, Mar. 2007.
71. Chen, H.H., Chan, S.C., and Ho, K.L., "Adaptive beamforming using frequency invariant uniform concentric circular arrays," *IEEE Trans. Circ. Sys.*, vol. 54, nr. 9, pp. 1938–1949, Sep. 2007.
72. Russell, D.J., and Palmer, R.D., "Application of APES to adaptive arrays on the CDMA reverse channel," *IEEE Trans.* Veh. Tech.., vol. 53, nr. 1, pp. 3–17, Jan. 2004.
73. Zheng, Y.R., Goubran, R.A., and El-Tanany, M. "A nested sensor array focusing on near field targets," in *Proc. IEEE Sensors*, 2003, pp. 843–848, Apr. 2004.
74. Zheng, Y.R., Goubran, R.A., and El-Tanany, M., "Robust near-field adaptive beamforming with distance discrimination," *IEEE Trans. Speech. Audio Proc.*, vol. 12, nr. 5, pp. 478–488, Sep. 2004.
75. Karlsson, R., and Gustafsson, F., "Recursive Bayesian estimation: bearings-only applications," *IEE Proc.-Radar Sonar Nav.*, vol. 152, nr. 5, pp. 305–313, Oct. 2005.

76. Yang, H.-L., and Wu, W.-R., "A novel adaptive antenna array for DS/CDMA code acquisition," *IEEE Trans. Sig. Proc.*, vol. 55, nr. 9, pp. 4567–4580, Sep. 2007.
77. Lapierre, F.D., Verly, J.G., Himed, B., Klemm, R., and Lesturgie, M., "Radar space-time adaptive processing," *EURASIP J. Appl. Signal Processing*, vol. 2006, Article ID 93805, pp 1–4, 2006.
78. White, G., Falletti, E., Xu, Z., Borio, D., Sellone, F., Zakharov, Y., Presti, L.L., and Daneshgaran, F., "Report on adaptive beamforming algorithms for advanced antenna types for aerial platform and ground terminals," FP6-IST-2003-506745 CAPANINA, pp. 175, Jan. 2006.
79. De Lorenzo, D.S., "Navigation accuracy and interference rejection for GPS adaptive antenna arrays," *Ph.D. Dissertation, Department of Aeronautics and Astronautics*, Stanford University, Paolo Alto, CA, 125 pp., August 2007.
80. Coskun, C., "Robust adaptive beamforming," Technical University of Denmark, Institute for Informatics and Mathematical Modeling, 94 pp, Sep. 30, 2008.
81. Cardoso, J.F., and Souloumiac, A., "Blind beamforming for non-Gaussian signals," *IEE Proc.* vol. 140, nr. 6, pp. 362–370, Dec. 1993.
82. Filho, D.Z., Cavalcante, C.C., Resende, L.S., and Romano, J.M.T., "On adaptive LCMV beamforming for multiuser processing in wireless systems," in *Proc. SBMO/IEEE MTT-S International Microwave & Optoelectronics Conference (IMOC 2007)*, pp. 521–525, 2007.
83. Berefelt, F., Boberg, B., Eklöf, F., Malmström, J., Pääjärvi, L., Strömbäck, P., Wirkander, S-L., "INS/GPS integration with adaptive beamforming," in *Proc. ION-GPS 2003*, Portland, OR, pp. 1096–1106, Sep. 2003.
84. Preisig, J.C., "Adaptive matched field processing in an uncertain propagation environment," *RLE Technical Report No. 567, Research Laboratory of Electronics*, MIT, 174 pp., Jan. 1992.
85. Liu, W., Weissand, S., Hanzo, L., "A subband-selective broadband GSC with cosine-modulated blocking matrix," *IEEE. Trans. Ant. Prop.*, vol. 52, nr. 3, pp. 813–820, Mar. 2004.
86. More, T.D., "Analytic study of space-time and space-frequency adaptive processing for radio frequency interference suppression," *Ph.D. Dissertation*, Ohio University, 226 pp, 2002.
87. Bevelacqua, P.J., and Balanis, C.A., "Optimizing antenna array geometry for interference suppression," *IEEE Trans. Ant. Prop.*, vol. 55, nr. 3, pp. 637–641, Mar. 2007.
88. Chae, C.-B., Mazzarese, D., Jindal, N., and Heath, R.W. Jr., "Coordinated beamforming with limited feedback in the MIMO broadcast channel," *IEEE. J. Sel. Area Comm.*, vol. 26, nr. 8, pp. 1505–1515, Oct. 2008.
89. Comisso, M., "Beamforming techniques for wireless communications in low-rank channels: analytical models and synthesis algorithms," Department of Electro-technique, Electronic, and Informatics, *Ph.D. Dissertation*, University of Trieste, 232 pp., 2009.
90. Nicoli, M., Sala, M., Simeone, O., Sampietro, L., and Santacesaria, C., "Adaptive array processing for time-varying interference mitigation in IEEE 802.16 systems," in *Proc. PIMRC 2006*, Helsinki, pp. 1–4, Sept. 2006.
91. Hayashi, K., and Hara, S., "A new spatio-temporal equalization method using estimated channel impulse response," in *Proc. WTC-2000*, Tokyo, 2000.
92. Montalbano, G., Ghauri, I., and Slock, D.T.M., "Spatio-temporal array processing for aperiodic CDMA downlink transmission," in *Proc. ASILOMAR*, vol.2, pp. 912–916, Oct. 1999.
93. Rotman, R., Raz, O., Barzilay, S., Rotman, S., and Tur, M., "Wideband antenna patterns and impulse response of broadband RF phased arrays with RF and photonic beamforming," *IEEE Trans. Ant. Prop.*, vol. 55, nr. 1, pp. 36–44, Jan. 2007.
94. Li, Y., "Broadband beamforming and direction finding using concentric ring array," *Ph.D. Dissertation*, University of Missouri, July 2005.
95. Fabrizio, G.A., Gershman, A.B., and Turley, M.D., "Robust adaptive beamforming for HF surface wave over-the-horizon radar," *IEEE Trans. Aero. Elec. Syst.*, vol. 40, nr. 2, pp. 510–525, Apr. 2004.

96. Fuchs, J.-J., "The global matched filter: the application to DOA estimation with an uniform circular array," in *Proc. IEEE Sensor Array Multi. Sig. Proc. Work.*, pp. 469–473, Mar. 2000.
97. Bell, K.L., Ephraim, Y., and Van Trees, H.L., "A Bayesian approach to robust adaptive beamforming," *IEEE Trans. Sig. Proc.*, vol. 48, nr. 2, pp. 386–398, Feb., 2000.
98. Liu, W., and Langley, R., "Adaptive wideband beamforming with combined spatial/temporal subband decomposition," *PIERS Online J.*, vol. 3, nr. 7, pp. 991–996, Jul. 2007.
99. Rong, Z., "Simulation of adaptive array algorithms for CDMA systems," *MS Thesis, Virginia Polytechnic Institute and State University*, Virginia, 143 pp., 1996.
100. Guerci, J.R., "Introduction to STAP and knowledge-aided processing," Tutorial 3, *IEEE RadarCon 2009*, Pasadena, CA, 4 May 2009.
101. Goldstein, M., Picciolo, M., and Griesbach, J., "Advanced STAP," Tutorial 16, *IEEE RadarCon 2009*, Pasadena, CA, 4 May 2009.

Appendix A. RF Signals Simulink Models

RF Signals for Indoor GRFS Systems

RF Signals for Wireless Networks

Figure A.1 includes a description of a baseband model of an IEEE® 802.11a physical layer WLAN [1]. The model supports all mandatory and optional data rates: 6, 9, 12, 18, 24, 36, 48, and 54 Mb/s. The demo also illustrates adaptive modulation and coding over a dispersive multipath fading channel, whereby the simulation varies the data rate dynamically. Note that the model uses an artificially high channel fading rate to make the data rate change more quickly and thus make the visualization more animated and instructive [1].

The demonstration contains components that model the essential features of the WLAN 802.11a standard. The top row of block contains the WLAN 802.11 transmitter components as illustrated in Fig. A.2; while the bottom row contains the receiver components as depicted in Fig. A.2 [1]. Further details about this block can be obtained in [1].

Figure A.4 illustrates Simulink simulation results of the Simulink Block diagram of Fig. A.1. Starting from top to bottom and from left to right we have:

1. *TX Data*: the transmitter binary data stream.
2. *Un-equalized signal*: the I and Q of the unequalized received signal.
3. *RX power spectrum (dB)*: the double sided RX power spectrum in (dB).
4. *SNR (dB)*: the signal-to-noise ratio at the input of the receiver in (dB).
5. *Equalized signal*: equalized I and Q symbols. Current plot in Fig. A.4 shows 64 QAM modulation. (Other forms of modulation are BPSK, QPSK, 16 QAM, 64 QAM as shown in Fig. A.3).
6. *Equalized power spectrum*: equalized power spectrum after the equalization on the receiver side.
7. *Bit rate (Mb/s)*: variable bit rate of the WLAN. Current plot in Fig. A.4 shows bit rates on 24, 36, 48 Mb/s.

I. Progri, *Geolocation of RF Signals*,
DOI 10.1007/978-1-4419-7952-0, © Springer Science+Business Media, LLC 2011

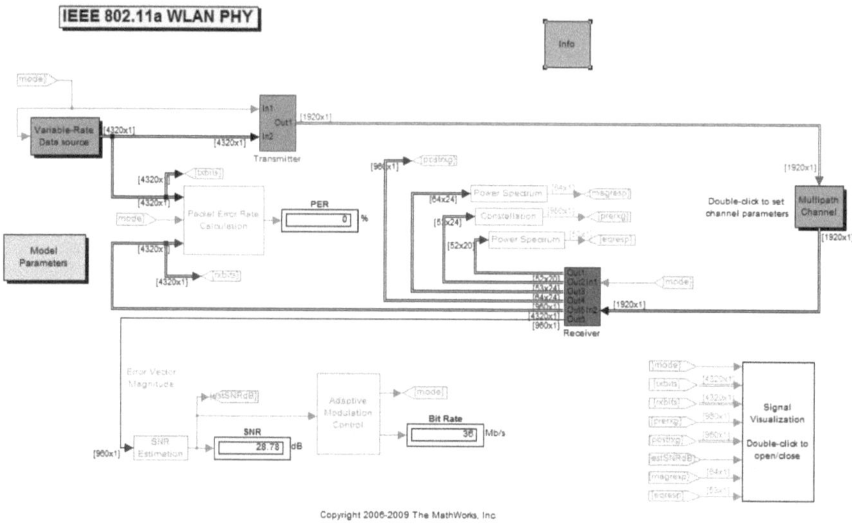

Fig. A.1 A description of the Simulink block diagram of IEEE® 802.11a WLAN Physical Layer. Reprinted with permission copyright © 2006–2009 The MathWorks, Inc. [1]

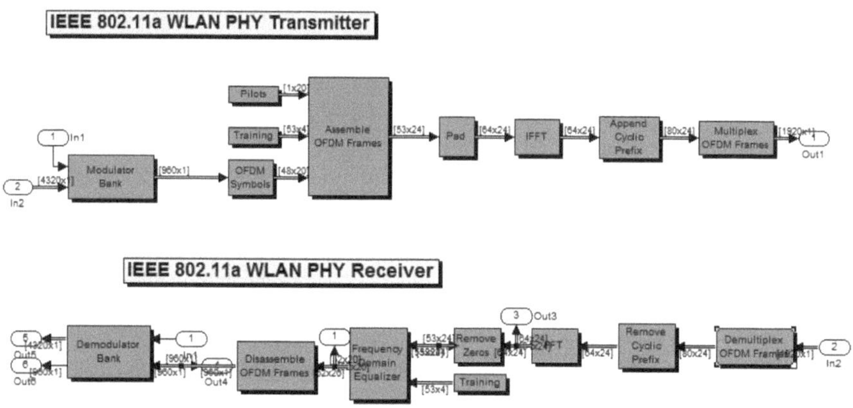

Fig. A.2 A description of the Simulink block diagram of IEEE® 802.11a WLAN Physical Layer Transmitter and Receiver. Reprinted with permission copyright © 2006–2009 The MathWorks, Inc. [1]

Figure A.5 provides a description of the baseband Simulink block diagram of IEEE® 802.11b WLAN Physical Layer.

Figure A.6 shows a description of the baseband Simulink block diagram of IEEE® 802.11b WLAN Physical Layer Transmitter (Top) and Receiver (Bottom). Figure A.7 presents a description of the baseband Simulink block diagram of IEEE® 802.11b WLAN Physical Layer instrumentation (left) and instruments

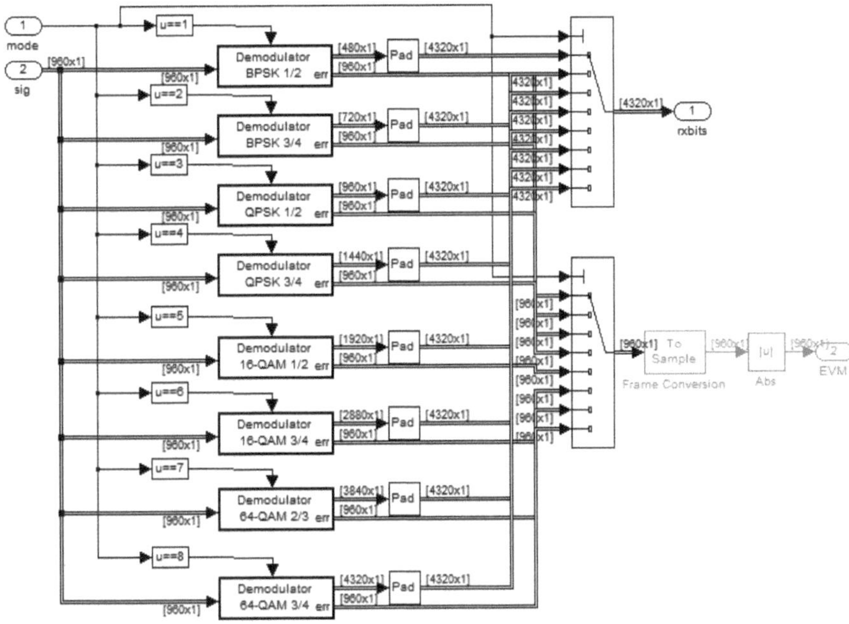

Fig. A.3 A description of the Simulink block diagram of IEEE® 802.11a WLAN Physical Layer Receiver Demodulator showing BPSK, QPSK, 16-QAM, and 64-QAM demodulation. Reprinted with permission copyright © 2006–2009 The MathWorks, Inc. [1]

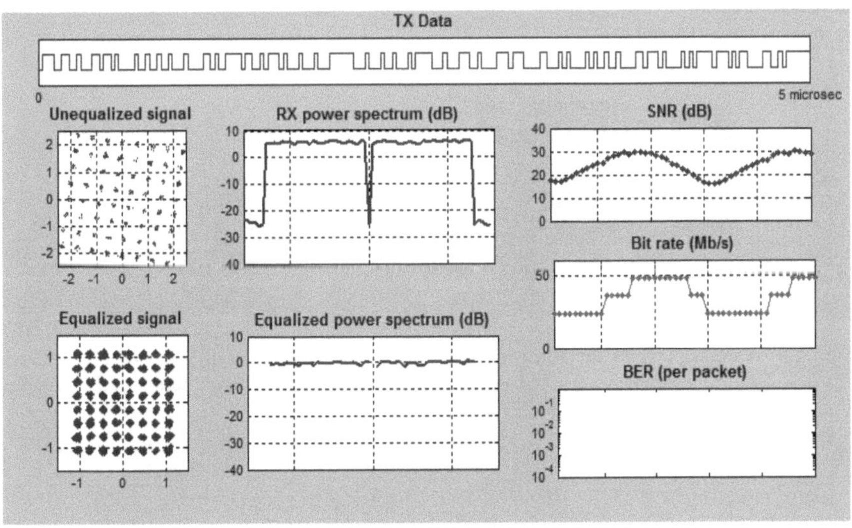

Fig. A.4 A description of Simulink simulation of the IEEE® 802.11a WLAN Physical Layer. Reprinted with permission copyright © 2006–2009 The MathWorks, Inc. [1]

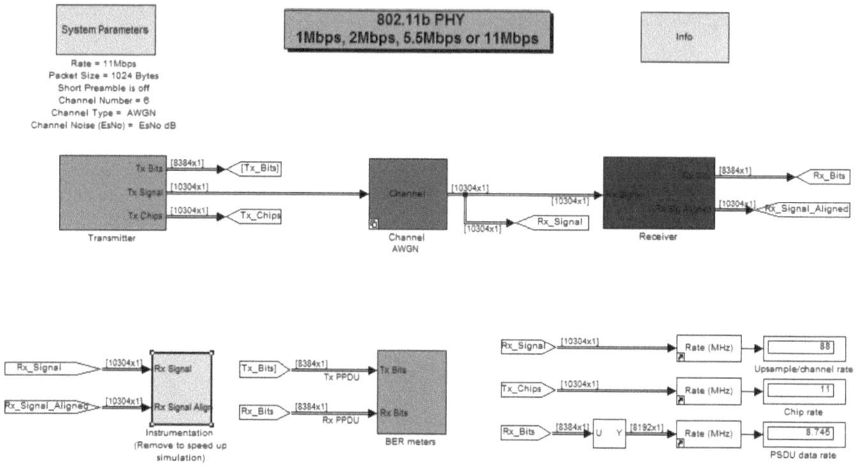

Fig. A.5 A description of the baseband Simulink block diagram of IEEE® 802.11b WLAN Physical Layer. Reprinted with permission copyright © 2006–2009 The MathWorks, Inc. [2]

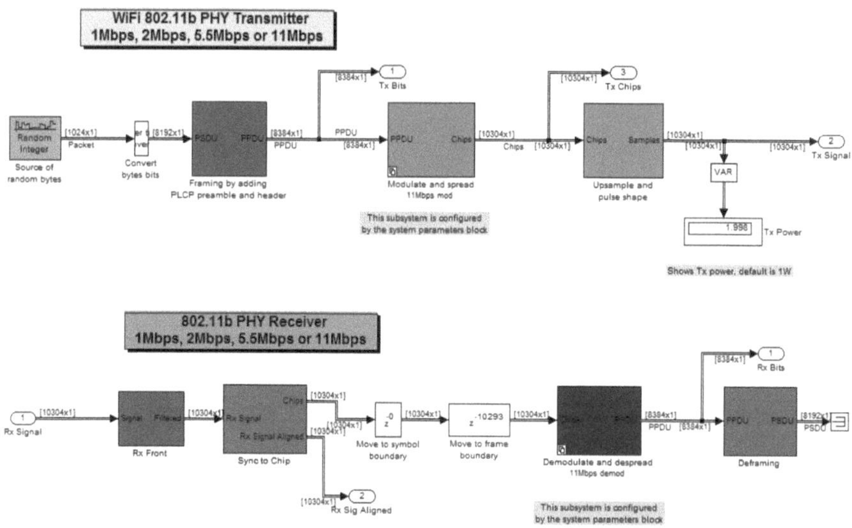

Fig. A.6 A description of the baseband Simulink block diagram of IEEE® 802.11b WLAN Physical Layer Transmitter (*top*) and Receiver (*bottom*). Reprinted with permission copyright © 2006–2009 The MathWorks, Inc. [2]

(right). Figure A.8 offers a description of the baseband Simulink block diagram of IEEE® 802.11b WLAN Physical Layer BER meters and Fig. A.9 summarizes a description of Simulink simulation results of the baseband IEEE® 802.11b

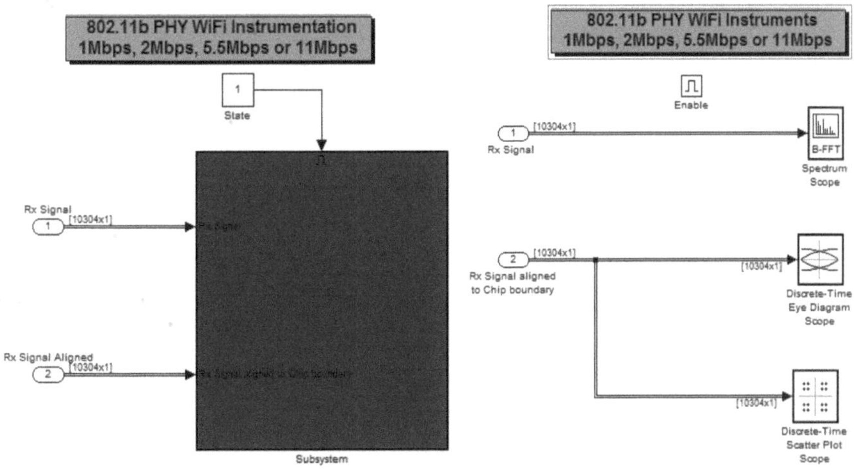

Fig. A.7 A description of the baseband Simulink block diagram of IEEE® 802.11b WLAN Physical Layer instrumentation (left) and instruments (right). Reprinted with permission copyright © 2006–2009 The MathWorks, Inc. [2]

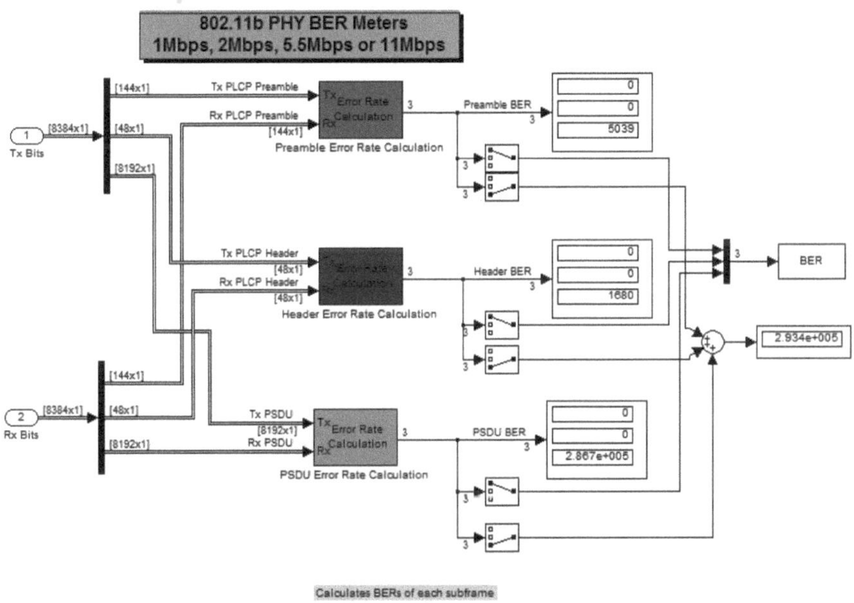

Fig. A.8 A description of the baseband Simulink block diagram of IEEE® 802.11b WLAN Physical Layer BER meters. Reprinted with permission copyright © 2006–2009 The MathWorks, Inc. [2]

WLAN Physical Layer. Other WLAN Simulink forms of the IEEE 802.11 are very similar with 802.11a and 802.11b; therefore, we leave them as an exercise for the reader.

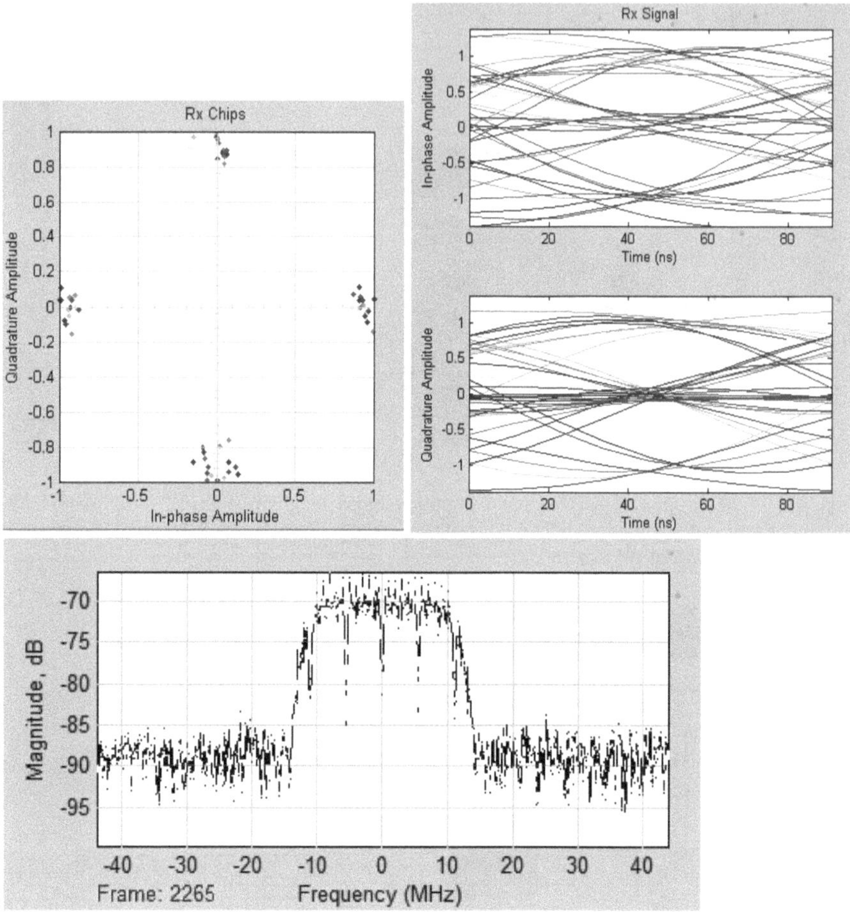

Fig. A.9 A description of Simulink simulation results of the baseband IEEE® 802.11b WLAN Physical Layer. Reprinted with permission copyright © 2006–2009 The MathWorks, Inc. [2]

Figure A.10 shows a description of the baseband Simulink block diagram of IEEE® 802.15.3 UWB Multiband OFDM Physical Layer. Figure A.11 illustrates a description of the baseband Simulink block diagram of IEEE® 802.15.3 UWB Multiband OFDM Physical Layer Transmitter (top) and Receiver (bottom). And Fig. A.12 depicts a description of Simulink simulation results of the baseband IEEE® 802.15.3 UWB Multiband OFDM Physical Layer.

The Simulink design shown in Fig. A.10 only considers the QPSK modulation. One can redesign the Simulink to take into consideration other forms of modulation such as trellis coded QPSK and 16/32/64-QAM which will result in a very similar Simulink implementation as the one shown in Fig. A.3. We will leave this as an exercise for the reader. Again, I would like to stress that my main objective in this book is to provide a broad and detailed description of the RF signals and workable Simulink demos that an experienced designer can go ahead and build more

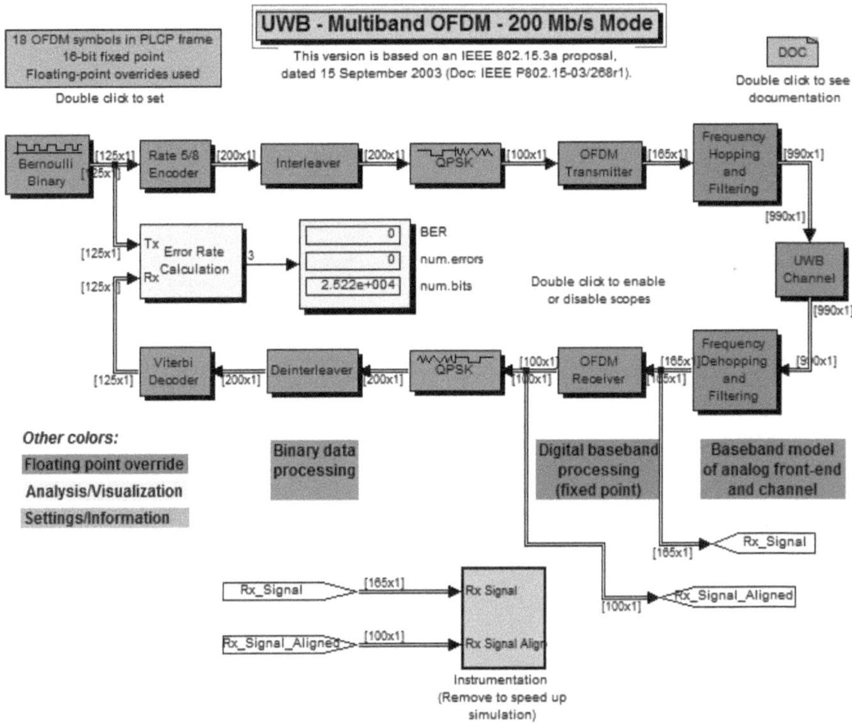

Fig. A.10 A description of the baseband Simulink block diagram of IEEE® 802.15.3 UWB Multiband OFDM Physical Layer. Reprinted with permission copyright © 2006–2009 The MathWorks, Inc. [3]

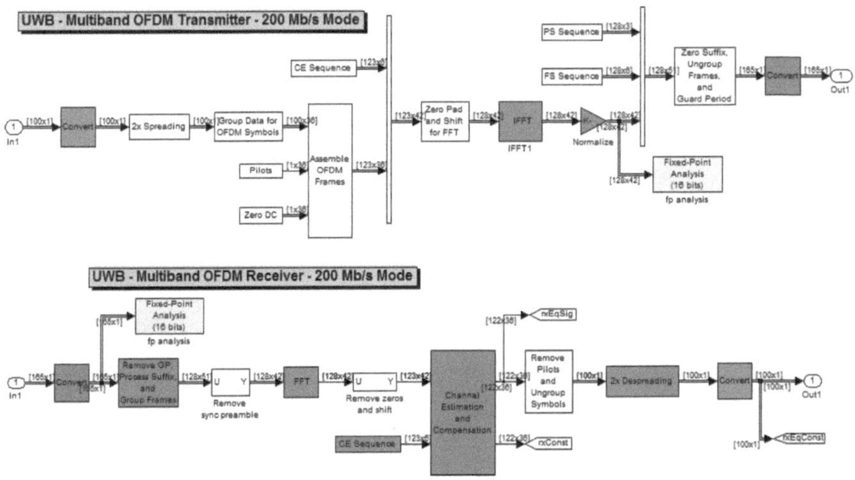

Fig. A.11 A description of the baseband Simulink block diagram of IEEE® 802.15.3 UWB Multiband OFDM Physical Layer Transmitter (*top*) and Receiver (*bottom*). Reprinted with permission copyright © 2006–2009 The MathWorks, Inc. [3]

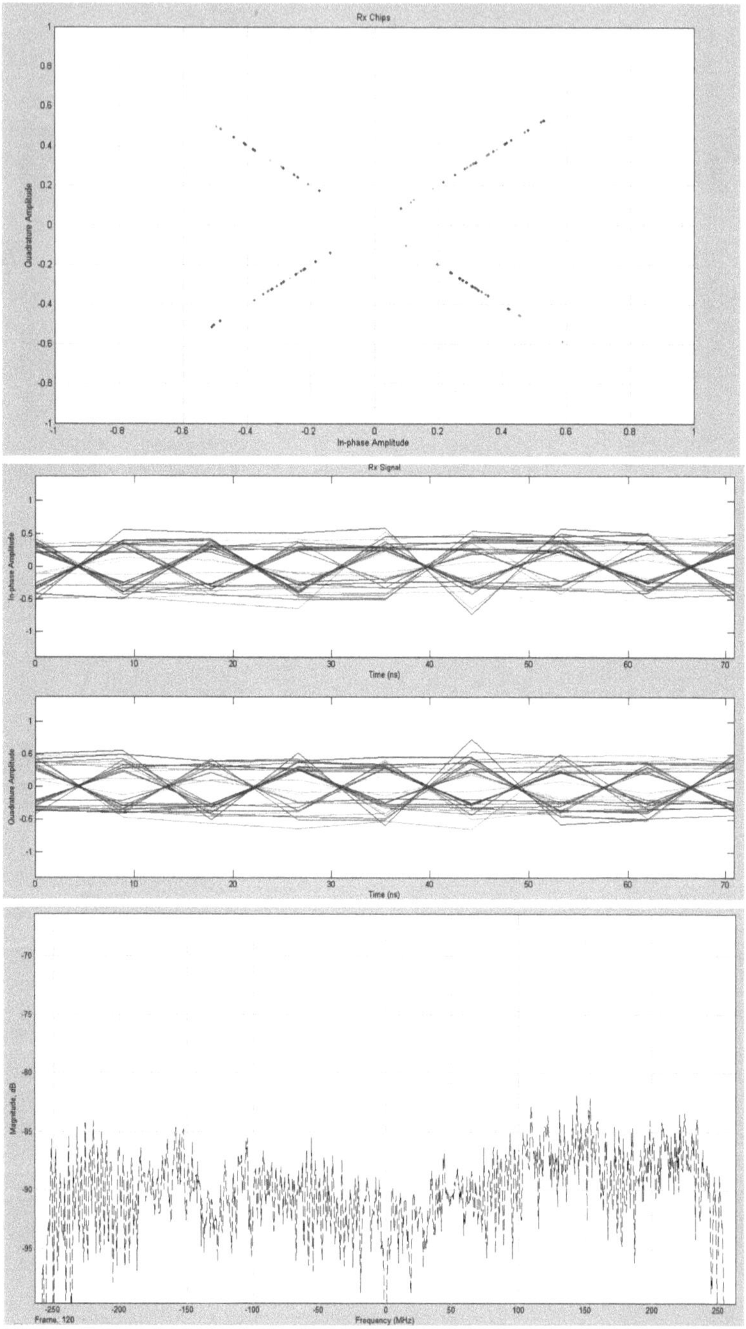

Fig. A.12 A description of Simulink simulation results of the baseband IEEE® 802.15.3 UWB Multiband OFDM Physical Layer. Reprinted with permission copyright © 2006–2009 The MathWorks, Inc. [3]

sophisticated and more realistic Simulink models and run more accurate simulation results which are as close to the real life as possible.

RF Signals for Urban GRFS Systems

RF signals for urban GRFS systems include: (1) RF signals for mobile systems and metropolitan area networks (MAN) in Sect. A.2.1.

RF Signals for Mobile Systems and Metropolitan Area Networks

Figure A.13 depicts a description of the baseband Simulink block diagram of CDMA2000 Physical Layer. Figure A.14 shows a description of the baseband Simulink block diagram of CDMA2000 Physical Layer Transmitter (first two top plots) and Receiver (bottom two top plots). Figure A.15 illustrates a description of the baseband Simulink simulation results block diagram of CDMA2000 Physical Layer [4].

Figure A.16 indicates a description of the baseband Simulink block diagram of IEEE 802.16-2004 OFDM Physical Layer Including Space–Time Block Coding. Figure A.17 presents a description of the baseband Simulink block diagram of IEEE 802.16-2004 OFDM Physical Layer Including Space–Time Block Coding Transmitter (top) and Receiver (bottom). And Fig. A.18 shows a description of the

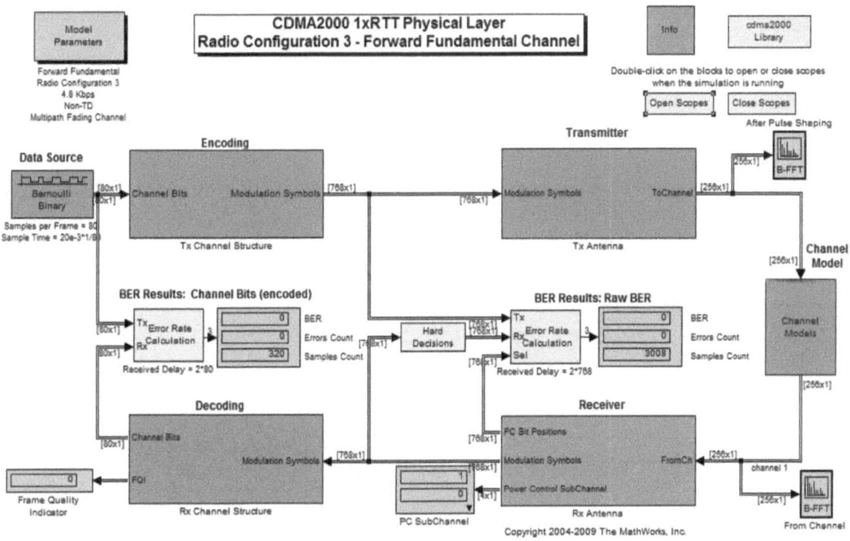

Fig. A.13 A description of the baseband Simulink block diagram of CDMA2000 Physical Layer. Reprinted with permission copyright © 2006–2009 The MathWorks, Inc. [4]

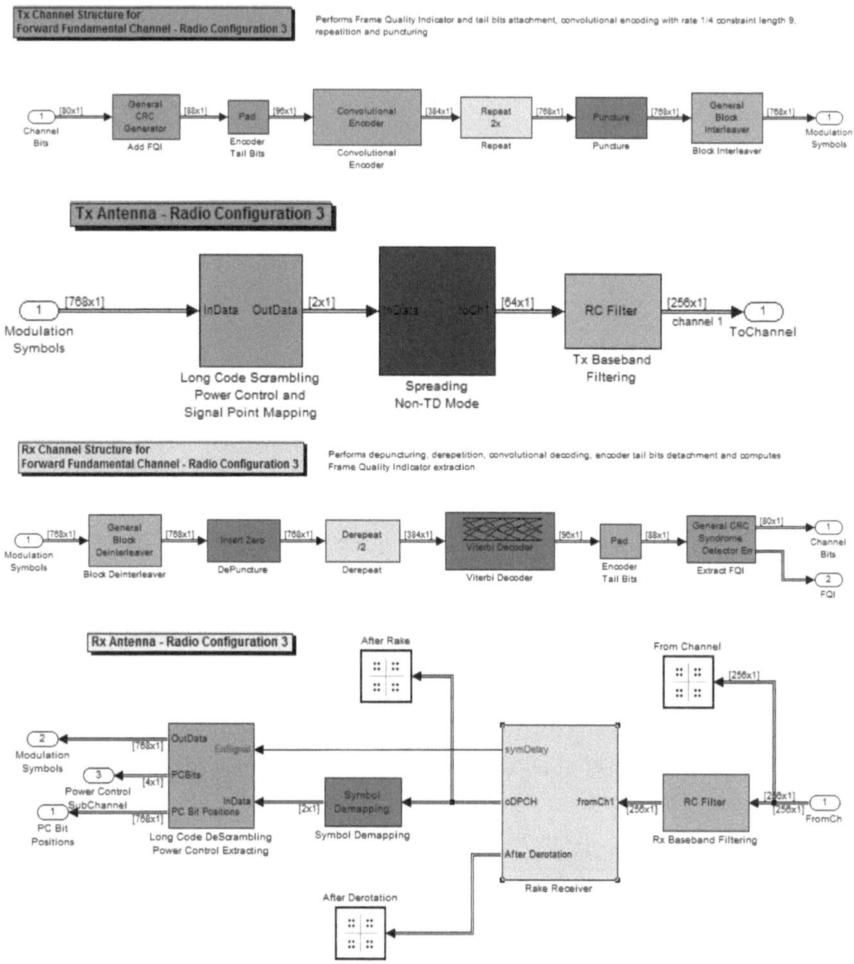

Fig. A.14 A description of the baseband Simulink block diagram of CDMA2000 Physical Layer Transmitter (*first two top plots*) and Receiver (*bottom two top plots*). Reprinted with permission copyright © 2006–2009 The MathWorks, Inc. [4]

baseband Simulink simulation results of IEEE 802.16-2004 OFDM Physical Layer Including Space–Time Block Coding [5].

RF Signals for Satellite GRFS Systems

For the purpose of this book and for the purpose of this chapter, the satellite signals of interests are those used as part of RF signals for Global Navigation Satellite Systems (GNSS); (2) communications connectivity for voice, data, video, and

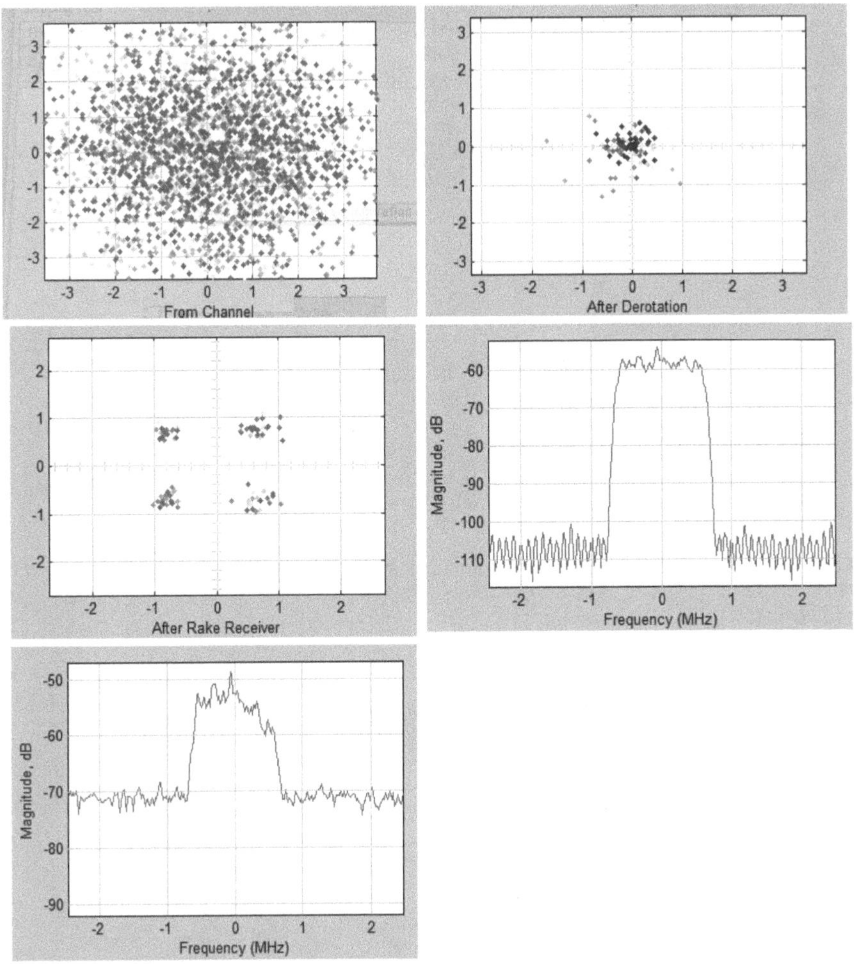

Fig. A.15 A description of the baseband Simulink simulation results block diagram of CDMA2000 Physical Layer. Reprinted with permission copyright © 2006–2009 The MathWorks, Inc. [4]

picture as treated in Sect. A.3.1 and part of the RF signals for satellite television technology (STT).

RF Signals for Satellite Television Technology

Figure A.19 depicts a description of a baseband Simulink block diagram of the RF Satellite Link [6] which starts with: (1) a satellite downlink transmitter (see Fig. A.20 (top)); (2) the downlink path (free space path loss) and Doppler and

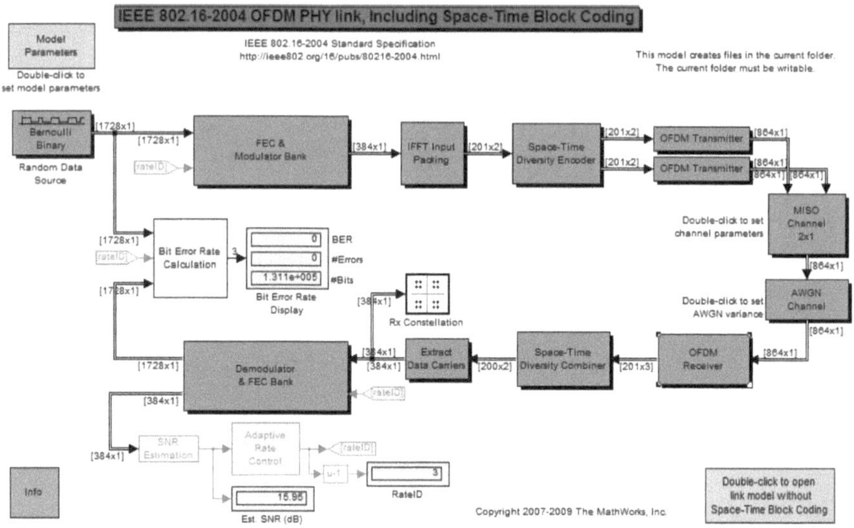

Fig. A.16 A description of the baseband Simulink block diagram of IEEE 802.16-2004 OFDM Physical Layer Including Space–Time Block Coding. Reprinted with permission copyright © 2006–2009 The MathWorks, Inc. [5]

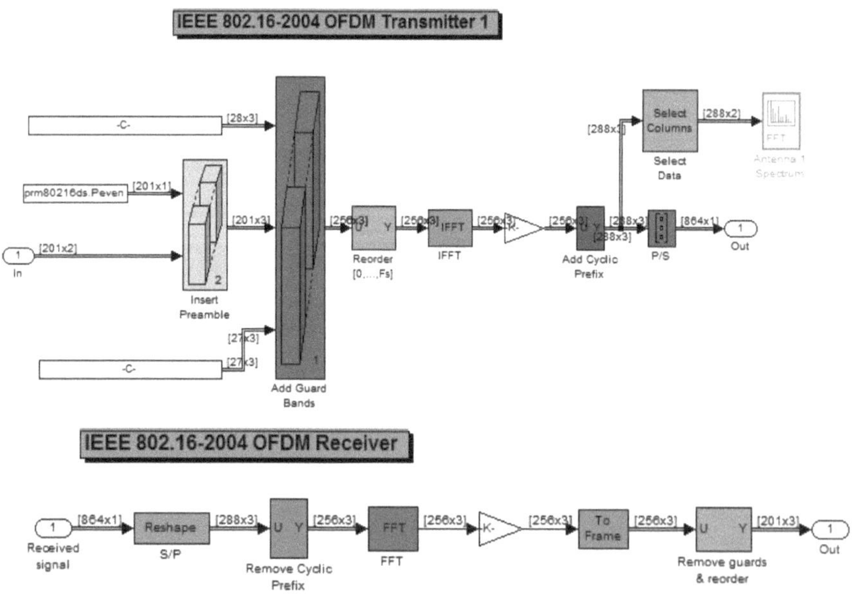

Fig. A.17 A description of the baseband Simulink block diagram of IEEE 802.16-2004 OFDM Physical Layer Including Space–Time Block Coding Transmitter (*top*) and Receiver (*bottom*). Reprinted with permission copyright © 2006–2009 The MathWorks, Inc. [5]

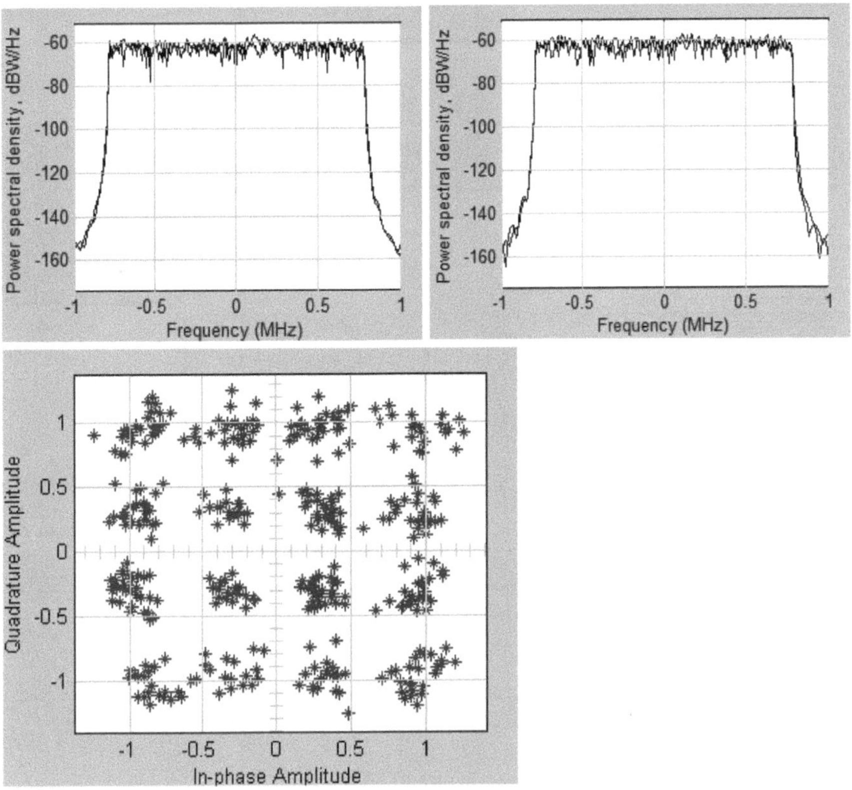

Fig. A.18 A description of the baseband Simulink simulation results of IEEE 802.16-2004 OFDM Physical Layer Including Space–Time Block Coding. Reprinted with permission copyright © 2006–2009 The MathWorks, Inc. [5]

phase error (phase and frequency offset); (3) Ground station downlink receiver (see Fig. A.20 (bottom)). Figure A.20 shows a description of a baseband Simulink block diagram of the RF satellite link transmitter (top) and receiver (bottom) [6]. As shown in Fig. A.20 (top), the Satellite downlink transmitter block diagram contains a random integer generator, a rectangular 16-QAM modulator, a square root raised cosine filter, a high power amplifier, and Tx dish antenna. In Fig. A.20 (bottom), the ground station downlink receiver shows the Rx dish antenna gain, the phase noise, the I/Q imbalance, DC removal, magnitude AGC, Doppler and phase compensation, raised cosine receive filter, and the rectangular 16-QAM. Figure A.21 presents simulation results of a baseband Simulink block diagram of the RF Satellite Link [6]. The top plot shows the Tx and Rx spectrum in (dB) versus the frequency (Hz). In the pass-band, (40 kHz double side band centered at the 0 Hz line) both the Tx and Rx spectrum overlap with each other; however, in the stop-band, the Tx spectrum is below the Rx spectrum due to noise and other channel impairments such as Doppler and Phase rotation, I/Q imbalance etc. in the Rx signal. Next, we have the constellations before and after high power amplifier in

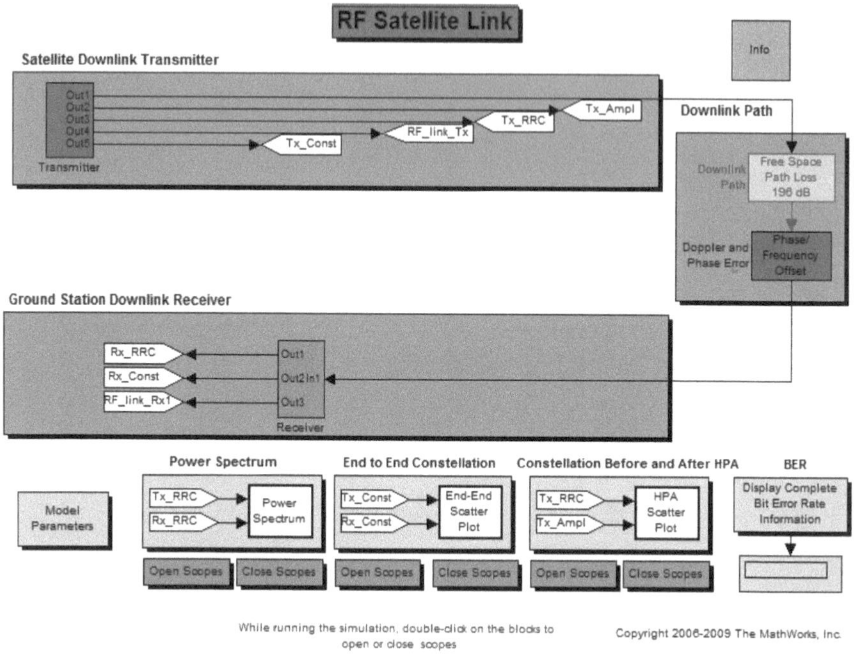

Fig. A.19 A description of a baseband Simulink block diagram of the RF Satellite Link. Reprinted with permission copyright © 2006–2009 The MathWorks, Inc. [6]

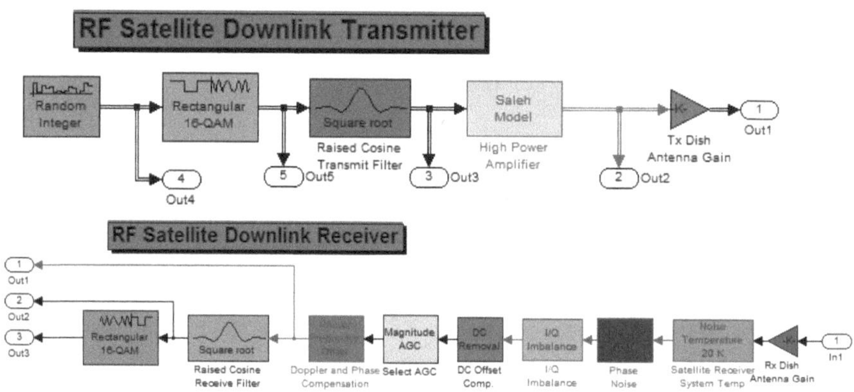

Fig. A.20 A description of a baseband Simulink block diagram of the RF Satellite Link Transmitter (*top*) and Receiver (*bottom*). Reprinted with permission copyright © 2006–2009 The MathWorks, Inc. [6]

Fig. A.21 (plots 2 and 3 from the top). The last two plots of Fig. A.21 are the end-to-end constellation scatter plot which clearly indicates that the 16-symbol-signals as shown in Ref [6]. This concludes the example of a voice satellite radio RF link Simulink demo and all the other Simulink demos in the book.

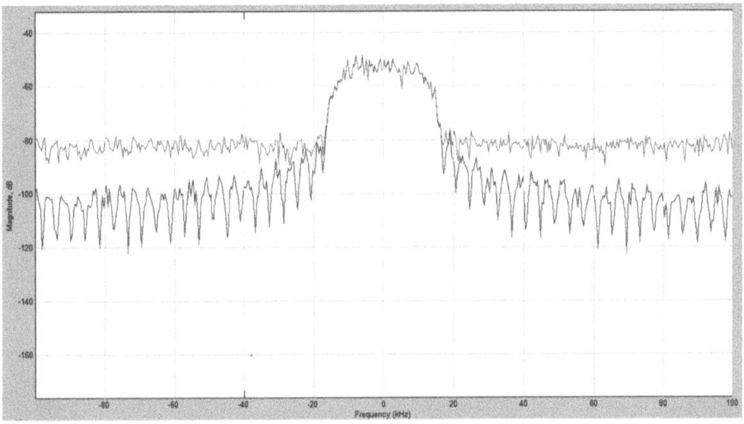
Magnitude, dB
Frequency (kHz)

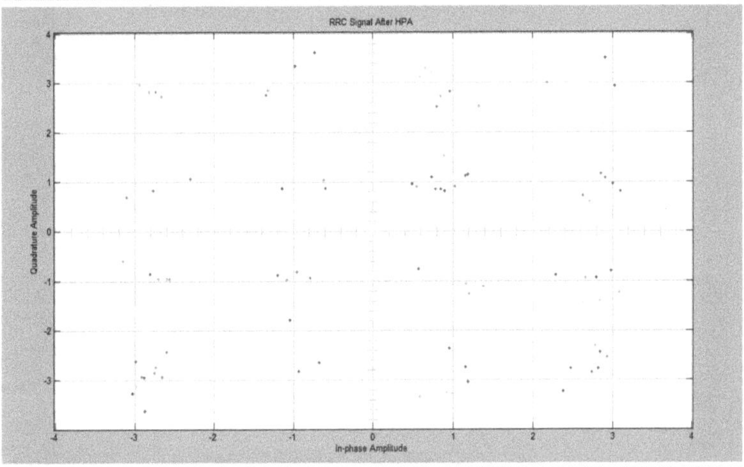
RRC Signal After HPA
Quadrature Amplitude
In-phase Amplitude

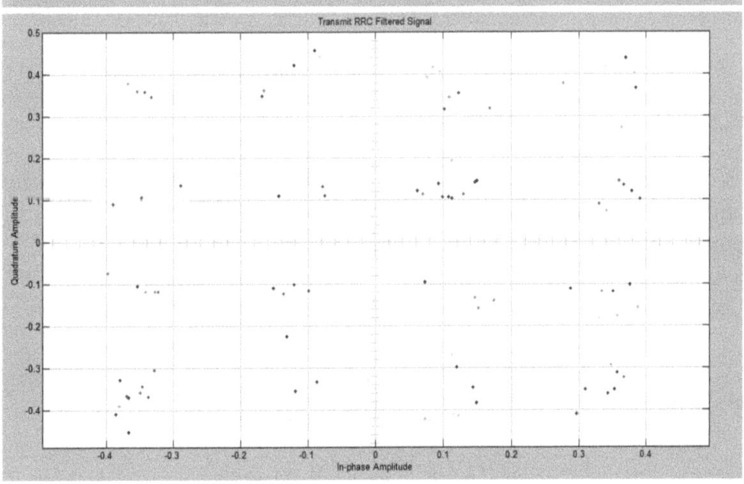
Transmit RRC Filtered Signal
Quadrature Amplitude
In-phase Amplitude

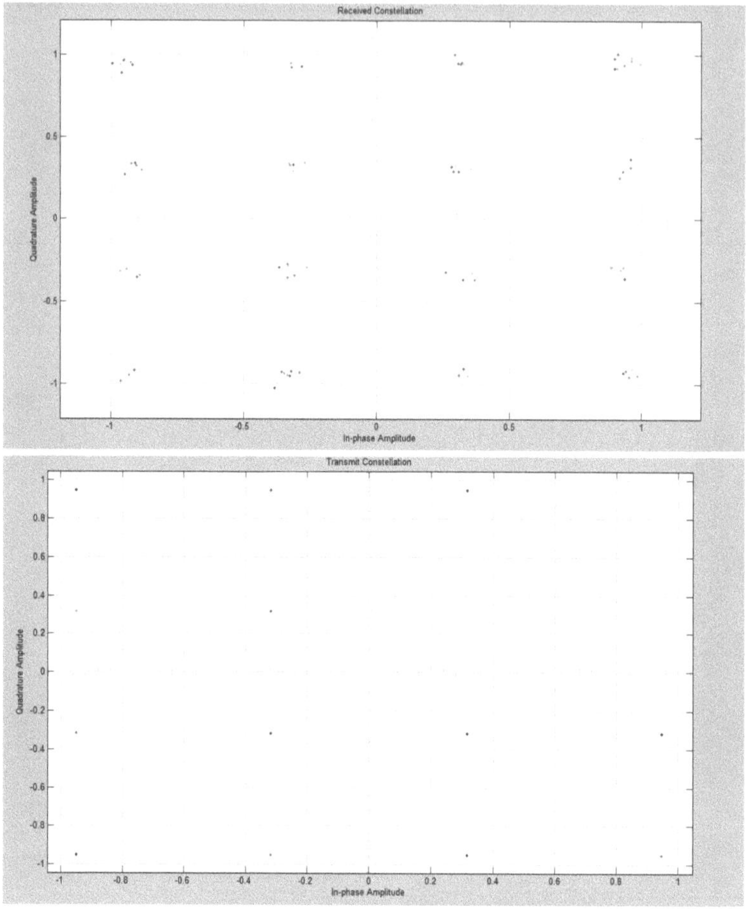

Fig. A.21 A description of a baseband Simulink simulation results of IEEE 802.16-2004 OFDM Physical Layer Including Space–Time Block Coding. Reprinted with permission copyright © 2006–2009 The MathWorks, Inc. [6]

This concludes the Simulink demo case studies of this first edition because we have provided enough case studies to illustrate some of the hottest signal designs in the communications world. Other signal designs such as Satellite TV, Video Broadcasting, GPS etc. can be illustrated in the same manner as these which we might include them either in separate publications or in future editions of this book.

References

1. Demo of an end-to-end baseband model of the physical layer of a wireless local area network (WLAN) according to the IEEE® 802.11a standard. The MathWorks, Inc., Copyright 2006–2009, MATLAB and Simulink 2009b.

2. Demo of an end-to-end baseband model of the physical layer of a wireless local area network (WLAN) according to the IEEE® 802.11b standard. The MathWorks, Inc., Copyright 2006–2009, MATLAB and Simulink 2009b.
3. Demo of an end-to-end baseband model of the physical layer of a Ultra Wide Band (UWB) Multiband OFDM according to the IEEE® 802.15.3a standard The MathWorks, Inc., Copyright 2006–2009, MATLAB and Simulink 2009b.
4. Demo of an end-to-end baseband model of the physical layer of the CDMA2000 standard. The MathWorks, Inc., Copyright 2006–2009, MATLAB and Simulink 2009b.
5. Demo of an end-to-end baseband model of the physical layer of the IEEE 802.16-2004 OFDM including Space-Time Block Coding. The MathWorks, Inc., Copyright 2006–2009, MATLAB and Simulink 2009b.
6. Demo of an end-to-end baseband model of the physical layer of the RF Satellite Link. The MathWorks, Inc., Copyright 2006–2009, MATLAB and Simulink 2009b.

Index

About the Author

Dr. Ilir F. Progri is a world-renowned leading authority in Geolocation, GPS, GNSS, and Wireless Communications in all aspects of RF, IF, and baseband signal specifications, simulation, software development, and implementation of significant new capabilities in GPS, GNSS, and Wireless Communications and indoor geolocation systems. As principle investigator, he has led research and development engineering projects for over 14 years. He has over 55 published papers, three world class tutorials, and one patent in all aspects of geolocation, RF geolocation, geointelligence, and georeference systems. Dr. Progri's work has been recognized by over 67 publications from experts, researchers, and scientists of US, Canada, Australia, UK, Finland, France, Germany, Croatia, Russia, China, Singapore, Japan, Korea, etc.

Dr. Progri is currently the Permanent Chairman of the Board of Directors and the President and CEO of Giftet, Inc. (http://www.giftet.com).

Dr. Progri received the 2010 VIP Presidential Who's Who Award recognized among outstanding business and professional achievers. He also received the 2008 America's Registry Roundtable Award of America's Registry of Outstanding Professionals as an elite group of 50 professionals dedicated to exceptional achievement and accomplishment in his chosen field. He received the 2008 Presidential Who's Who Award recognized among business and professional achievers.

Dr. Progri is the 2010 IEEE Worcester County Section (WCS) ComSoc Chapter chairman. He is the Senior Member Elevation Selection Committee Chairman at the IEEE WCS. He was also the 2009 and 2008 IEEE Los Angeles Council Vice-Chair and the 2009 IEEE Foothill Section Vice-Chair. He was the Local Arrangements Chair of the Organizing Committee of IEEE Globecom 2012, 2009; member of the Organizing Committee of the IEEE RadarCon 2009 and Program Co-Chair for WTS 2006 and 2005. Dr. Progri is an Associate Editor and the Editor of Position Navigation and Timing of IEEE Systems Magazine. Dr. Progri has been a Reviewer for the *IEEE Transactions on Wireless Communications*, 2010 and *IEE*

Proceedings for Radar, Sonar, and Navigation 2010, 2009. Dr. Progri has attended for the first time the IEEE Board of Directors Meeting Series in West Los Angeles, CA, June 2009. Dr. Progri has been a National Science Foundation (NSF) SBIR/STTR Phase II Review Panelist Member 2009; Reviewer for the NSF SBIR/STTR Phase I 2008; Reviewer *IEEE WHNC 2008*, 2008; *IEEE Transactions on Aerospace and Electronic Systems*, 2007, 2008; Sadiku, M.N.O. Elements of Electromagnetics, 4th ed., *New York, NY: Oxford University Press*, 2006; Session Chair and Co-Chair for the ION 2004, 2005; WTS, 2004, 205, 2005; IEEE Communications Magazine 2002; and IEEE ICC 2002.

He is a Fellow Member of RIN, Senior Member of the IEEE, Com Soc, AESS, and GRSS, and Member of ION, AIG, and Sigma Xi (the Scientific Research Society). He received his Ph.D. and MS degrees in Electrical Engineering from WPI, Worcester, Massachusetts in May 2003 and in May 1997, respectively. He received his Diploma of Engineer Degree in Electrical Engineering from the Polytechnic University of Tirana (PUT), Albania in July 1994.

FSC
www.fsc.org
MIX
Papier | Fördert
gute Waldnutzung
FSC® C083411

Kolibri 360 GmbH
Ferdinand-Jühlke-Straße 7
99095 Erfurt, Deutschland
produktsicherheit@kolibri360.de